STATISTICS

A Gentle Introduction

Second Edition

Frederick L. Coolidge

University of Colorado at Colorado Springs

SAGE Publications
Thousand Oaks ▪ London ▪ New Delhi

For information:

Sage Publications, Inc.
2455 Teller Road
Thousand Oaks, California 91320
E-mail: order@sagepub.com

Sage Publications Ltd.
1 Oliver's Yard
55 City Road
London EC1Y 1SP
United Kingdom

Sage Publications India Pvt. Ltd.
B-42, Panchsheel Enclave
Post Box 4109
New Delhi 110 017 India

Printed in the United States of America

Library of Congress Cataloging-in-Publication Data

Coolidge, Frederick L. (Frederick Lawrence), 1948-
Statistics: A gentle introduction / Frederick L. Coolidge.—2nd ed.
p. cm.
Includes bibliographical references and index.
ISBN 978-1-4129-2494-8 (pbk. : acid-free paper)
1. Mathematical statistics—Textbooks. I. Title. QA276.12.C67 2006
519.5—dc22

2005022221

This book is printed on acid-free paper.

08 09 10 10 9 8 7 6 5 4 3 2

Acquisitions Editor: Lisa Cuevas Shaw
Editorial Assistant: Karen Gia Wong
Production Editor: Melanie Birdsall
Copy Editor: Gillian Dickens
Typesetter: C&M Digitals (P) Ltd.
Proofreader: Liann Lech
Cover Designer: Candice Harman

Contents

Acknowledgments xvii

1. A Gentle Introduction 1
   - How Much Math Do I Need to Do Statistics? 2
   - The General Purpose of Statistics: Understanding the World 2
   - Another Purpose of Statistics: Making an Argument or a Decision 3
   - What Is a Statistician? 3
     - One Role: The Curious Detective 3
     - Another Role: The Honest Attorney 4
     - A Final Role: A Good Storyteller 5
   - Liberal and Conservative Statisticians 6
   - Descriptive and Inferential Statistics 7
   - Experiments Are Designed to Test Theories and Hypotheses 8
   - Oddball Theories 9
   - Bad Science and Myths 10
   - Eight Essential Questions of Any Survey or Study 11
     - 1. Who Was Surveyed or Studied? 11
     - 2. Why Did the People Participate in the Study? 12
     - 3. Was There a Control Group and Did the Control Group Receive a Placebo? 12
     - 4. How Many People Participated in the Study? 13
     - 5. How Were the Questions Worded to the Participants in the Study? 14
     - 6. Was Causation Assumed From a Correlational Study? 15
     - 7. Who Paid for the Study? 16
     - 8. Was the Study Published in a Peer-Reviewed Journal? 16
   - On Making Samples Representative of the Population 17
   - Experimental Design and Statistical Analysis as Controls 18
   - The Language of Statistics 19
   - On Conducting Scientific Experiments 20
   - The Dependent Variable and Measurement 21
   - Operational Definitions 21

Measurement Error 21
Measurement Scales: The Difference Between Continuous and Discrete Variables 22
Types of Measurement Scales 23
Nominal Scales 23
Ordinal Scales 23
Interval Scales 24
Ratio Scales 25
Rounding Numbers and Rounding Error 25
Statistical Symbols 26
Summary 27
History Trivia: Achenwall to Nightingale 28
Key Terms, Symbols, and Definitions 29
Chapter 1 Practice Problems 31
Chapter 1 Test Questions 33

2. Descriptive Statistics: Understanding Distributions of Numbers 37
The Purpose of Graphs and Tables: Making Arguments and Decisions 38
How a Good Graph Stopped a Cholera Epidemic 39
How Bad Graphs and Tables Contributed to the Space Shuttle *Challenger* Explosion 40
How a Poor PowerPoint Presentation Contributed to the Space Shuttle *Columbia* Disaster 41
A Summary of the Purpose of Graphs and Tables 41
1. Document the Sources of Statistical Data and Their Characteristics 42
2. Make Appropriate Comparisons 42
3. Demonstrate the Mechanisms of Cause and Effect and Express the Mechanisms Quantitatively 43
4. Recognize the Inherent Multivariate Nature of Analytic Problems 43
5. Inspect and Evaluate Alternative Hypotheses 44
Graphical Cautions 44
Frequency Distributions 46
Shapes of Frequency Distributions 49
Grouping Data Into Intervals 49
Advice on Grouping Data Into Intervals 50
1. Choose Interval Widths That Reduce Your Data to 5 to 10 Intervals 50
2. Choose the Size of Your Interval Widths Based on Understandable Units, for Example, Multiples of 5 or 10 52
3. Make Sure That Your Chosen Intervals Do Not Overlap 52
The Cumulative Frequency Distribution 52

Cumulative Percentages, Percentiles, and Quartiles 53
Stem-and-Leaf Plot 54
Nonnormal Frequency Distributions 55
On the Importance of the Shapes of Distributions 56
Additional Thoughts About Good Graphs
Versus Bad Graphs 57
Low-Density Graphs 57
Chart Junk 57
Changing Scales Midstream (or Mid-Axis) 57
Labeling the Graph Badly 58
The Multicolored Graph 58
PowerPoint Graphs and Presentations 58
History Trivia: De Moivre to Tukey 59
Key Terms and Definitions 60
Chapter 2 Practice Problems 61
Chapter 2 Test Questions 62

3. Statistical Parameters:
Measures of Central Tendency and Variation 67
Measures of Central Tendency 67
The Mean 68
The Median 69
The Mode 71
Choosing Between Measures of Central Tendency 72
Klinkers and Outliers 73
Uncertain or Equivocal Results 74
Measures of Variation 75
The Range 75
The Standard Deviation 76
Correcting for Bias in the Sample Standard Deviation 77
How the Square Root of x^2 Is Almost
Equivalent to Taking the Absolute Value of x 78
The Computational Formula for Standard Deviation 78
The Variance 79
The Sampling Distribution of Means, the Central
Limit Theorem, and the Standard Error of the Mean 80
The Use of the Standard Deviation for Prediction 80
Practical Uses of the Empirical Rule:
As a Definition of an Outlier 82
Practical Uses of the Empirical Rule:
Prediction and IQ Tests 82
Some Further Comments 82
History Trivia: Fisher to Eels 83
Key Terms, Symbols, and Definitions 83
Chapter 3 Practice Problems 84
Chapter 3 Test Questions 85

4. Standard Scores, the z Distribution, and Hypothesis Testing 89
Standard Scores 89
The Classic Standard Score: The z Score and the z Distribution 91
Calculating z Scores 92
More Practice on Converting Raw Data Into z Scores 92
Converting From z Scores to Other Types of Standard Scores 94
The z Distribution 95
Interpreting Negative z Scores 96
Testing the Predictions of the Empirical Rule With the z Distribution 96
Why Is the z Distribution So Important? 97
How We Use the z Distribution to Test Experimental Hypotheses 98
More Practice With the z Distribution and T Scores 99
Example 1: Finding the area in a z distribution that falls above a known score where the known score is above the mean 99
Example 2: Finding the area in a z distribution that falls below a known score where the known score is above the mean 100
Example 3: Finding the area in a z distribution that falls below a known score where the known score is below the mean 102
Example 4: Finding the area in a z distribution that falls above a known score where the known score is below the mean 103
Example 5: Finding the area in a z distribution that falls between two known scores where both known scores are above the mean 105
Example 6: Finding the area in a z distribution that falls between two known scores where one known score is above the mean and one is below the mean 107
Example 7: Finding the area in a z distribution that falls between two known scores where both known scores are below the mean 108
Summarizing Scores Through Percentiles 110
History Trivia: Karl Pearson to Egon Pearson 112
Key Terms and Definitions 113
Chapter 4 Practice Problems 113
Chapter 4 Test Questions 114

5. Inferential Statistics: The Controlled Experiment, Hypothesis Testing, and the z Distribution 119
Hypothesis Testing in the Controlled Experiment 121
Hypothesis Testing: The Big Decision 122
How the Big Decision Is Made: Back to the z Distribution 122

The Parameter of Major Interest in Hypothesis Testing: The Mean 124
Nondirectional and Directional Alternative Hypotheses 125
A Debate: Retain the Null Hypothesis or Fail to Reject the Null Hypothesis 126
The Null Hypothesis as a Nonconservative Beginning 127
The Four Possible Outcomes in Hypothesis Testing 127
1. Correct Decision: Retain H_0, When H_0 Is Actually True 127
2. Type I Error: Reject H_0, When H_0 Is Actually True 128
3. Correct Decision: Reject H_0, When H_0 Is Actually False 128
4. Type II Error: Retain H_0, When H_0 Is Actually False 128
Significance Levels 129
Significant and Nonsignificant Findings 129
Trends, and Does God Really Love the .05 Level of Significance More Than the .06 Level? 130
Directional or Nondirectional Alternative Hypotheses: Advantages and Disadvantages 130
Did Nuclear Fusion Occur? 131
Baloney Detection 132
How Reliable Is the Source of the Claim? 132
Does This Source Often Make Similar Claims? 133
Have the Claims Been Verified by Another Source? 133
How Does the Claim Fit With Known Natural Scientific Laws? 134
Can the Claim Be Disproven or Has Only Supportive Evidence Been Sought? 135
Do the Claimants' Personal Beliefs and Biases Drive Their Conclusions or Vice Versa? 136
Conclusions About Science and Pseudoscience 137
The Most Critical Elements in the Detection of Baloney in Suspicious Studies and Fraudulent Claims 137
Can Statistics Solve Every Problem? 138
Probability 139
The Lady Tasting Tea 139
The Definition of the Probability of an Event 140
The Multiplication Theorem of Probability 140
Combinations Theorem of Probability 141
Permutations Theorem of Probability 142
Gambler's Fallacy 145
Coda 146
History Trivia: Egon Pearson to Karl Pearson 146
Key Terms, Symbols, and Definitions 147

Chapter 5 Practice Problems 148
Chapter 5 Test Questions 149

6. An Introduction to Correlation and Regression 153
Correlation: Use and Abuse 155
A Warning: Correlation Does Not Imply Causation 157
1. Marijuana Use and Heroin Use Are Positively Correlated 158
2. Milk Use Is Positively Correlated to Cancer Rates 158
3. Weekly Church Attendance Is Negatively Correlated With Drug Abuse 158
4. Lead Levels Are Positively Correlated With Antisocial Behavior 158
5. The Risk of Getting Alzheimer's Dementia Is Negatively Correlated With Smoking Cigarettes 159
6. Sexual Activity Is Negatively Correlated With Increases in Education 159
7. An Active Sex Life Is Positively Correlated With Longevity 160
8. Coffee Drinking Is Negatively Correlated With Suicidal Risk 160
9. Excessive Drinking and Smoking Causes Women to Be Abused 160
Another Warning: Chance Is Lumpy 161
Correlation and Prediction 161
The Four Common Types of Correlation 161
The Pearson Product-Moment Correlation Coefficient 162
Testing for the Significance of a Correlation Coefficient 164
Obtaining the Critical Values of the *t* Distribution 165
Step 1: Choose a One-Tailed or Two-Tailed Test of Significance 166
Step 2: Choose the Level of Significance 166
Step 3: Determine the Degrees of Freedom (*df*) 166
Step 4: Determine Whether the *t* From the Formula (Called the Derived *t*) Exceeds the Tabled Critical Values From the *t* Distribution 166
If the Null Hypothesis Is Rejected 167
Representing the Pearson Correlation Graphically: The Scatterplot 167
Fitting the Points With a Straight Line: The Assumption of a Linear Relationship 167
Interpretation of the Slope of the Best-Fitting Line 169
The Assumption of Homoscedasticity 172
The Coefficient of Determination: How Much One Variable Accounts for Variation in Another Variable: The Interpretation of r^2 172

Quirks in the Interpretation of Significant and Nonsignificant Correlation Coefficients 173
Linear Regression 174
Reading the Regression Line 175
R 177
R-Square 178
Adjusted *R*-Square 178
Final Thoughts About Regression Analyses 180
Spearman's Correlation 180
Significance Test for Spearman's *r* 182
Ties in Ranks 182
Point-Biserial Correlation 184
Testing for the Significance of the Point-Biserial Correlation Coefficient 186
Phi (φ) Correlation 187
Testing for the Significance of Phi 188
History Trivia: Galton to Fisher 188
Key Terms, Symbols, and Definitions 190
Chapter 6 Practice Problems 191
Chapter 6 Test Questions 192

7. The *t* Test for Independent Groups 197
The Statistical Analysis of the Controlled Experiment 197
One *t* Test But Two Designs 198
Assumptions of the Independent *t* Test 199
Independent Groups 199
Normality of the Dependent Variable 200
Homogeneity of Variance 200
The Formula for the Independent *t* Test 200
You Must Remember This! An Overview of Hypothesis Testing With the *t* Test 201
What Does the *t* Test Do? Components of the *t* Test Formula 201
What If the Two Variances Are Radically Different From One Another? 202
A Computational Example 202
Steps in the *t* Test Formula 203
Steps in Determining Significance 205
The Power of a Statistical Test 207
Effect Size 208
The Correlation Coefficient of Effect Size 208
Confidence Intervals 209
Estimating the Standard Error 212
History Trivia: Gosset and Guinness Brewery 213
Key Terms and Definitions 214
Chapter 7 Practice Problems 215
Chapter 7 Test Questions 215

8. The *t* Test for Dependent Groups 219
Variations on the Controlled Experiment 219
Design 1 220
Design 2 220
Design 3 221
Assumptions of the Dependent *t* Test 221
Why the Dependent *t* Test May Be More Powerful Than the Independent *t* Test 222
How to Increase the Power of a *t* Test 222
Drawbacks of the Dependent *t* Test Designs 223
One-Tailed or Two-Tailed Tests of Significance 223
Hypothesis Testing and the Dependent *t* Test: Design 1 224
Design 1 (Same Participants or Repeated Measures): A Computational Example 225
Determination of Effect Size 227
Design 2 (Matched Pairs): A Computational Example 228
Determination of Effect Size 231
Design 3 (Same Participants and Balanced Presentation): A Computational Example 231
Determination of Effect Size 234
History Trivia: Fisher to Pearson 234
Key Terms and Definitions 235
Chapter 8 Practice Problems 235
Chapter 8 Test Questions 236

9. Analysis of Variance: One-Factor Completely Randomized Design 241
A Limitation of Multiple *t* Tests and a Solution 241
The Equally Unacceptable Bonferroni Solution 242
The Acceptable Solution: An Analysis of Variance 242
The Null and Alternative Hypotheses in Analysis of Variance 243
The Beauty and Elegance of the *F* Test Statistic 244
The *F* Ratio 245
How Can There Be Two Different Estimates of Within-Groups Variance? 245
ANOVA Designs 247
ANOVA Assumptions 248
Pragmatic Overview 248
What a Significant ANOVA Indicates 249
A Computational Example 249
Degrees of Freedom for the Numerator 252
Degrees of Freedom for the Denominator 252
Determining Effect Size in ANOVA 253
History Trivia: Gosset to Fisher 254
Key Terms and Definitions 256

Chapter 9 Practice Problems 257
Chapter 9 Test Questions 258

10. After a Significant Analysis of Variance: Multiple Comparison Tests 263
Conceptual Overview of Tukey's Test 264
Computation of Tukey's HSD Test 264
What to Do If the Error Degrees of Freedom Are Not Listed in the Table of Tukey's *q* Values 266
Determining What It All Means 266
On the Importance of Nonsignificant Mean Differences 268
Final Results of ANOVA 268
Tukey's With Unequal *N*s 268
Key Terms, Symbols, and Definitions 269
Chapter 10 Practice Problems 269
Chapter 10 Test Questions 269

11. Analysis of Variance: One-Factor Repeated-Measures Design 273
The Repeated-Measures ANOVA 273
Assumptions of the One-Factor Repeated-Measures ANOVA 274
Computational Example 274
Determining Effect Size in ANOVA 278
Key Terms and Definitions 279
Chapter 11 Practice Problems 279
Chapter 11 Test Questions 280

12. Analysis of Variance: Two-Factor Completely Randomized Design 283
Factorial Designs 283
The Most Important Feature of a Factorial Design: The Interaction 284
Fixed and Random Effects and In Situ Designs 284
The Null Hypotheses in a Two-Factor ANOVA 285
Assumptions and Unequal Numbers of Participants 285
Computational Example 286
Computation of the First Main Effect 287
Computation of the Second Main Effect 287
Computation of the Interaction Between the Two Main Effects 288
Interpretation of the Results 290
Key Terms and Definitions 291
Chapter 12 Practice Problems 291
Chapter 12 Test Questions 292

13. Post Hoc Analysis of Factorial ANOVA 297
Main Effect Interpretation: Gender 297

Why a Multiple Comparison Test Is Unnecessary for a Two-Level Main Effect, and When Is a Multiple Comparison Test Necessary? 298
Main Effect: Age Levels 298
Multiple Comparison Test for the Main Effect for Age 299
Warning: Limit Your Main Effect Conclusions When the Interaction Is Significant 301
Multiple Comparison Tests 302
Interpretation of Interaction Effect 302
For the Males 304
For the Females 305
Males Versus Females 305
Final Summary 305
Writing Up the Results Journal Style 305
Language to Avoid 306
Exploring the Possible Outcomes in a Two-Factor ANOVA 306
Determining Effect Size in a Two-Factor ANOVA 308
History Trivia: Fisher and Smoking 309
Key Terms, Symbols, and Definitions 310
Chapter 13 Practice Problems 310
Chapter 13 Test Questions 311

14. Factorial Analysis of Variance: Additional Designs 315
The Split-Plot Design 315
Overview of the Split-Plot ANOVA 316
Computational Example 316
Main Effect: Social Facilitation 321
Main Effect: Trials 321
Interaction: Social Facilitation × Trials 322
Two-Factor ANOVA: Repeated Measures on Both Factors Design 322
Overview of the Repeated-Measures ANOVA 322
Computational Example 323
Key Terms and Definitions 330
Chapter 14 Practice Problems 330
Chapter 14 Test Questions 331

15. Nonparametric Statistics: The Chi-Square Test 335
Overview of the Purpose of Chi-Square 336
Overview of Chi-Square Designs 337
Chi-Square Test: Two-Cell Design (Equal Probabilities Type) 337
Computation of the Two-Cell Design 338
The Chi-Square Distribution 339
Assumptions of the Chi-Square Test 340
Chi-Square Test: Two-Cell Design (Different Probabilities Type) 340

Computation of the Two-Cell Design 341
Interpreting a Significant Chi-Square Test for a Newspaper 342
Chi-Square Test: Three-Cell Experiment
(Equal Probabilities Type) 343
Computation of the Three-Cell Design 343
Chi-Square Test: Two-by-Two Design 344
Computation of the
Chi-Square Test: Two-by-Two Design 345
What to Do After a Chi-Square Test Is Significant 348
When Cell Frequencies Are Less Than 5 Revisited 349
History Trivia: Pearson and *Biometrika* 350
Key Terms, Symbols, and Definitions 350
Chapter 15 Practice Problems 350
Chapter 15 Test Questions 351

16. Other Statistical Parameters and Tests 355
Health Science Statistics 356
Test Characteristics 356
Risk Assessment 359
Parameters of Mortality and Morbidity 361
Analysis of Covariance 363
Multivariate Analysis of Variance 363
Multivariate Analysis of Covariance 364
Factor Analysis 365
Multiple Regression 366
Canonical Correlation 366
Linear Discriminant Function Analysis 367
Cluster Analysis 368
A Summary of Multivariate Statistics 368
Coda 369
Key Terms and Definitions 369
Chapter 16 Practice Problems 371
Chapter 16 Test Questions 371

Appendix A: *z* Distribution 375

Appendix B: *t* Distribution 381

Appendix C: Spearman's Correlation 383

Appendix D: The Chi-Square Distribution 385

Appendix E: *F* Distribution 387

Appendix F: Tukey's Table 389

References 391

Index 393

About the Author 397

Acknowledgments

I want to thank my undergraduate students over the years for their curiosity and questions that have helped shape this book. Over the years, many graduate students have also helped me revise and correct its various editions, and I wish to thank them as well.

1 A Gentle Introduction

Chapter 1 Goals

- Understand the general purposes of statistics
- Understand the various roles of a statistician
- Learn the differences and similarities between descriptive and inferential statistics
- Understand how the discipline of statistics allows the creation of principled arguments
- Learn eight essential questions of any survey or study
- Understand the differences between experimental control and statistical control
- Develop an appreciation for good and bad experimental designs
- Learn four common types of measurement scales
- Learn to recognize and use common statistical symbols

The words *DON'T PANIC* appear on the cover of the book *A Hitchhiker's Guide to the Galaxy*. Perhaps it may be appropriate for some of you to invoke these words as you embark on a course in statistics. However, be assured that this course and materials are prepared with a different philosophy in mind, which is that statistics should help you understand the world around you and help you make better informed decisions. Unfortunately, statistics may be legendary for driving sane people mad or, less dramatically, causing undue anxiety in the hearts of countless college students. However, statistics is simply the science of the organization and conceptual understanding of groups of numbers. By converting our world to numbers, statistics helps us to understand our world in a quick and efficient way. It also helps us to make conceptual sense so that we might be able to communicate this information about our world to others. More important, just as in the practice of logic, statistics allows us to make arguments based on established and rational principles, and thus, it can promote wise and informed decision making. As in any other scientific discipline, statistics has its own language, and it will be important for you to learn many new terms and symbols and to see how some common words that you already know may have very different meanings in a statistical context.

How Much Math Do I Need to Do Statistics?

Can you add, subtract, multiply, and divide with the help of a calculator? If you answered yes (even if you are slow at the calculations), then you can handle statistics. If you answered no, then you should brush up on your basic math and simple algebra skills. Statistics is not really about math. In fact, some mathematicians secretly revel in the fact that their science may have little relevance to the real world. The science of statistics, on the other hand, is actually based on decision-making processes, and so statistics must make conceptual sense. Most of the statistical procedures presented in this course can also be performed through specially designed computer software programs such as Statistical Packages for the Social Sciences (SPSS; www.spss.com) and GB-Stat (www.gbstat.com). Programs such as these make it easy to perform statistical calculations, and they do so with blazing speed. However, these programs provide very little or no explanation for the subsequent output, and that is at least one of the purposes of the present course.

The General Purpose of Statistics: Understanding the World

A group of numbers in mathematics is called a set, but in statistics, this group is more frequently called *data* (a single number is called a *datum*). Typically, the numbers themselves represent scores on some test, or they might represent the number of people who show up at some event. It is the purpose of statistics to take all of these numbers or data and to present them in a more efficient way. An even more important use of statistics, contrary to some people's beliefs, is to present these data in a more comprehensible way. People who obfuscate (to bewilder or confuse) with statistics are not really representative of the typical and ethical statistician. This course will also provide you with some protection against the attempts of some people and their statisticians who try to convince you that a product works when it really does not or the superiority of a product over another when it really is not superior.

Another Purpose of Statistics: Making an Argument or a Decision

People often use statistics to support their opinion. People concerned with reducing the incidence of lung cancer use statistics to argue that cigarettes increase the likelihood of lung cancer. Car and truck makers use statistics to show the reliability of their cars or trucks. It has been said that being a statistician is like being an honest lawyer. One can use statistics to sell a product, advance an argument, support an opinion, elect a candidate, improve societal conditions, and so on. Because clearly presented statistical arguments can be so powerful, say, compared to someone else's simple unsupported opinion, the science of statistics can become a very important component of societal change.

What Is a Statistician?

Although many of you would probably rather sit on a tack than become statisticians, imagine combining the best aspects of the careers of a curious detective, an honest attorney, and a good storyteller. Well, that is often the job of a statistician (see Abelson, 1995). Let us examine the critical elements of these roles in detail.

One Role: The Curious Detective

The curious detective knows a crime has been committed and examines clues or evidence at the scene of the crime. Based on this evidence, the detective develops a suspicion about a suspect who may have committed the crime. In a parallel way, a statistician develops suspicions about suspects or causative agents, like which product is best, what causes Alzheimer's disease, or how music may or may not affect purchasing behavior. In the case of Alzheimer's disease, the patients' lives are quantified into numbers (data). The data become the statistician's clues or evidence, and the experimental design (how the data were collected) is the crime scene. As you already know, evidence without a crime scene is virtually useless, and equally useless are data without knowing how they were collected. Health professionals who can analyze these data statistically can subsequently make decisions about causative agents (for example, does aluminum in food products cause Alzheimer's disease?) and choose appropriate interventions and treatments.

A good detective is also a skeptic. When other detectives initially share their suspicions about a suspect, good detectives typically reserve judgment until they have reviewed the evidence and observed the scene of the crime. Statisticians are similar in that they are not swayed by popular opinion. They should not be swayed by potential profits or losses. Statisticians examine the data and experimental design and develop their own hypotheses (educated guesses) about the effectiveness or lack of effectiveness of some procedure or product. Statisticians also have the full capability of developing their own research designs (based on established procedures) and testing their own hypotheses. Ethical statisticians are clearly a majority. It is unusual to find cases of clear fraud in the world of statistics, although it sometimes occurs. More often, statisticians are like regular people: They can sometimes be unconsciously swayed by fame, money, loyalties, and prior beliefs. Therefore, it is important to learn thoroughly the fundamental statistical principles in this course. As noted earlier, even if you do not become a producer of statistics, you and your family, friends, and relatives will always be consumers of statistics (like when you purchase any product or prescription drug), and thus it is important to understand how statistics are created and how they may be manipulated either intentionally, unintentionally, or fraudulently. There is the cliché that if something seems too good to be true, it may not be true. This cliché also holds in the world of statistics. Interestingly, however, it may not be the resulting statistics that make something appear too good to be true, but it may be how the statistics were gathered, also known as the experimental design. This course will present both the principles of statistics and their accompanying experimental designs. And as you will soon learn, the power of the experimental design is greater than the power of the statistical analysis.

Another Role: The Honest Attorney

An honest attorney takes the facts of a case and creates a legal argument before a judge and jury. The attorney becomes an advocate for a particular position or a most likely scenario. Frequently, the facts may not form a coherent whole, or the facts may have alternative explanations. Statisticians are similar in that they examine data and try to come up with a reasonable or likely explanation for why the data occurred. Ideally, statisticians are not passive people but active theoreticians (i.e., scientists). Scientists are curious. They have an idea about the nature of life or reality. They wonder about relationships among variables, for example, what causes a particular disease, why people buy this brand as opposed to that brand, or whether a product is helpful or harmful. These hypotheses are tested through experiments or surveys. The results of an experiment or survey are quantified (turned into numbers). Statisticians then become attorneys when they honestly determine whether they feel the data support their original suspicions or hypotheses. If the data do support the original **hypothesis,** then statisticians argue their case (study) on behalf of their hypothesis before a judge (journal editor, administrator, etc.). In situations where statisticians wish to publish the results of their findings, they select an appropriate journal and send the paper and a letter of justification to the editor of the journal. The journal editor then sends the paper to experts in that field of study (also known as peer review).

The reviewers suggest changes or modifications if necessary. They are also often asked to decide whether the study is worthy of publication. If the researcher has convinced the editor and peer reviewers that this hypothesis is the most likely explanation for the data, then the study may be published. The time between submission, acceptance, and the actual publication of the article will take between 6 months and 2 years, and this time period is known as publication lag.

There are, of course, unscrupulous or naive attorneys, and sadly, too, as noted earlier, there are unscrupulous and naive statisticians. These types either consciously or unconsciously force the facts or data to fit their hypothesis. In the worst cases, they may ignore other facts not flattering to their case or even make up their data. In science, although there are a few outright cases of fraud, it is more often that we see data forced into a particular interpretation because of a strong prior belief. In these cases, the role of a skeptical detective comes into play. We may ask ourselves, are these data too good to be true? Are there alternate explanations? Fortunately in science, and this is where we so strongly differ from a courtroom, our hypothesis will not be decided by one simple study. It is said that we do not "prove" the truth or falsity of any hypothesis. It takes a series of studies, called **replication,** to show the usefulness of a hypothesis. A series of studies that fails to support a hypothesis will have the effect of making the hypothesis fall into disuse. There was an old psychological theory that body type (fat, skinny, or muscular) was associated with specific personality traits (happy, anxious, or assertive), but a vast series of studies found very little support for the original hypothesis. The hypothesis was not disproven in any absolute sense, but it fell into disuse among scientists and in their scientific journals. Ironically, this did not "kill" the scientifically discredited body-type theory for it still lives in popular but unscientific monthly magazines.

A Final Role: A Good Storyteller

Storytelling is an art. When we love or hate a book or movie, we are frequently responding to how well the story was told. I once read a story about the making of steel. Now, steel making is not high (or even medium) on my list of interesting topics, but the writer unwound such an interesting and dramatic story that my attention was completely riveted. By parallel, it is not enough for a statistician to be a curious detective and an honest attorney. One must also be a good storyteller. A statistician's hypothesis may be the real one, but statisticians must state their case clearly and in convincing style. Thus, a successful statistician must be able to articulate what was found in an experiment, why this finding is important and to whom, and what the experiment may mean for the future of the human race (OK, I may be exaggerating on this latter point). Articulation (good storytelling) may be one of the most critical aspects of being a good statistician and scientist.

In fact, good storytelling may be one of the most important roles in the history of humankind (see Sugiyama, 2001).

There are many examples of good storytelling throughout science. The origin of the universe makes a very fascinating story. Currently, there are at least two somewhat rival theories (theories are bigger and grander than hypotheses, but essentially theories, at their hearts, are no more than educated guesses): In one story, a supreme being created the universe from nothingness in 6 days, and in the other story, the Big Bang, the universe started as a small egg that exploded to create the universe. Notice that each theory has a fascinating story associated with it. One problem with both stories is that we are trying to explain how something came from nothing, which is a logical contradiction. It has been suggested that the object of a myth is to provide some logical explanation for overcoming a contradiction; however, this may be an impossible task when the contradiction is real. Still, both of these theories remain very popular among laypeople and scientists alike, in part because they make interesting, provocative, and fascinating stories.

Science is replete with interesting stories. Even the reproductive behavior of planaria or the making of steel can be told in an interesting and convincing manner (i.e., any topic has the potential to make a good story).

Liberal and Conservative Statisticians

As we proceed with this course in statistics, you may come to realize that there is considerable leeway in the way data are investigated (also called the experimental methodology or experimental design) and in the way data are tabulated and reported (called statistical analyses). There appear to be two camps, the liberals and the conservatives. Just as in politics, neither position is entirely correct as both philosophies have their advantages and disadvantages.

Scientists as a whole are generally conservative, and because statisticians are scientists, they too are conservative as a general rule. Conservative statisticians stick with the tried and true. They prefer conventional rules and regulations. They design experiments in the same historical way, and they interpret and report their interpretations in the same conventional fashion. This position is not as stodgy as it may first appear. The conservative position has the advantage of being more readily accepted by the scientific community (including journal editors and peer reviewers). If one sticks to the rules and accepted statistical conventions, and one argues successfully according to these same rules and conventions, then there is typically a much greater likelihood that the findings will be accepted and published, and receive attention from the scientific community. Conservative statisticians are very careful in their interpretation of their data. They guard against chance playing a role in their findings by rejecting any findings or treatment effects that are small in nature. Their findings must be very clear or their treatment effect (like from a new drug) must be very large for them to conclude that their data are real, and consequently, there is a very low probability that pure chance could account for their findings.

The disadvantages of the conservative statistical position are that new investigative research methods, creative statistical analyses, and radical conclusions are avoided. For example, in the real world, sometimes new drug treatments are somewhat effective but not on everyone, or the new drug treatment may work on nearly everyone, but the improvement is modest or marginal. By always guarding so strongly against chance, conservative statisticians frequently end up in the position of "throwing out the baby with the bath water." They may end up concluding that their findings are simply due to chance when in reality something is actually happening in the data that is not due to chance.

Liberal statisticians are in a freer position. They may apply exciting new methods to investigate a hypothesis and apply new methods of statistically analyzing their data. Liberal statisticians are not afraid to flout the accepted scientific statistical conventions. The drawback to this position is that scientists, as a whole, are like people, in general. Many of us initially tend to fear new ways of doing things. Thus, liberal statisticians may have difficulty getting their results published in standard scientific journals. They will often be criticized on the sole grounds of investigating something a different way and not for their actual results or conclusions. In addition, there are other real dangers from being a statistical liberal. Inherent in this position is that they are more willing than conservative statisticians to view small improvements as real treatment effects. In this way, liberal statisticians may be more likely to discover a new and effective treatment. However, the danger is that they are more likely to call a chance finding a real finding. If scientists are too hasty or too readily jump to conclusions, the consequences of their actions can be deadly or even worse. What can be worse than deadly? Well, consider a tranquilizer called thalidomide in the 1960s. Although there were no consequences for men, more than 10,000 babies were born to women who took thalidomide during pregnancy, and the babies were born alive but without hands or feet. Thus, scientific liberalism has its advantages (new, innovative, creative) and its disadvantages (it may be perceived as scary, flashy, or bizarre, or have results that are deadly or even worse). Neither position is a completely comfortable one for statisticians. This book will teach you the conservative rules and conventions. It will also encourage you to think of alternative ways of exploring your data. But remember, as in life, no position *is* a position, and any position involves consequences.

Descriptive and Inferential Statistics

The most crucial aspect of applying statistics consists of analyzing the data in such a way as to obtain a more efficient and comprehensive summary of the overall results. To achieve these goals, statistics is divided into two areas, descriptive and inferential statistics. At the outset, do not worry about the distinction between them too much: The areas they cover overlap, and descriptive statistics may be viewed as building blocks for the more complicated inferential statistics.

Descriptive statistics is historically the older of these two areas. It involves measuring data using graphs, tables, and basic descriptions of numbers such as averages or means. These universally accepted descriptions of numbers are called **parameters**, and the most popular and important of the parameters are the mean and standard deviation (which we will discuss in detail in Chapter 3).

Inferential statistics is a relatively newer area, which involves making guesses (inferences) about a large group of data (called the **population**) from a smaller group of data (called the **sample**). The population is defined as the entire collection or set of objects, people, or events that we are interested in studying. Interestingly, statisticians rarely deal with the population because typically, the number of objects that meet the criterion for being in the population is too large (e.g., taxpayers in the United States), it is financially unfeasible to measure such a group, and/or we wish to apply the results to people who are not yet in the population but will be in the future. Thus, statisticians use a sample drawn from the population. For our inferences about our sample to be representative of the population, statisticians have two suggestions. First, the sample should be randomly drawn from the population. The concept of a random sample means that every datum or person in the population has an equal chance of being chosen for the sample. Second, the sample should be large relative to the population. We will see later that the latter suggestion for sample size will change depending on whether we are conducting survey research or controlled experiments.

Statistical studies are often reported as being based on **stratified samples.** A stratified sample means that objects are included in the sample in proportion to their frequency in the subgroups that make up the population. For example, if it was reported that a study was conducted on a ethnically stratified sample of the U.S. population, then the ratios of ethnic groups in the sample would resemble the ratios in the population. Currently, the population of the United States is approximately 291 million people. Approximately 197 million identify themselves as non-Hispanic White or Caucasian, 40 million as Hispanic, 36 million as Black or African American, 12 million as Asian, 2 million as American Indian, and 4 million as being of two or more races. Thus, a stratified sample might be 68% White or Caucasian, 14% Hispanic, 12% Black or African American, 4% Asian, 0.6% American Indian, and 1.4% biracial or multiracial. Again, large samples, random samples, and stratified samples help us to make inferences that are more than likely to be true about the population that we are studying.

Experiments Are Designed to Test Theories and Hypotheses

A theory can be considered a group of general propositions that attempts to explain some phenomenon. Typically, theories are also grand; that is, they tend to account for something major or important, such as theories of supply and demand, how children acquire language, or how the universe began. A good theory should provoke people into thinking. A good theory should also be able to generate testable propositions. These propositions are actually guesses about the ways things should be if the theory is correct. Hypotheses are specifically stated propositions created from and consistent with the theory, which are then tested through experimental research. Theories whose specific hypotheses are frequently supported by research tend to be regarded as useful. Theories whose hypotheses fail to receive support tend to be ignored. It is important to remember that theories and hypotheses are never really proven or disproved in any absolute sense. They are either supported or fail to receive support from research. In the real world, it is also frequently the case that some theories and hypotheses receive mixed results. Sometimes, research findings support the propositions, and other experiments fail to support them. Scientists and statisticians tend to have critical attitudes about theories and hypotheses because, as noted earlier, the repercussions in science for a bad theory can be deadly or even worse. Thus, statisticians like well-designed and well-thought-out research, such that the findings and conclusions are clear, compelling, and unambiguous.

Oddball Theories

Science is not simply a collection of rigid rules. Scientific knowledge does not always advance smoothly but typically moves along in sputters, stops and starts, dead ends, and controversy. Carl Sagan, the late astronomer, said that science requires the mating of two contradictions: a willingness to think about new, unique, strange, or even bizarre explanations coupled with rigorous skepticism and hard evidence. If scientists only published exactly what they thought they would find, new discoveries would be exceedingly rare. Thus, scientists must be willing to take risks and dare to be wrong. The physicist Wolfgang Pauli wrote that being "not even wrong" is even worse than being wrong because that would imply that one's theory is not even worth contradiction or dispute. The Nobel Prize for medicine was awarded to an American neurologist, Stanley Prusiner, who proposed that infectious particles called prions do not contain any genes or genetic material yet reproduce and cause "mad cow disease" (a dementia-like disease caused by eating infected beef). For years, Prusiner's ideas were considered revolutionary or even heretical; however, most research now supports his more than two decades of experimental work. His "oddball" theory was provocative, generated many testable hypotheses, and was supported by his own rigorous testing and skepticism, and scientific knowledge has benefited greatly from his theory. However, remember there is no shortage of oddballish people creating oddball theories. A good theory is not only interesting and provocative but must also be supported by testable hypotheses and solid evidence.

Bad Science and Myths

In my Sunday newspaper, there is a "Fitness Column," and a recent question concerned adhesive strips that go across the bridge of the nose. These are the same strips that seem to be so popular among professional athletes. "Do they work?" asks the letter writer. The magazine's expert says they reduce airflow resistance "as much as 30%." However, because the "expert" says it is true, does that make it true? Of course not! Nevertheless, we are bombarded daily by a plethora of statistics, such as one in four college-age women have been raped, marriage reduces drug abuse, classical music boosts IQ, people use only 10% of their brains, and left-handers die younger. We even pass down many myths as truths—such as that Eskimos have 30 words for snow or that one should change the oil in one's car every 3,000 miles. The Internet has even sped up the process of passing this information. Some of these common myths are also called "urban legends." Of course, anyone can post any belief or idea on the Internet, and the mere posting or retelling of a "story" does not make it true. For example, Eskimos apparently have just as many descriptions of snow as other people. Because most of us have been raised to tell the truth, we typically do not challenge the facts or statistics fed to us daily. However, let us be realistic. Many studies are poorly designed, statistics can be misused or manipulated, and many people in our society are not interested in the truth, but they are interested in money and power (that brings more money). Recently, the magazine *Consumer Reports* empirically investigated the myth of the 3,000-mile oil change and consistently found no benefits compared to much longer intervals. Interestingly also, no reputable scientist ever published the idea that people use only 10% of their brains. If it were true, I would very much hope that my 10% are the ones involved in heart rate, respiration, and blood pressure. A "motivational speaker" probably created this myth, but there's not a shed of scientific evidence for its truthfulness. It may have persisted because the consequences of the belief (that humans are capable of doing more than they do) are not harmful. But now, let us return to the "nose strips."

Ultimately, why did someone create nose strips? Probably to make money. Helping people breathe right was perhaps, if we are very lucky, secondary. In fact, it is easy to imagine that many products in the market are not even remotely designed to do what they claim. Their sole purpose is to make money. However, let us give the nose strip creator the benefit of the doubt. Why should we use them? Well, the nose strip propaganda says they have scientific evidence that they work, and they present the following "facts": Breathing takes 10% of our total energy, and their nose strips reduce air resistance by up to 30%. With regards to this first fact, I think it would be fair to know how total energy was measured. Isn't it rather difficult to come up with a single measure of energy expended? And even if we could come up with an acceptable measure, did they measure this 30% expenditure when a person was exercising or at rest? If he or she was exercising (because I assume that's when we'd want to use nose strips), how did they measure total energy expended when the person was running around? Did they remotely or telemetrically send the energy expenditure information? I am not sure that this is even possible. It's easier for me to imagine the person at rest when he or she was measured for energy expenditure. Therefore, at-rest measures may not be appropriate to someone who is actually running about. Furthermore, who was the person tested? What was that person's age? Will we be able to generalize to the average person? Is it fair to generalize from just one test subject to all people? Was the participant paid? Some people will say anything for money. An even worse scenario is that some people will even deceive themselves about how well something works if they are paid enough money (curiously, Festinger, a social psychologist, found that some people will deceive themselves for too little money).

The same criticism appears to hold for the second "fact." How did they ever measure this 30% air resistance reduction? Was that 30% over a period of time? Was it a mean score for a large number of participants? Was it a single score? Who did the testing? Was it an independent testing group, or did the nose strip people conduct the research? If the nose strip people did the testing, I think they may have been consciously or unconsciously biased in favor of the product.

A second reason the nose strip people say that we should buy their product is that Jerry Rice, the great football player, wears them. Now, we are tempted to use them because of an expert opinion and the prestige associated with that expert. But does that mean they really work? Of course not, but we rarely challenge the opinion of an "expert." Many other factors are operating here, such as the powerful process of people identifying with the rich and famous. Some people think that if Jerry uses them, then they will be as great as Jerry if they use them. We can see as we closely examine our motivations that many are not based on rational or logical reasoning. They are based, however, on unconscious and powerful forces that cause us to believe, trust, imitate, and follow others, particularly people whom we perceive as having more power or status than we have. This willingness to believe on sheer faith, that a product or technique works, has been called the **placebo effect.**

In summary, when we closely examine many of our attitudes, beliefs, facts, and "studies," we find that they sorely lack any scientific validity. And if we blindly accept facts, figures, cures, and snake oil, we are putting ourselves in a very dangerous position. We may make useless changes in our lives or even dangerous ones. We may also unnecessarily subject ourselves to needless stress and worry.

Eight Essential Questions of Any Survey or Study

1. Who Was Surveyed or Studied?

Remember, most of the time, we hope that we can use the product or would benefit from a new treatment or technique. A majority of studies still use only men. Would we be able to generalize to women if only men are studied? Would we be able to generalize to children if only adults are studied? Would we be able to generalize to people if only animals are studied? Will the product be safe for people if we do not use animals in our study? If our moral values prevent us from using animals in research, what other methods are available to ensure the safety of a product?

For a sample to be representative of the larger group (the population), it should be randomly chosen. Did the study in question use random sampling? Random sampling implies that everyone in the population had an equal chance to be in the sample. If we telephone a random sample of voters in our county, is that a random sample? No, because not everyone has a telephone, and not everyone has an equal chance to answer the phone if we call at 11 a.m.

2. Why Did the People Participate in the Study?

Did the people volunteer, or were they paid? If they were paid, and paid a lot, they might skew the results in favor of the experimenter's hypothesis. Even if they were not paid, participants have been known to unconsciously bias an experiment in the experimenter's favor. Why? Because sometimes we are just fond of some people—we'd like to help them out, or we'd be embarrassed if they failed in front of us. We may not wish to share their humiliation. Or it is easy to imagine some participants who would like to see the experimenter fail because they enjoy other people's suffering and humiliation. Ultimately, it is in the experiment's best interest if the experimenter's hypothesis is kept a secret from the participants. Furthermore, the experimenter should not know who received the experimental treatment, so that the experimenter isn't biased when he or she is assessing the results of the experimental treatment or product being tested. If the experimenter doesn't know who was in what group (but someone important does) and the participants do not know what is being tested, the study is said to be a **double-blind experiment.**

3. Was There a Control Group and Did the Control Group Receive a Placebo?

Even if the participants were not paid and were blind to the experimenter's hypothesis, we sometimes wish to change so badly that we change even if the product or treatment in question does not really work. This is known as the placebo effect. Placebo effects are well known throughout the scientific world, and they are real. In other words, some people really do get better when they are given fake substances or sugar pills. Some people really do get better when we shake rattles in front of their faces or wave our fingers back and forth or light incense and chant. Our current scientific method calls for a control group if we are testing some drug, product, or new type of psychotherapy, and furthermore, the new drug or technique should be superior to the improvement we frequently see in the control group that receives only the placebo. Thus, the nose strips should have been tested against a group of participants who received some kind of placebo (like a similar piece of tape on their noses), and both groups should have received the same instructions. Interestingly, the word *placebo* means "I shall believe" in Latin. It is the first word in evening prayers said for the dead. In about the 12th century, these prayers became known as "placebos." People began to hire professional mourners who would recite the "placebos," and relatively quickly the word came to have an unflattering and pejorative meaning, which has persisted to modern times (see Brown, 1998, for a provocative discussion of the placebo effect).

4. How Many People Participated in the Study?

There is no firm set of rules of how many people should be studied. However, the number of people participating can have a profound effect on our conclusions. In surveys, for example, the more the merrier. The larger the number of people sampled, the more likely that the sample will be representative of the population. Thus, to some extent, the size of our sample will be determined by the size of the population to which we hope to generalize. For example, if we are interested in a study of taxpaying U.S. citizens, then a sample size of 500 might still be considered a small sample if there are 100 million taxpayers. On the other hand, 500 might be considered an unnecessarily large sample if we are attempting to study the population of Colorado Buddhists.

In studies where we employ the scientific method (also known as a controlled experiment) using an experimental group and a control group, large sample sizes may actually be harmful to the truth. This occurs because as we increase the number of participants in each group, there is a peculiar statistical artifact that increases the chances we say the product works when it really does not. Increasing the sample sizes unnecessarily is called an **abuse of power. Power** in this context is defined as the ability to detect a real difference in the two groups due to the treatment. Rarely if ever will the experimental group be equal to the control group even if the treatment does not work! This occurs because of simple chance differences between the groups. However, if an experimenter increases the sample sizes (to 100 or more in each group), then chance differences will be seen as real differences in most inferential statistical tests. The way to avoid an abuse of power is to perform a statistical test known as **power analysis,** which can help the experimenter choose an appropriate sample size. Power analyses are beyond the scope of this course, but consult an advanced text if you are interested.

There is also the opposite situation where a real difference between two groups might not be detected because of insufficient power or too few participants are used in each group. In this situation, the treatment may actually work, but we might conclude that it does not because we have not used enough people in each group. Again, a power analysis should help in this situation. Most experimenters use a simple rule of thumb: There should be at least 10 participants in each group, or the overall experiment should include at least 30 participants.

5. How Were the Questions Worded to the Participants in the Study?

Remember, I mentioned earlier the "fact" that one in four college women has reported being raped. How did the experimenter actually define the word *rape?* In this case, the experimenter asked 6,159 college students, "Have you ever had sexual intercourse when you didn't want to because a man gave you alcohol or drugs?" We might argue that this definition of "rape" is too broad. It is possible that because of a woman's religious beliefs or family values, she may be ambivalent about sex (some part of her didn't want to under ANY circumstances outside of marriage). The experimenter in this situation defended her broad definition of rape because she said that rape has so many different meanings. She took the provocative position that if a student's report met the experimenter's definition of rape, then the student was raped, and "whether she realizes it or not is irrelevant."

How about the "fact" stated on milk cartons that more than 1 million children have been abducted? Here, the debatable issue is the definition of *abducted.* Some missing children organizations defined *abducted* as any child not living with his or her legal custodian parent. Thus, it might be argued that some of these abducted children are not really missing if they are living with one of their biological parents, however, just not the parent who has legal custody. While this may still be a serious matter, it might not reach the same level of importance as a child who has been taken by someone unknown to the family, when the whereabouts of the child are a complete mystery. If we consult police and FBI records, we would find that there are fewer than 1,000 children who meet the latter criterion. The problem with the broad definition of abduction is that if too many children meet the definition of abducted, then we will not have the time and resources to hunt for the truly missing children.

The sexual abuse of children is certainly an important issue and highly worthy of our attention. Recently, it was stated that 25% of all males have reported the tendency to sexually abuse children. This statistic might be highly alarming except that the men surveyed were asked, "If you knew you would never get caught or you knew that no one would ever find out, have you ever had the fantasy of sexually fondling a girl under the age of 14?" You might still consider the 25% of men who said yes to this question to be highly despicable, perverts, or sick, yet it could be argued that the question had the word *fantasy* in it. By definition, a fantasy is something not occurring in reality. It seems unfair to ask someone about a fantasy and then assume that it reflects a true tendency in reality. Furthermore, perhaps the men were recalling an old fantasy, one they had when they were under the age of 14 themselves. What would have happened if we had asked the question like this: "Would you ever consider having sexual intercourse with a child if you knew you would receive the death penalty, there was a very high probability that you could get caught, your mother would be the first person to know about it, and she would then commit suicide because you broke her heart?"

A related issue in measurement is how clearly the experimenter has defined the problem he or she is trying to study or solve. Have you heard the "fact" that one in three children goes to bed hungry? How clearly did the experimenter define what he or she meant by hungry? Because I go to bed late every night, I nearly always go to bed hungry even after a big meal earlier in the evening. How did the experimenter define the word *hungry?* Did he or she ask all the children in the world whether they were hungry before they went to bed? I doubt it. And what if the experimenter defined *hungry* by asking a child, "Would you like anything else to eat, like a cookie or candy, before you go to sleep?" It is easy to imagine that one definition of *hungry* is whether a child asks for more to eat, but then, many overweight children would also ask for more to eat and meet the definition of going to bed hungry. Perhaps you can perceive the monumental task involved in clearly defining any variable, even words as seemingly simple as *hungry*. Children who are really starving is a terrible reality. However, blatant lying, conscious exaggeration, and abusing statistical methods are also abhorrent, even in the cause of ending world hunger.

6. Was Causation Assumed From a Correlational Study?

Probably the single most pervasive abuse of statistics occurs when people infer causation from a correlational study. A correlational study is one in which two factors or variables are measured together in a setting. In correlational studies, there is no random assignment of the participants to two groups, there is no treatment applied, and there is no control group. In the correlational study, if the two variables do appear to be related to each other, many people unfortunately assume that one of these variables *causes* the other. While it is possible that there is a causative relationship, a correlational study should never be interpreted as evidence for causation. For example, a correlational study of marriage and drug use found that drug use declined after marriage. The headlines in the newspapers read, "Responsibilities of marriage and family often curb drug use." One of the problems in this interpretation is that the study was a correlative one, and we must always guard against the inference of causation. It would be equally wrong, but no less absurd, to assume that drug use curbs marriage. A second problem with correlational studies is that many other factors or variables are also operating and may contribute as a cause. If a scientist does not measure these other variables, then the scientist cannot assume that it is only a single variable such as marriage curbing drug use. In fact, an overwhelming majority of behaviors have multiple and complex causes, and to assume that a single variable such as alcohol use causes crime is not only statistically wrong but wrong in theory as well.

7. Who Paid for the Study?

In an ideal world, the investigators in a study should not have any financial ties or interest to the outcome of a study. However, a few years ago, a study reported that balding men were more likely to have heart attacks than nonbalding men. Who financed the research? A drug company that manufactures a popular hair-growing product! While this financial interest in the study does not automatically invalidate the study, it does tend to cast suspicion on the outcome of the study or the motives of the investigators.

More recently, a controversy began when a study linked a high blood pressure medicine to an increased risk of heart attacks. It was noted that 70 scientists, doctors, or researchers came to the defense of this particular blood pressure medicine in journal articles, reports, or medical publication commentaries. When the 70 were questioned about their defense, 67 of the 70 admitted they had financial relationships with the manufacturer of the drug, which included travel grants, public speaking fees, research and educational grants, and consulting fees and contracts.

It has been suggested that scientists have been naive about the extent to which financial relationships may affect research. Again, in an ideal world, there should be no financial complications between the investigators and the outcomes of their studies. However, short of the ideal, perhaps scientists should report any appearances of a conflict of interest at the outset of their study, either to the journal editor to which the study is submitted or in a footnote in the published version of the study. In this way, the readers of the articles can determine for themselves to what extent they feel the studies may be biased. For example, recently, a prominent medical journal published a meta-analysis (a study summarizing many other studies) of hair-growing medications, and the author concluded that hair-growing medications do indeed work. Only in a subsequent issue of the journal was it revealed that the author had been supported for years by grants from companies that manufacture hair-growing medications.

8. Was the Study Published in a Peer-Reviewed Journal?

Nearly all scientific journals use a peer-review procedure where any article submitted to an editor is sent out to other scientists for their opinion on whether the study should be accepted for publication. Some journals reject as many as 90% of all the articles submitted, while some of the "easier" journals may reject only 50%. Even if an article is accepted for publication (a process that could take years), it is rarely accepted without any changes.

Typically, a journal editor will summarize the criticisms of the peer reviewers and forward them to the authors. The peer reviewers may request additional information about the literature surrounding the issue, more information about the participants, more information about the tests employed or the procedures, additional statistical analyses, or modifications of the conclusions or implications of the study. The first study that ever produced evidence for cold fusion, a process whereby nearly limitless energy could be obtained from a special kind of water, was not published in a peer-reviewed journal. It was presented by teleconference! Thus, the specific procedures were not revealed in detail, and the study was not subject to any acceptable form of peer review. Is it any wonder that the issue of whether those scientists actually created cold fusion is in question?

What about books? Do they undergo peer review? Typically not. It does vary from publisher to publisher, but there are some publishing companies where anything you write can be published in a book (for a fee). Many scientists will present the results of their research at national or international conventions. Are convention presentations peer-reviewed? Most convention presentations do have some form of peer review, but the standards for convention presentations are much more lax than for journal articles, and the acceptance rates can vary from 50% to 100%. Thus, published journal articles meet the highest standards of scientific scrutiny, convention presentations are a distant second, and some books will meet absolutely no scientific standards. We must consider paid advertisements in newspapers and magazines as well as infomercials on television as the least likely venues for truth or justice.

What about information on the Internet? The Internet is a dicey proposition. There are very legitimate sources of information on the Internet, and reputable scientific journals are increasingly making their studies available on the Internet. There are also quite reputable journals that exclusively publish on the Internet. However, anyone can post anything he or she likes on the Internet. Thus, one must be able to discern between legitimate Internet sources and the illegitimate or questionable ones. I have personally noticed many scientific and pseudo-scientific articles posted on the Internet by their authors. The veracity of such studies is completely unknown. These studies may be of value, or they may have been rejected by legitimate journals and may be completely fraudulent. The moral of the Internet story is, let the surfer beware!

On Making Samples Representative of the Population

To ensure that the hypotheses we make from the sample about the population are good ones, remember the two requirements: First, the sample should be randomly drawn from the population, and second, the sample should be relatively large. However, these two guidelines do not guarantee that a sample will be representative of the population. There will always exist the possibility that a sample, although large and randomly drawn, may still lack some important characteristics that are present in the population. Perhaps this is why single experiments can rarely ever prove a hypothesis or a theory. It takes repeated experiments (replication) by different experimenters before scientists begin to accept that a particular hypothesis may or may not be useful in explaining some phenomenon. In fact, I tell my students that Father Guido Sarducci (of *Saturday Night Live* fame) created a 5-minute university education because that is all students will remember 5 years after they graduate. Thus, if my students will only remember one word from my entire statistics course, then I ask them to remember the importance of the word *replication.* Before trusting in a finding, we should see whether the finding has been replicated over a series of studies by different scientists.

Experimental Design and Statistical Analysis as Controls

To obtain useful and meaningful information from both descriptive and inferential statistics, it is necessary for the data to be collected on the basis of an explicit and appropriate design called the **experimental design.** Essentially, an experimental design is a blueprint for how the data collection will be conducted. At the very beginning of this data collection or experiment, there are two forms of control or power that we have at our disposal: *experimental control* and *statistical control.* Experimental control concerns the design of the experiment—for example, how many people will be used, how many groups there will be, what kind of people will be used, how they will be tested or measured, and so on. The second form of control or power is statistical control, and this is employed after the data collection or the experiment is complete. It is important to remember that although statistical techniques are powerful tools for describing data, experimental control is the more important of the two. One reason for this is that statistics can become meaningless and will communicate no knowledge whatsoever unless we adhere to some basic experimental principles.

For example, a recent TV poll asked viewers to call two different phone numbers. The first phone number was to be called if the viewer thought that a particular university football team should be ranked as the best team in the country, and the second number was to be called if the viewer thought that the same team should not be top-ranked. This plan for the experiment is called the experimental design. The statistical analysis consisted of comparing the number of calls for the two phones and seeing which was higher. In this case, the experimental design is so poor that the statistical analysis is rendered meaningless. One of the many problems with this design is that the sample of TV viewers was not random. Therefore, any conclusions that the results reflect popular opinion will be false. Another problem is that a viewer could call in more than once and thus unduly influence the outcome. Recently,

ESPN.com conducted an online survey about which team had the best sports uniform. The Denver Broncos rallied to win the poll with 137,257 votes to the University of Michigan's 88,743 votes. Upon further review, it was revealed that 71,465 Bronco votes came from a single Internet address. Thus, the experimental design is so poor that it is not meaningful to make any conclusions whatsoever about which team has the best sports uniforms.

The statistical analysis cannot take into account this problem or variable. As previously noted, these uncontrolled problem variables in experimental designs are called **confounding variables.** In summary, the major experimental design problem with TV, Internet, and call-in polls is that some people are allowed to participate more than once, and the sample of viewers will not be random. As noted earlier, such studies result in futile, meaningless, and highly untrustworthy data. Often, these polls will cost the viewer some amount of money to participate, and the true purpose of the poll is solely to generate money, not to determine the voters' opinions. However, the experimental design is often so poor that the only truly ethical conclusion would be that 137,257 is a larger number than 88,743. It cannot be concluded that more people thought a particular team had a better uniform than another. Any conclusion that gave the implication that a particular team's uniform was the choice of most people would be completely unethical. Subsequently, in the interest of "fairness," the ESPN.com pollsters disqualified all votes for the Broncos and designated the University of Michigan as the winner. In this case, ESPN.com is making a post hoc (after-the-fact) adjustment to the experimental design, further calling its results into question. Again, the experimental design is so poor that no amount of statistical power or analysis can correct the problem. Why do unscientific and unethical telephone surveys persist on TV, the Internet, and newspapers? Money! These polls, although completely unscientific, have entertainment value, and entertainment value means money. Also, at 50 cents or more per vote in some telephone polls, the owners of these polls are making money.

In summary, poor TV, Internet, newspaper, and other surveys demonstrate how you must think about designing your experiment in terms of your statistical analysis. Although statistical control is powerful, it cannot save or make interpretable a shoddy, poorly designed experiment.

The Language of Statistics

Learning about statistics is much like learning a new language. Many of the terms used in statistics may be completely new to you, such as *bimodal.* Also, some words you already know will be used in new combinations, such as *standard deviation.* Finally, some words will have a new meaning when used in the context of statistics. For example, the word **significance** has a different meaning in a statistical context. When used outside of statistics, significance is used as a value judgment. If something is said to be significant, then we usually mean that it is important or of consequence. The opposite of significant is usually insignificant. However, in statistics, the word *significance* has a different meaning. Significance refers to an effect that has occurred that is not likely due to chance. In statistics, the opposite of significance is nonsignificance, and this means that an effect is likely due to chance. It will be important for you to remember that the word *insignificant* is a value judgment, and it typically has no place in your statistical language.

On Conducting Scientific Experiments

Scientific experiments are generally performed to test some hypothesis. Remember, a theory is some grand idea about the way nature (at any level) seems to work. A theory generally should make some predictions in the forms of hypotheses. The hypotheses are then tested through scientific experiments. The typical experiment is a two-group experimental design where there is an experimental group and a control group. Most typically, the experimental group gets or receives some special treatment. The control group usually does not get this treatment. Then, the two groups are measured in some way. These two levels of treatment are aspects of the **independent variable.** The independent variable is the variable that the experimenter manipulates. The experimenter wishes to see whether the independent variable affects the dependent variable. The test that the two groups are measured on is called the **dependent variable** or response variable. Sometimes, it is very important that the participants in each group do not know whether they are receiving some special treatment. If this experiment was to determine whether Vitamin C affects colds, then the control group participants should also be given the same treatment as the experimental group, with the exception of actually receiving Vitamin C. Typically, control groups receive a placebo, which is a pill indistinguishable from the pills that contain the Vitamin C that the experimental groups receive. Remember, the participants should not be aware of who is actually receiving Vitamin C and who is getting the placebo. Also, the experimenter or statistician, who will evaluate the two groups after treatment, should not know who is receiving Vitamin C or the placebo until after the groups have been evaluated because it is a well-known phenomenon that experimenters can have an unconscious role in making their experiments come out in favor of their hypotheses. As previously noted, procedure is said to be a double-blind experiment. An example of this unconscious influence occurred in a proposed new treatment of childhood autism (an early and severe psychotic disturbance). The treatment was called facilitated communication. The therapists (facilitators) would help the highly uncommunicative autistic children form their answers to questions on a keyboard. An experiment was designed to control for the experimenter's influence, whereby the questions were either the same or different for the autistic children and the facilitator. Despite having received millions of dollars in grants to develop and train people to become facilitators, sadly, in no case could the children answer a question correctly when the questions were different between the children and their "facilitators." It appeared that the facilitators were simply consciously or unconsciously answering the questions correctly on behalf of the autistic children.

The Dependent Variable and Measurement

Remember that the test or whatever we use to measure the participants is called the dependent variable. In the Vitamin C experiments, there are a number of different measures that we might use as dependent variables. Some experimenters have focused on the number of new colds in a 1-year period. We could also measure the number of days sick with a cold. The choice of the dependent variable could be very important with regard to the conclusions of our study. For example, some studies have shown Vitamin C to be ineffective in preventing colds; however, there is some evidence that Vitamin C may reduce the severity or the duration of a cold.

Operational Definitions

If we are going to count the number of colds each participant gets in a 1-year period, then each participant would somehow then be checked for how many colds he or she got. Is there some subjectivity in how you decide whether a person has a cold? Of course there is! To minimize this subjectivity, perhaps the same person (like a medical doctor) would rate all of the participants to determine the presence or absence of a cold. However, doctors might vary among themselves about whether a person actually has a cold. Perhaps the diagnoses might vary as a function of what each doctor defines as a cold. For example, one doctor's definition of a cold might be whether his or her patient claims to have a cold. Another doctor's definition might require the presence of congested sinuses and elevated white blood cells. There is usually no single correct definition; however, whatever definition you do choose should be stated clearly according to some criteria. When you list the criteria clearly in your study that all of the participants met, then those criteria are said to be your **operational definition.**

Measurement Error

No dependent variable will ever be perfectly measured for each participant. The variation in the dependent variable that is not due to the independent variable is ironically called **measurement error** in statistics. The individual participants contribute the most error to the experiment and to the dependent variable. Other sources of error may come from subtle variations in the testing condition such as time of day, temperature, extraneous noise levels, humidity, equipment problems, and a plethora of other minor factors. In our Vitamin C study, it is possible that some participants never get colds anyway, and some may frequently get sick. The experimental error associated with the participants is hopefully balanced out when we randomly assign the participants to the two conditions. After all, we could not, for example, assign all our friends to the experimental group and all others to the control group. Random assignment of the participants to the groups helps to balance out the effects of the participants as a source of error. Using a large number of participants in each group also helps to balance out this source of error.

When you hear about or read about experiments or surveys in any form of media, remember to look at the dependent variable closely. Sometimes, the entire interpretation of an experiment may hinge upon an adequate dependent variable. How about the results of a study that some city, say "city x," is the best place to live. One critical variable in this study is how they measure "best city." Is it in terms of the scenery? What happens if you like or hate ocean views? Typically, these studies try to be "scientific" so their judges rate the cities on a number of different variables. However, many of these questions may relate to economic growth. Therefore, if a city is not growing rapidly, it may not fare well in the rating system. Would you consider a fast-growing city the best place in which to live? It becomes obvious there are many intangibles in developing a survey to rate the best place in which to live, and any survey that claims to be able to rate the best city is highly questionable. Be sure to examine the operational definitions of critical variables in experimental studies. In this latter example, it would be interesting to see the specific operational definition of "best city." If the authors do not provide an operational definition, then the study is virtually uninterpretable.

Measurement Scales: The Difference Between Continuous and Discrete Variables

A variable is typically anything that can change in value, and a variable usually takes on some numeric value. Statisticians most commonly speak of continuous and discrete variables. A **continuous variable** can be measured along a line scale, which varies from a small number to a large number. For example, a continuous variable would be the time in seconds in an experiment for a participant to complete a task. A **discrete variable,** for statisticians, typically means that the values are unique and are qualitatively distinct from one another. Speculations are not made between the values. For example, gender could be considered a discrete variable. If the category "male" is assigned a value of "0" and "female" is assigned a value of "1," then interpretations will not be appropriate for values between 0 and 1. It should also be noted that when there are only two values of a discrete variable, it is also referred to as a **dichotomous variable.**

Types of Measurement Scales

There are various kinds of measurement scales. In any statistical analysis, the type of scale must be identified first, so the appropriate statistical test can be chosen. Some measurement scales supply minimal information about their respective participants, and thus the statistical analysis may be limited. Other types of scales supply a great deal of information, and consequently, the statistical possibilities are enhanced. The following are four common types of measurement scales:

Nominal Scales

Nominal scales assign people or objects to qualitatively different categories. Nominal scales are also referred to as *categorical scales* or *qualitative scales*. One example is the assignment of people to one of the two categories of gender. Thus, when measured on a nominal scale, all of the people or objects in a category are the same on some particular value. Notice also that the people in the category are all considered equal with respect to that value. For example, all the males who fit the male category are considered equal with respect to the category. Thus, membership in a category does not imply magnitude. Some males in the category are not considered more "male" than other males in the same category. The frequency of each category can be analyzed, however. For example, it might be noted that 37 males and 44 females participated in a study. Another example of a nominal scale would be a survey question that required the answer yes or no, or yes, no, or undecided.

It is also important to note that there are no intermittent values possible on a nominal scale. This means that if, for statistical purposes, a value of 0 is assigned to the male category and 1 is assigned to the female category, there are no values allowed or assumed between the 0 and 1. It is, of course, physiologically possible to have a person who has a mixed biological gender. It is also psychologically possible to have a mixed gender identity. However, it is not possible to have intermittent values if gender is measured on a nominal scale.

Ordinal Scales

Ordinal scales involve ranking people on some variable. The person or object that has the highest value on the variable is ranked "number 1" and so on. The ordinal scale, therefore, requires classification (how much of the value does an individual have) and ranking (where the individual stands relative to all other members of the group).

The ordinal scale has one major limitation—the differences between rankings may appear equal when in reality it is known that they are not. For example, if we rank athletes after a race, the difference in times between the first- and second-place athletes may be huge, while the difference in times between second and third place may be very small. Nevertheless, with an ordinal scale, the appearance is given that the differences between the first, second, and third rankings are all equal.

This limitation is not necessarily bad. Ordinal scales may be sufficient if it is known that the classification variable possesses some arbitrariness. Thus, it may be useful to rank all 50 participants in an experiment with respect to some variable and then compare on some other variables or tests the top 5 ranked participants with the bottom 5 ranked participants.

It has also been assumed in the previous discussion that the people or objects ranked received their respective rankings by a single classification variable. Of course, it is also possible to rank people or objects through more than one classification variable or even nebulous or hazy criteria. For example, movie critics are frequently known for presenting their rankings for the 10 best movies or the 10 worst movies. What are their classification variables? In this case, there are probably many classification or criteria variables, and some of them may even be inexplicable or unconscious. At the very least, this may mean that some types of ordinal scales may be suspect.

Interval Scales

Interval scales probably receive the most statistical attention in all sciences. Interval scales give information about people or objects with respect to their ranking on some classification variable, and interpretations can be made with regard to how far apart the people or objects are on the variable.

With an interval scale, it is assumed that the difference of a particular size is the same along all points of the scale. For example, on an attitude survey, the difference between the scores of 40 and 41 is the same as the difference between the scores of 10 and 11. On an interval scale, it would also be assumed that a difference of 10 points between two scores would represent the same subjective difference anywhere along the scale. For example, it is assumed that the difference between scores of 40 and 50 on an attitude survey is the same degree of subjective magnitude as between 5 and 15. Obviously, this is a crucial and difficult assumption to meet on any interval scale. Nevertheless, most measurement scales, particularly in the social sciences, are assumed to be interval scales. Examples of interval scales are scores on intelligence tests, scores on attitude surveys, and most personality and psychopathology tests. However, there is some debate about whether these tests should be considered interval scales because it is questionable whether these scales meet the equal magnitude assumption. In practice, however, most statisticians agree that the purported interval scales at least have the property of ordinal scales, and they may at least approximate interval scales. In a more practical evaluation, it is well documented that these purported interval scales yield a plethora of interpretable findings.

Ratio Scales

Ratio scales have the properties of interval scales and, in addition, have some rational zero point. This means that a zero point on a ratio scale has some conceptual meaning. It must be noted that in most social disciplines such as psychology, sociology, or education, ratio scales are rarely used. In economics or business, an income variable could be measured on a ratio scale because it makes sense to talk of "zero" income, while it makes no sense to talk about "zero" intelligence.

Although ratio scales may be thought of as the most sophisticated of the types of scales, they are not necessary to conduct research. Most types of statistical analyses and tests are designed to be used with interval scales. Indeed, one of the primary purposes of the science of statistics is to organize and understand groups of numbers and to make inferences about the nature of the world. Its purpose is not to create the perfect measuring device.

Rounding Numbers and Rounding Error

The general rule for rounding decimal places is if the number to be dropped is 5 or greater, then the remaining number is rounded up. If the number to be dropped is less than 5, then leave the remaining number unchanged. For example, when rounding to the nearest tenth (or rounding to one decimal place), look at just the tenth and hundredth decimal places and ignore any places beyond (like thousandths):

10.977 would round to 11.0

125.63 would round to 125.6

100.059 would round to 100.1

6.555 would round to 6.6

6.5499 would round to 6.5

Anytime a number is rounded off, rounding error is introduced. Income tax forms often ask for rounding to whole numbers so only the first decimal place would be used to round. For example, when rounding to whole numbers, the following would occur:

$10.98 would round to $11

$125.63 would round to $126

$100.06 would round to $100

$6.55 would round to $7

$6.49 would round to $6

Statistical Symbols

Statisticians use symbols to represent various concepts. It will also be important for you to learn most of the common symbols they use. This should not be a very difficult task because the symbols are used over and over again, so you will have many opportunities to commit them to memory. Two of the most common symbols are $\bar{x}$ (pronounced x bar), which stands for the mean or average of a sample of numbers, and the Greek capital letter Σ (**sigma**), which is used to indicate the sum of a group of numbers. Let us practice with these two symbols. Find the average or $\bar{x}$ for the following group of numbers:

5, 9, 10

Intuitively, you probably know how to add the numbers up and divide by 3. Therefore, the arithmetic mean (which is commonly used instead of the word *average*) is 8. However, let us use the word *set* instead of group. When you added each of the numbers together, you were taking the sum. Each number in the set of numbers can be represented by x. Therefore, in statistics, the formula for the mean would look like this:

$$\bar{x} = \frac{\sum x}{N}$$

where $\bar{x}$ = the mean or average of the set of numbers, Σ = the sum of all the numbers in the set, and N = how many numbers are in the set:

$$\bar{x} = \frac{5 + 9 + 10}{3}$$
$$\bar{x} = \frac{24}{3}$$
$$\bar{x} = 8$$

Another common symbol is $\Sigma\, x^2$, which indicates that each number in the set is individually squared and then the products are added together. So, for the original set of numbers 5, 9, 10, obtain $\Sigma\, x^2$.

$$\sum x^2 = 5^2 + 9^2 + 10^2$$
$$\sum x^2 = 25 + 81 + 100$$
$$\sum x^2 = 206$$

What does $(\Sigma x)^2$ indicate? Remember, in mathematics, you must simplify what is in the parentheses first before you begin any other operations. So for this same set of numbers, add the numbers up first and then square this value.

$$\left(\sum x\right)^2 = (5+9+10)^2$$
$$\left(\sum x\right)^2 = (24)^2$$
$$\left(\sum x\right)^2 = 576$$

Does Σx^2 equal $(\Sigma x)^2$? By using the same numbers, you can verify that they are not equal! $\Sigma x^2 = 5^2 + 9^2 + 10^2$ or $\Sigma x^2 = 25 + 81 + 100$, which means $\Sigma x^2 = 206$. On the other hand, $(\Sigma x)^2 = (5 + 9 + 10)^2$ or $(24)^2$, which means $(\Sigma x)^2 = 576$. Thus, you can see the resulting values are very different, and they cannot be used interchangeably. For those of you who may be weak in mathematics, it may be useful for you to note the following formulas carefully because you will need them later in this book.

Summary

For the set of numbers 5, 9, 10:

Sum of the set	$\sum x = 5+9+10$ $\sum x = 24$
Mean	$\bar{x} = \frac{\sum x}{N}$ $\bar{x} = \frac{5+9+10}{3}$ $\bar{x} = 8$
Sum of each number in the set squared	$\sum x^2 = 5^2+9^2+10^2$ $\sum x^2 = 25+81+100$ $\sum x^2 = 206$
Squared sum of the set	$\left(\sum x\right)^2 = (5+9+10)^2$ $\left(\sum x\right)^2 = (24)^2$ $\left(\sum x\right)^2 = 576$

Sometimes the symbol x_i (x sub i) appears. x_i refers to the ith number in the set; thus, x_1 is the first number in the set. For the previous set of numbers (5, 9, 10), $x_1 = 5$, $x_2 = 9$, and $x_3 = 10$. Sometimes you will see x_n (x sub n), which stands for the nth number in the set or the last number in the set. Almost as commonly, you will see x_k (x sub k), which also stands for the last number in the set. In this book, a shorthand notation is used for $\Sigma\, x$. In reality, the formal notation is $\sum_{i=1}^{k} x_i$, where $i = 1$ below the Σ and k above the Σ form a programming loop for the x_i. Thus, x_i starts with x_1 and moves progressively through the second number in the set, third number in the set, and so on, through x_k (or the last number in the set).

History Trivia

Achenwall to Nightingale

Although the history of mathematics is ancient, the science of statistics has a much more recent history. It is claimed that the first use of the word *statistics* was by German Professor Gottfried Achenwall (1719–1792) in 1749, and he implied that statistics meant the use of mathematics in the service of the nation. For Achenwall, *statistics* was another word for state or political arithmetic, which might include counting the population, the size of the army, or the amount and rate of taxation.

One of the earliest uses of the word *statistics* in the English language was by John Sinclair (1754–1835), a Scottish lawyer and writer who published a 21-volume survey of Scotland from 1791 to 1799. In its preface, Sinclair acknowledged the political nature of statistics in Germany; however, he emphasized the use of statistics as a method of inquiry into the psychological state of a country, for example, to ascertain the level of happiness of its inhabitants and a means by which its future improvements could be made. He specifically thought a new word from the German language, *statistics,* might attract greater public attention.

Concurrent with these developments, mathematicians were making contributions in the area of probability theory that would ultimately form the foundations of inferential statistics. Pierre Simon, the Marquis de LaPlace (1749–1827), better known as "LaPlace," was a French mathematician who contributed much to early probability theory. Karl Friedrich Gauss (1755–1855) was a German mathematician and astronomer who also made valuable contributions to the foundations of both descriptive and inferential statistics.

In the middle to late 1800s, the application of statistics to social problems became more prominent. Adolph Quetelet (1796–1874), a Belgian mathematician and astronomer, extended some of Gauss's ideas to the analysis of crime in society. Francis Galton (1822–1911), cousin of Charles Darwin, was an English explorer, meteorologist, and scientist. His book, *Natural Inheritance,* published in 1889, has been recognized as the start of the first great wave of modern statistics. He also profoundly influenced another English person, Karl Pearson (1857–1936), who has been called the founder of the modern science of statistics. Pearson's intellectually provocative book, *The Grammar of Science,* was published in 1892. In it, he stressed the importance of the scientific method to society and knowledge, and he believed statistical procedures were fundamental to the scientific method.

Florence Nightingale (1820–1910) was a contemporary of Karl Pearson and a friend of Francis Galton. She is typically remembered as a nurse and hospital reformer. However, she might just as well be remembered as the mother of descriptive statistics. She trained to become a nurse during the 1850s, and it was her strict observance to sanitation in hospitals that dropped death rates dramatically. She performed not only her nursing duties but administrative duties as well. She also founded a school for the training of nurses and established nursing homes in England.

In the course of her administrative work, she developed a uniform procedure for hospitals to report statistical information about their patients. She is credited with developing the pie chart, which represents portions of the whole as pieces of a pie. She also argued to get statistics in the curriculum of higher education. She had suggested to Galton that a professorship be established for the statistical investigation of societal problems, and she pointed out issues to Galton that should be studied under the auspices of this professorship. These issues included crime, education, health, and social services. In 1911, the University of London finally established a Department of Applied Statistics and appointed Karl Pearson its head with the title, Galton Professor.

Through the graphic representation of data, called a frequency histogram, Florence Nightingale convinced the queen and the prime minister of England to establish a commission on the health and the care of the British army. She did this by showing clearly with graphs that the rate of deaths for the military while at home in England was almost double the rates of equivalent nonmilitary English males.

One of her biographers has argued that Florence Nightingale's interest in statistics transcended her interest in health care and was closely related to her strong religious convictions. She felt the laws governing social phenomena were also laws of moral progress; therefore, they were God's laws and could be revealed by the use of statistics.

Key Terms, Symbols, and Definitions

Abuse of power—The use of more participants than necessary in a controlled experiment.

Confounding variable—A variable that was not accounted for in the experimental design, varies systematically with the dependent variable, and prevents a clear interpretation of the effect of the independent variable on the dependent variable. Also called a nuisance variable.

Continuous variable—A measurement scale where an individual measurement can be made at any point along the range of the scale.

Dependent variable—In an experiment, a measure expected to vary across different levels of the independent variable. It is also called the response variable.

Descriptive statistics—A group of techniques used to describe data in a straightforward, abbreviated manner and may include means, pie charts, graphs, and tables.

Dichotomous variable—A measurement scale where an individual measurement can only fall into two discrete categories.

Discrete variable—A variable that has values that are unique and qualitatively distinct from one another, where values between the discrete variables are not meaningful (e.g., levels of gender).

Double-blind experiment—An experiment where the participants are kept unaware of the experimenter's hypothesis, and the experimenter is kept unaware of the participants' group affiliation (experimental or placebo group) until after the dependent variable has been measured.

Experimental design—The framework, blueprint, or structure for how an experimental study will be conducted (e.g., will it be a survey or a controlled experiment, how many groups will be studied, how many participants will be in each group, who will receive the experimental treatment, will a placebo be necessary, etc.).

Hypothesis—An educated guess that guides research.

Independent variable—In an experiment, the variable that the experimenter manipulates. It may also be called the treatment variable or predictor variable.

Inferential statistics—Techniques that are used on samples to make inferences about population values.

Interval scale—A measurement scale in which the units of measurement are equal along the length of the scale, but there is no rational zero point.

Measurement error—The difference between an obtained value of the dependent variable and what would be the theoretically true value.

Nominal scale—A measurement scale in which the data are simply named, labeled, or categorized, such as males/females.

Operational definition—The definition of a concept by means of the operations or procedures that were used to make, produce, or measure it.

Ordinal scale—A measurement scale in which the data are rank ordered according to the trait.

Parameters—Common and conventionally accepted ways of measuring data characteristics, such as means.

Placebo effect—The belief of the participant in an experiment that the independent variable will affect the participant's behavior. A placebo can also refer to an inert substance given to participants in the control group to control for placebo effects.

Population—Most often a theoretical group of all possible scores with the same trait or traits.

Power—The ability in a controlled experiment to detect whether a treatment or procedure really works.

Power analysis—A statistical analysis designed to help the researcher determine the minimum sample size that will help determine whether a treatment or procedure really works.

Ratio scale—A measurement scale in which the units of measurement are equal, and there is a rational zero point, such as the Kelvin temperature scale, with absolute zero.

Replication—The repeated ability to duplicate the results of a scientific experiment by different experimenters, which helps establish a hypothesis's usefulness (or nonusefulness in cases where the finding cannot be replicated).

Sample—A smaller group of scores selected from the population of scores.

Sigma ($\sum_{i=1}^{k} x_i$)—A symbol used to indicate that the complete set of numbers should be added together.

Significance—A treatment effect in an experiment that is not likely due to chance. Its opposite in statistical language is *not significant* or *nonsignificant. Insignificant* is considered a value judgment that typically has no place in statistics.

Stratified sample—A sample in which people or objects are included in the sample in the same proportion to their frequency in the subgroups that make up the population.

Chapter 1 Practice Problems

1. Distinguish between a theory and a hypothesis.
2. Name two requirements to help make samples representative of the population.
3. Define *dependent* and *independent variable*. Give examples of each and explain how they relate.
4. Based on the following list, identify whether each item represents experimental control or statistical control.
   a. adding an additional group to an experiment
   b. increasing the number of participants in the study
   c. analyzing the data with a different statistical test
   d. using a double-blind experiment
5. Which of the following are continuous variables?
   a. stock price
   b. eye color
   c. country of origin
   d. gender
   e. blood pressure
   f. height
6. Identify the type of measurement scale that each of the following represents (i.e., nominal, ordinal, interval, or ratio).
   a. weight
   b. distance
   c. birthplace
   d. heart rate
   e. IQ score
   f. first-, second-, and third-place winners in a race
   g. race
   h. eye color
   i. 10 best nursing programs in America

7. Round each number to a whole number.

   a. 10.999 c. 10.55
   b. 10.09 d. 11.399

8. Round each number to the nearest tenth.

   a. 12.988 c. 5.555
   b. 110.74 d. 55.549

9. Round each number to the nearest hundredth.

   a. 12.999 c. 6.055
   b. 225.433 d. 90.107

10. Find $\bar{x}$ for the following sets of numbers.

    a. 1, 2, 4, 5
    b. 8, 10, 15
    c. 7, 14, 22, 35, 40

11. Obtain $\sum x^2$ for the following sets of numbers.

    a. 6, 9, 12
    b. 61, 75, 84, 85
    c. 100, 105, 110, 120

12. Obtain $(\sum x)^2$ for the following sets of numbers.

    a. 7, 9, 11
    b. 30, 41, 52, 63
    c. 225, 245, 255, 275

13. A researcher is interested in determining whether a new "anti-anxiety" drug relieves anxiety in college students studying statistics. Fifteen students were assigned to one of three conditions: five students received a placebo, five received 50 mg of the new drug, and five received 100 mg of the drug. The students were then placed in a quiet room and were given 15 statistics problems to do. They were administered by the researcher's assistant. After completing the problems, the papers were scored, and the number of errors made by each student was used to determine their levels of anxiety.

    Based on the information provided above:

    a. Identify the dependent and independent variables.
    b. Determine the type of measurement scale used (interval, ratio, ordinal, nominal).
    c. Explain why the sample used is or is not representative of the population.
    d. Using only the information above, give an operational definition of anxiety.
    e. Define confounding variable and cite some examples that could create problems in this experiment.

14. The governor of the state of Georgia recently proposed that the state should provide the parents of every Georgia newborn a classical music CD or cassette to raise the intelligence of the child. The governor cited studies that have shown that college students who listen to classical music have higher IQ scores. What is the most essential experimental design problem in the governor's proposal? What are some potential confounding variables in the studies of college students?

Chapter 1 Test Questions

1. Which of the following is NOT a role of the statistician according to Abelson?
   a. curious detective
   b. honest attorney
   c. good storyteller
   d. moral advocate

2. A series of studies is used to demonstrate the usefulness of a theory. This process is called
   a. replication
   b. duplication
   c. inferential testing
   d. demonstrativeness

3. True or false: Theories are often proven or disproven in science.
   a. true
   b. false

4. Historically, which is the oldest form of statistics?
   a. descriptive
   b. inferential
   c. parametric
   d. nonparametric

5. Which of the following is NOT a characteristic of a good theory?
   a. provable
   b. testable hypotheses
   c. provocative
   d. grandness
   e. all of the above

6. Who wrote that "not [ever] being wrong" was even worse than being wrong in science?
   a. Nightingale
   b. Pauli
   c. Prusiner
   d. Abelson

7. Anxiety was measured in a group of anxious elderly patients who were divided into two groups. One group received a new anti-anxiety drug and the other a placebo. The operational definition of the dependent variable is
   a. anxiety
   b. not specified
   c. anxious elderly patients
   d. all of the above
   e. none of the above

8. The word *placebo* comes from Latin and means
   a. untrue
   b. fake belief
   c. I shall believe
   d. placid

9. Some patients in a control group will get well because of their belief that they are receiving the real treatment. This phenomenon is known as the
   a. experimenter effect
   b. placebo effect
   c. Hawthorne effect
   d. belief effect
   e. all of the above

10. If we randomly choose names from the phone book, call them, and determine whether they are registered voters, does this process result in a random sample of voters?
    a. yes
    b. no

11. Which of the following results in an abuse of statistical power?
    a. too large a sample
    b. too small a sample
    c. a nonrandom sample
    d. a correlative experimental design
    e. all of the above

12. Which of the following was NOT presented as one of the eight questions of any study or survey?
    a. Who was surveyed?
    b. Were the participants paid?
    c. How many people participated?
    d. Did the survey have too many questions or items?
    e. Was causation assumed from a correlational study?

13. A study linked a high blood pressure medicine to an increased risk of heart attacks. How many of the 70 published defenders of the medicine had financial ties to the manufacturer of the drug?
    a. 24
    b. 47
    c. 51
    d. 67
    e. all of them

14. Which of the following orders is the correct one for the level of scientific scrutiny of a research study (from highest to lowest scrutiny)?
    a. journal article, convention paper, book
    b. book, convention paper, journal article
    c. convention paper, book, journal article
    d. newspaper, convention paper, book
    e. none has the correct order

15. Which of the following is the most important and powerful form of scientific control in an experiment?
    a. the design of the experiment
    b. the statistical analyses
    c. a correlational design
    d. a large sample

16. After a normative study of blood pressure in people older than age 50, the researchers realized that they did not record the people's ages. The lack of knowledge about their ages is considered a potential
    a. confounding variable
    b. dependent variable
    c. independent variable
    d. response variable
    e. diagnostic variable

17. Which of the following words typically has no place in a statistical report?
    a. significant
    b. insignificant
    c. nonsignificant

18. When the experimenter evaluates the outcome of a study without knowing which participants were in the experimental and control groups and the participants are also unaware, then the experiment is said to be
    a. retrospective
    b. archival
    c. retroblind
    d. archivally blind
    e. double blind

19. In a study of autistic children being taught facilitated communication, what percentage of the autistic kids could answer the questions correctly without any help from the facilitators after their training?
    a. 0%
    b. 25%
    c. 55%
    d. 92.6%

20. An experimenter defines "consumer assertion" as the total number of products that a consumer has returned for his or her money back in a 1-year period. This definition in statistics is
    a. a confounding variable
    b. a definitional periodicity
    c. an operational definition
    d. a continuous but discrete definition
    e. measurement "error"

21. Which of the following scales requires a rational zero point?
    a. ordinal
    b. nominal

c. ratio
d. rational
e. interdenominational

22. The capital Greek letter Σ indicates that the reader should _______ a set of numbers.
    a. add
    b. subtract
    c. multiply
    d. divide
    e. calculate

23. For the set of numbers 2, 3, 5, the value of $\Sigma\, x^2$ would be equal to
    a. 38
    b. 100
    c. 10

24. For the set of numbers 4, 5, 6, 9, the value of $(\Sigma\, x)^2$ would be equal to
    a. 24
    b. 158
    c. 576

25. For the set of numbers 2, 3, 5, 6, 9, the value of $\sum x^2 - \frac{(\sum x)^2}{N}$ would be equal to
    a. –94
    b. 30
    c. 94
    d. 125

2 Descriptive Statistics

Understanding Distributions of Numbers

Chapter 2 Goals

- Learn the purposes of graphs and tables
- Learn how a good graph stopped a cholera epidemic
- Learn how bad graphs, tables, and presentations contributed to the space shuttle *Challenger* explosion and the space shuttle *Columbia* disaster
- Learn how to make graphs and tables
- Understand how to avoid chart junk
- Learn how to make a frequency distribution
- Understand the essential characteristics of frequency distributions

Back in the 1850s, when Karl Marx was just beginning to work on *Das Kapital,* the formal discipline of statistics and its use in making decisions were virtually unknown. One of the largest and most successful manufacturing companies at that time was a Manchester, England, cotton mill, which employed slightly fewer than about 300 people. Ironically, it was owned by one of Karl Marx's friends, Friedrich Engels. In his mill, there were no managers. There were only "charge hands" who were also workers but were involved in maintaining discipline over their fellow workers. No doubt, business decisions were the sole responsibility of the owner. Because these owners did not have, at that time, the advantages of making decisions based on statistical analyses, they must have had to rely strictly on their own intuitions. If they were right, the businesses prospered. If they were wrong, they floundered. In this chapter, you will learn how effectively made graphs and tables can help guide any decision, personal or organizational—and may save lives or even cost them.

By the later 1800s, the field of descriptive statistics was becoming fairly well established, although it consisted mostly of tables of numbers and some use of graphs, usually representing people's life span and other actuarial data. Today, the presentation of numbers in graphs and tables is still very popular because people can still get a good and quick conceptual picture of a large group of numbers. This conceptual picture then can be used to make informed decisions.

As noted previously, in the late 1800s, Florence Nightingale impressed the queen and the prime minister of England with her graph of death rates of British men versus British soldiers. Her graph, part of which is presented in Figure 2.1, is called a **bar graph**, and it is typically used with data based on nominal or ordinal scales. Florence Nightingale's nominal categories consisted of the British men and the British soldiers. The difference in their death rates can be seen by the differences in lengths between the two lines or bars.

The Purpose of Graphs and Tables: Making Arguments and Decisions

> Making decisions based on evidence requires the appropriate display of that evidence. Good displays of data help to reveal knowledge relevant to understanding mechanism, process and dynamics, cause and effect. That is, displays of statistical data should directly serve the analytic task at hand.
>
> —Edward Tufte (1997, p. 27)

Based on her bar graph, Florence Nightingale was able to effectively argue to the queen of England that unsanitary conditions in the English army led to higher death rates and that a national health commission should be established to improve living conditions. Her bar graph looks relatively simple and straightforward; however, it has a couple of deceptively powerful features. First, notice that she did not actually present any evidence of the actual unsanitary practices of the British army yet she was able to convince the queen that it was true. She made an effective argument for cause and effect by graphing an appropriate variable (death rates per 1,000 men) relevant to her case. She had a hypothesis (an educated guess) that unsanitary conditions led to sickness and death. It is possible that had she chosen another variable such as sickness rates, her argument may not have been as effective. Thus, choosing a variable relevant to her argument was one of her excellent decisions in preparing her graph. The second positive aspect of her simple bar graph

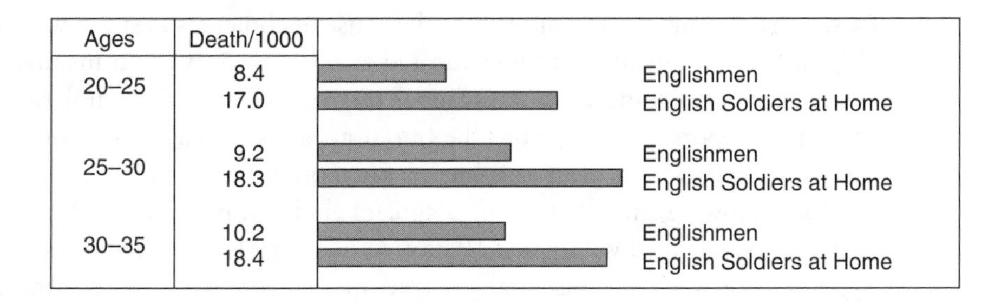

Figure 2.1 Relative Mortality Rates

was that she showed that death rates in the army were higher than those of men the same ages but not in the army. While this comparison may appear simple and obvious, it is a powerful lesson in making a clear argument. Florence Nightingale thought of two other explanations for the excessive death rates in the English army besides sanitation. Can you figure out what they were by examining her graphs?

The most obvious objection (alternative hypothesis) she faced was that army life is inherently dangerous: Wars kill people. It would be no wonder if death rates were higher for the British army. However, her bar graph very effectively dispelled this alternative hypothesis by making a relevant comparison. She graphed the death rates of English army men at home and not at war compared to typical Englishmen. By making this relevant comparison, she was able to show that it was not warlike conditions accounting for the higher death rates. Notice that this comparison did not directly prove her argument for cause and effect, but it did dispel a major rival explanation.

A second alternative hypothesis might have been that the age of the soldiers might have been the cause of the high death rates in the army. Perhaps English soldiers were simply older than the typical Englishman and thus died at greater rates while still in England and not at war. Notice that there is some evidence for this argument if we examine her three bars for the typical Englishman. The death rates do appear to rise as age rises. However, Florence Nightingale countered the age hypothesis by showing that English soldiers at home had higher death rates than Englishmen while comparing three different age groups, 20 to 25, 25 to 30, and 30 to 35. Again, Nightingale made a more effective argument for her hypothesis (unsanitary conditions) by making relevant comparisons and by controlling for alternative hypotheses.

How a Good Graph Stopped a Cholera Epidemic

During this same era in England, scores of people frequently died of cholera epidemics. Cholera is a disease still prevalent and deadly in the world today that comes from drinking water or eating food that has been contaminated by sewage. During the middle 1800s, the cause of cholera was still unknown, although there were at least two educated guesses, air or water transmission. There were also some more fantastical theories such as that cholera was caused by vapors escaping from the burial grounds of plague victims (even though they had been dead and buried for more than 200 years). In 1854, Dr. John Snow began investigating a cholera epidemic in London when more than 500 people died within just 10 days in one neighborhood. Snow's initial hypothesis proved to be essentially correct that cholera was caused by contaminated water. To make an effective argument for his hypothesis, Snow gathered evidence and made a graphic display. He accomplished this by getting a list of 83 officially recorded deaths within a short period of time. He plotted where these victims had lived on a map and discovered that a very large percentage lived near one particular well. After interviewing most of the families of the victims, he found that they did indeed get their drinking water from the Broad Street well. However, a rival argument was still plausible because not all of the 83 victims lived near the Broad Street well. Snow was able to fortify his hypothesis (and dispel a rival one) by his interviews because he found that, of the victims who lived nearer other wells, those victims had preferred the water from the Broad Street well or went to a school that got its water from the Broad Street well. Within about a week, Snow presented his hypothesis and his graphic map display to the water authorities. They removed the pump handle from the Broad Street well, and the cholera epidemic quickly ended (see Tufte, 1997, for the complete story).

How Bad Graphs and Tables Contributed to the Space Shuttle *Challenger* Explosion

Poorly conceived graphs and tables can also weaken arguments. The night of January 27, 1986, the makers of the space shuttle *Challenger* had a hypothesis that cold weather might make the rocket engine seals ineffective. It was predicted that the launch time temperature the next day would be about 27°F. The average temperature of 24 previous launches was 70°F. The lowest temperature of the previous 24 launches was 53°F, and that launch had five serious mishaps related to seal failures, which was far more than any other launching. The shuttle manufacturers prepared 13 graphs and tables in a few hours that evening to support their hypothesis and faxed them to NASA. However, their 13 graphs and tables did not present their argument clearly. In one chart listing all of the prior rocket seal mishaps, there was no information about temperature. In another chart, the same rocket was given three different names, making it difficult to determine which rocket had problems (yet there was only one rocket). Not one graph or table simply listed the number of mishaps as a function of temperature, yet this information was hidden in the data. The information was not effectively extracted and presented. NASA, based on the 13 graphs and tables and two follow-up telephone conversations later that evening, was unconvinced that lower temperatures might affect the function of the seals. The next day, the *Challenger* was launched, the rockets seals failed because of the cold weather, and the space shuttle blew up, killing all seven crew members (Tufte, 1997).

> *There are right ways and wrong ways to show data; there are displays that reveal the truth and displays that do not.* And if the matter is an important one, then getting the displays of evidence right or wrong can possibly have momentous consequences.
>
> —Edward Tufte (1997, p. 33)

How a Poor PowerPoint Presentation Contributed to the Space Shuttle *Columbia* Disaster

Sadly, this misrepresentation and misdirection in tables, graphs, and presentations appears to have continued with the space shuttle *Columbia* disaster that occurred on February 1, 2003. After its launch on January 16, a high-resolution film revealed that a piece of insulation foam (approximately 1,920 in^3) had struck the left wing of the shuttle. There were at least three requests from different shuttle management teams and engineers for high-resolution pictures of the shuttle's wing while in flight. Based on prior simulation studies of debris impacts up to 3 in^3 (in other words 640 times smaller!), a PowerPoint presentation was prepared by Boeing engineers (the makers of the shuttle) that minimized the potential damage. According to Yale Statistics Professor Emeritus Edward Tufte (2005), a key slide in this PowerPoint presentation, which resulted in the decision not to seek pictures while in flight, contained an unnecessary six different levels of hierarchy (big and little bullets, dashes, diamonds, and parentheses) to highlight just 11 sentences. The slide began with a title that already suggested that the shuttle managers had made the decision not to seek additional pictures. The slide also revealed that the engineers were relying on simulated data, not actual data, which was not even close to the size of the actual debris suspected. The latter aspect was noted in the slide, yet the executive summary statement absolutely minimized any potential damage.

A Summary of the Purpose of Graphs and Tables

Edward Tufte (1997) nicely summarizes the reasoning behind gathering statistical evidence and statistical graphs and tables.

1. Document the Sources of Statistical Data and Their Characteristics

Remember how Dr. Snow went to official death records? Not only was this method of gathering data more organized and official, but the data he obtained through this method also provided him with standardized and very essential information such as names, ages at death, and addresses where the victims lived. In the *Challenger* disaster, the rocket makers declined to put their individual names on the 13 charts and tables, so ultimate responsibility remained anonymous. It might have been useful for officials at NASA to have been able to talk directly to some of the engineers who had the hypothesis about seal failure in cold weather. Furthermore, at the bottom of each chart, the rocket makers placed a legal disclaimer that insinuated a kind of distrust for the charts, the chart makers, and any of the charts' viewers. The moral here is that if one is going to make an argument for a hypothesis, particularly an argument in favor of safety, one should state the argument as strongly and effectively as possible. Anonymity and legal disclaimers do not make for an effective argument.

2. Make Appropriate Comparisons

Remember how Florence Nightingale controlled for an alternative hypothesis by restricting her comparisons to Englishmen the same age as English soldiers? And she further strengthened her argument by using English soldiers not at war but at home. By making relevant comparisons, she eliminated doubt about rival explanations.

Recently, an "answer column" in the newspaper was asked if there was a gender effect in developing dementia. "Certainly," was the answer; women are affected at a rate of three to six times that men are. This evidence was gathered by going to nursing homes and counting the number of demented males and females. The "answer person" continued to speculate that since females have more of the hormone estrogen, perhaps it was estrogen that had a deleterious effect. One glaring problem with this reasoning is that women live longer than men. If we count the absolute numbers of men and women in nursing homes, we will always be able to count more women than men. The "answer person" failed to make a relevant comparison, that is, people at the same age.

Have you seen food store displays of 2% milk? What does the 2% represent? Do not feel badly if you are stumped. The 2% is supposed to represent the amount of fat in the milk. However, does 2% milk fat indicate that regular or whole milk has 100% milk fat or, in other words, does regular milk have 50 times the milk fat of 2% milk (100% ÷ 2% = 50 times greater rate)? The 2% milk advertisements are a good example of the problems in interpreting comparisons. Regular milk has 8 grams of total fat in a one-cup serving. There are 5 grams of total fat in 2% milk; thus, 2% milk has 62.5% of the total fat of regular milk. Therefore, 2% milk advertisements have been misleading, either intentionally or inadvertently, because few people would ever have imagined that 2%

milk had 62.5% of the fat of regular milk. This example of misleading advertising again shows us the value in making relevant comparisons.

3. Demonstrate the Mechanisms of Cause and Effect and Express the Mechanisms Quantitatively

Many times, it will not be sufficient to argue simply that we have discovered a real cause. The most effective argument for a causative hypothesis can be when we are able to demonstrate how varying the cause has a clear effect. Dr. Snow's hypothesis of causation was clearly supported when the water authorities removed the pump handle on the Broad Street well and death rates immediately declined. The demonstration of cause and effect had been clearly demonstrated through the mechanism of removing the pump handle.

In another highly visible display of the mechanism of cause and effect, a researcher proposed a few years ago that specific bacteria (*Helicobacter pylori*) were responsible for most ulcers. Despite some clinical evidence, there was much skepticism. To demonstrate a clear cause-and-effect relationship and in a highly visible display of the scientific method, the researcher (Australian Barry Marshall, who received the Nobel prize for this work in 2005) had himself injected with a bacterial extract from a patient with an ulcer. He quickly developed an ulcer and furthermore cured it with antibacterial drugs. In this example, the researcher demonstrated one mechanism of cause and effect by injecting himself with the suspected bacteria and getting an ulcer. But he also demonstrated another mechanism of cause and effect consistent with his original hypothesis when he was able to cure himself of ulcers by using antibacterial drugs.

More recently, a researcher claimed that HIV is not the cause of AIDS. With the same bravado, the researcher said he would put his controversial hypothesis to a similar test: He would inject himself with HIV. On the fateful day and before the media, the researcher did not show up.

4. Recognize the Inherent Multivariate Nature of Analytic Problems

Most problems in science have a **multivariate** nature (more than one cause). For example, while most ulcers may have a bacterial origin, some do not. Nor do all people exposed to the bacteria develop ulcers. In a recent newspaper headline, it was proposed that marriage tended to curb drug use. However, it is particularly true in psychology that there are multivariate causes. We are bombarded daily with overly simplistic explanations for behavior, such as crime is caused by drug and alcohol abuse. The implication is that removing the cause (drugs and alcohol) removes its effects (criminal behavior). However, it is extremely rare in the sciences that problems have a **univariate** nature (single cause). Eliminating drugs and alcohol from society will not decrease criminal behavior. In fact, there are some indications it might even increase. Forcing drug addicts to get married will not curb drug use. Severely addicted drug users will typically make terrible spouses and parents. What scientists can do, given the multivariate nature of most problems, is to argue clearly and effectively for some causal relationships while also remembering that nature is complex. Also, in many situations, varying causes may also vary in the strength of their contribution to a particular problem. Thus, criminal behavior may have a smaller contribution from heredity (they are born that way) and larger contributions from poverty, lack of education, and racial biases. Notice also that literally a hundred or more factors may be related to criminal behavior and that even when specifying a hundred factors, we still may not be able to predict accurately who will commit a crime. We continually read in the newspaper that *someone we would never suspect* has committed some heinous crime.

5. Inspect and Evaluate Alternative Hypotheses

We saw that Florence Nightingale evaluated at least two rival hypotheses to her contamination hypothesis: age and warlike conditions. By making relevant comparisons to English soldiers at home and at various age groupings, she was able to dismiss both of them as plausible alternatives. Many times in scientific articles, researchers cannot evaluate and test for rival hypotheses. However, Tufte's (1997) suggestion to at least inspect other ideas may be useful. Many published scientific papers will simply note rival hypotheses in the introduction or discussion sections of their papers. If researchers "save" the evaluation of rival hypotheses for the discussion section, they might do so by noting "while there remains Hypothesis A and Hypothesis B for the present findings. . . ." In this way science may be advanced, although the researcher has not formally evaluated the alternative hypotheses. Other researchers may then be able to generate research designs that may properly test rival ideas.

> When consistent with the substance and in harmony with the content, information displays should be documentary, comparative, causal and explanatory, quantified, multivariate, exploratory. . . . It also helps to have an endless commitment to finding, telling, and showing the truth.
>
> —Edward Tufte (1997, p. 48)

Graphical Cautions

A note of caution is in order. People can just as easily fool themselves and others by bar graphs. An example of this tomfoolery is shown in Figure 2.2.

In this example, School A's failure rate appears to be much lower than School B's or School C's. In reality, the difference among the failure rates for the three schools is very small (1%). However, by exaggerating the tiny differences with an inappropriate bar graph, School A's failure rate appears much better than the other two schools.

The following (Figure 2.3) graph tomfoolery occurred in a national truck advertisement.

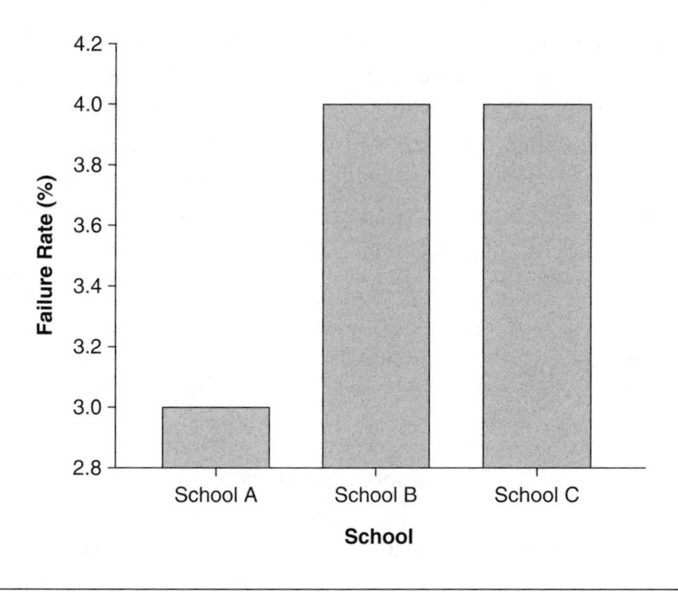

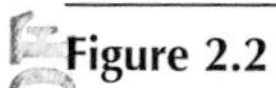

Figure 2.2

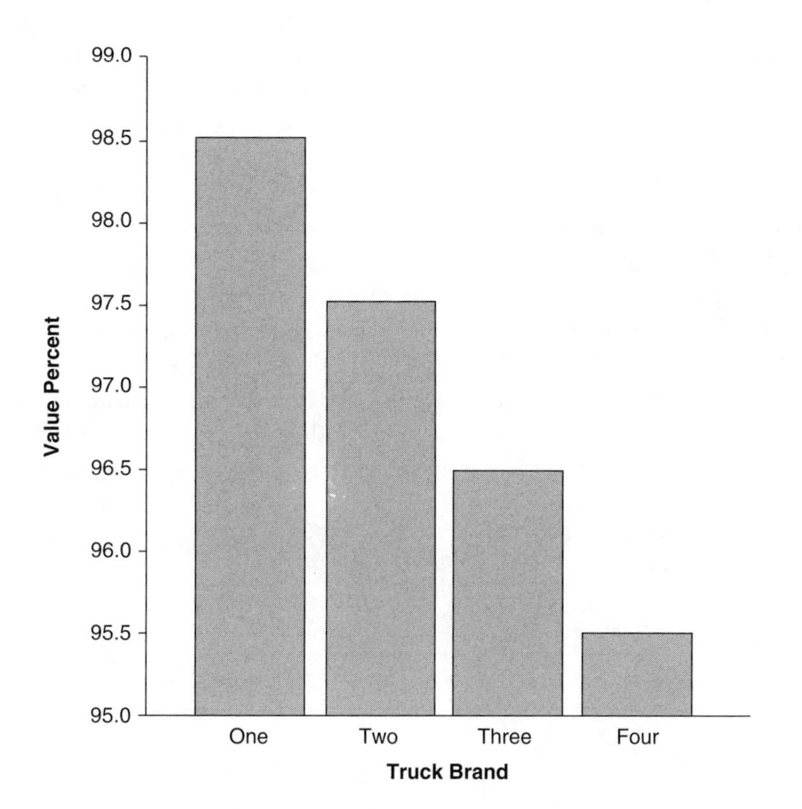

Figure 2.3 Percentage of Trucks on the Road After 10 Years

In this graph, the differences between the trucks' percentages have been magnified by cutting off the bottom 95% of the bars' heights. Although the graph makes it appear as if Truck A is much more reliable than any other truck (particularly that "terribly" unreliable Truck D), the actual difference in reliability is barely three percentage points. And despite the graphical appearance of a major difference between Trucks A and B, their reliability difference is less than one percentage point.

Frequency Distributions

When continuous line scales are used to measure the dependent variable as in interval or ratio scales, as is common in behavioral sciences, a **frequency distribution** may be constructed. The frequency distribution is one of the most important graphic presentations in modern statistics. For example, have you ever seen advertisements for a shoe sale? I once saw a large advertisement for an expensive name-brand shoe for sale at only $10! I raced down to the store to find two tables piled high with shoes. One table listed men's shoe sizes 4 to 6, and the other table had sizes 13 and greater. I left in disappointment, which later turned to disgust. If only, I thought, shoe storeowners were required, as part of their business license, to take a course in statistics.

So let us imagine a women's shoe storeowner who wishes to know what size shoes to order. The storeowner could poll every woman who entered the store for her shoe size over a period of time. The most inefficient way to present these data would be to write down all of the sizes in a single column. The problem with that approach is that the sizes are not organized in any coherent way. The frequency distribution, on the other hand, would give an immediate graphic or tabled picture of the shoe sizes. In addition, the frequency distribution can handle small or large samples. For the purpose of simplicity, let us suppose the survey of shoe sizes consisted of one size 7, three size 5s, one size 3, two size 6s, and two size 4s. The first step in constructing a frequency distribution would be to arrange the shoe sizes from low to high in a table with their corresponding frequencies (how many of each) beside them.

Note how easy it is now for the storeowner to figure out how many of each shoe he or she has on hand. You can also imagine that as the sample of

Table 2.1

Shoe Size	*Frequency*
3	1
4	2
5	3
6	2
7	1

shoes gets very large, this tabled frequency distribution will still be just as easy to understand.

Next, let us construct a graphic picture of this frequency distribution. The graph will consist of two continuous line scales at right angles to each other, which looks like this:

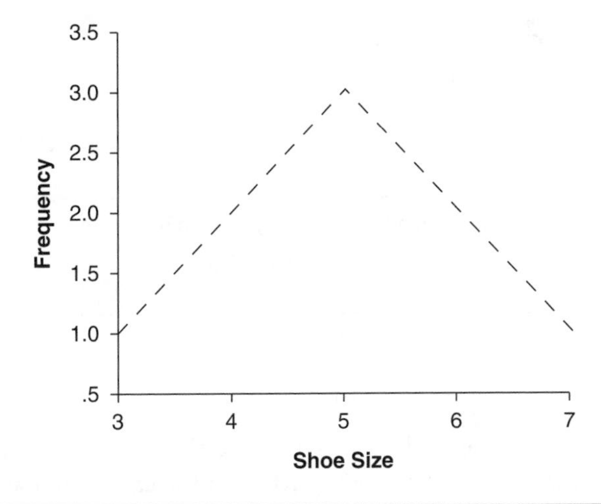

Figure 2.4

The horizontal axis most typically contains the line scale, which measures the dependent variable or the variable that we are measuring. In this case, the variable of interest that we are measuring is shoe size. The vertical axis usually measures the frequency or how many of each shoe size exist within the sample. *Note:* Each point in the graph represents an intersection of two lines drawn from each line scale. If you draw a line straight up from the lowest shoe size and draw a line straight across from the frequency of that shoe size, you will place a point at the intersection of these two lines. If you do this for each of the shoe sizes and you connect the points, it will generate a line, which represents the shape of the frequency distribution. When the points are directly connected to one another with straight lines, this graph is also called a **frequency polygon.** A polygon is a closed plane figure having three or more straight sides. If we had represented the frequencies with bars as we did in a bar graph, the result would be called a **frequency histogram.** The difference between a bar graph and a histogram is simple: Bar graphs have spaces between the bars and histograms do not. Figure 2.5 presents the shoe data as a frequency histogram.

The general shape of this frequency polygon or histogram represents one of the most significant concepts in statistics. The shape resembles what is called the **normal curve or bell-shaped curve.** It can also be referred to as a normally distributed frequency distribution. In simple terms, it means that when you are faced with a group of numbers representing most kinds of data, the resulting frequency distribution shows there are few cases that have a

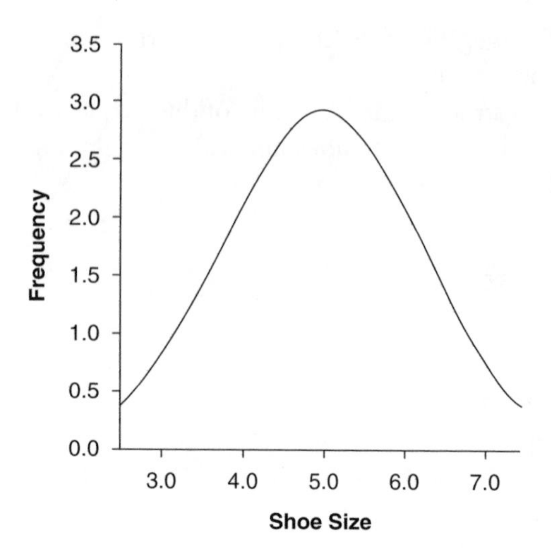

Figure 2.5

small amount of the dependent variable (the thing we are measuring, e.g., small shoe size, low IQ, light weight). Most of the cases will have a medium amount of the dependent variable, and finally, just a few cases will have the largest amount of the dependent variable (the largest shoe size, the highest IQ, the heaviest weight). Not all kinds of data will result in a normally distributed frequency distribution. However, it is interesting that many kinds of data, including behavioral and biological, will produce a bell-shaped curve, which approximates the normal distribution. The theoretical normal distribution is presented in Figure 2.6. The normal distribution also has special mathematical properties, which will be discussed simply (and kindly) later.

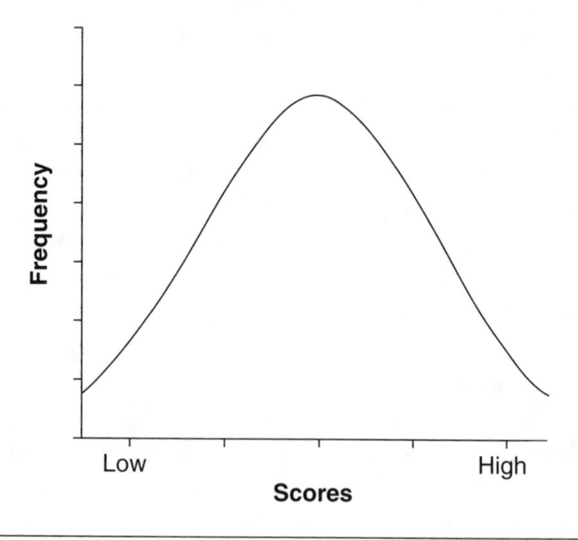

Figure 2.6

Shapes of Frequency Distributions

There are also common variations of the normal distribution. Sometimes, there are two different frequently occurring scores, as measured by the dependent variable. This curve results in a **bimodal distribution,** as presented in Figure 2.7A. If we use shoe size as an example, this would mean that there are a large number of people who have a shoe size around 5 and an equally large number of people with a shoe size around 8. There are also two kinds of distributions, which are variations on the normal distribution, and these are called **skewed distributions.** Figure 2.7B presents a **positively skewed** distribution (also called skewed right). Figure 2.7C presents a **negatively skewed** distribution (or skewed left).

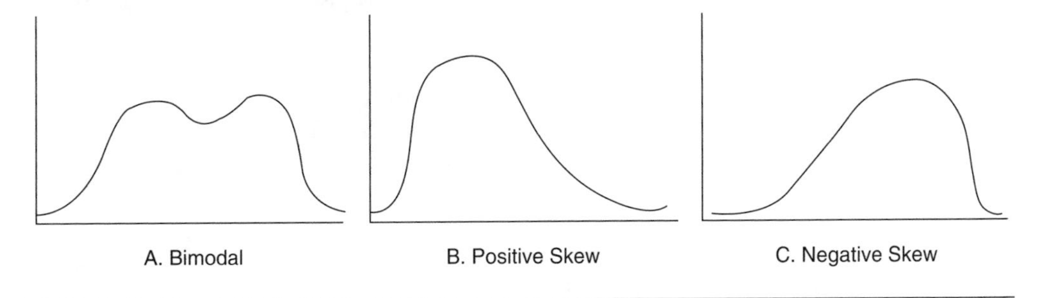

Figure 2.7

Here in Colorado, my students remember whether a curve is positively skewed or negatively skewed by looking at the distributions as snowy mountains. In Example B, if you were a "normal" skier, which side of the mountain would you ski down? You would ski to the right, so it is a distribution that is skewed to the right.

Many statistical software programs can calculate the skewness of a distribution. When you see a value of 0 for skewness, it means that the curve is not skewed, positive values indicate a right or positive skew, and negative values indicate a left or negative skew.

Grouping Data Into Intervals

When we are dealing with a large group of numbers, or a group of numbers that is spread out over a large range of the dependent variable, we may wish to group the individual scores into categories or intervals. For example, let us look at the following set of achievement scores by a third-grade class of children: 25, 27, 29, 30, 32, 36, 39, 44, 45, 47, 48, 48, 49, 52, 55, 56, 57, 63, 66, 67. First, let us table the scores in a frequency distribution. Table 2.2 presents the resulting frequency distribution.

Table 2.2

Achievement Score	*Frequency*	*Achievement Score*	*Frequency*
25	1	47	1
26	0	48	2
27	1	49	1
28	0	50	0
29	1	51	0
30	1	52	1
31	0	53	0
32	1	54	0
33	0	55	1
34	0	56	1
35	0	57	1
36	1	58	0
37	0	59	0
38	0	60	0
39	1	61	0
40	0	62	0
41	0	63	1
42	0	64	0
43	0	65	0
44	1	66	1
45	1	67	1
46	0		

Note that because the scores are spread out across the values of the dependent variable, and because many of the possible achievement scores have a frequency of 0, a table of the data, which simply lists frequency, is relatively meaningless. This is also the case if a graph of the raw data is presented (with many scores with a frequency of either 0 or 1, the graph is often called a sawtooth distribution; see Figure 2.8). Therefore, it may be better to group the scores together into intervals. Look at Table 2.3. The scores in the previous set of data have been grouped into intervals of 10.

By categorizing the data into intervals, we are now able to get a more meaningful picture. A graph of the grouped data appears in Figure 2.9.

Now that we have grouped our data into intervals, the graphic presentation of the frequency distribution looks approximately mound-shaped.

Advice on Grouping Data Into Intervals

1. Choose Interval Widths That Reduce Your Data to 5 to 10 Intervals

For example, if you have too few intervals, such as 2 or 3, then you may be crunching up your data too much. However, too many intervals may

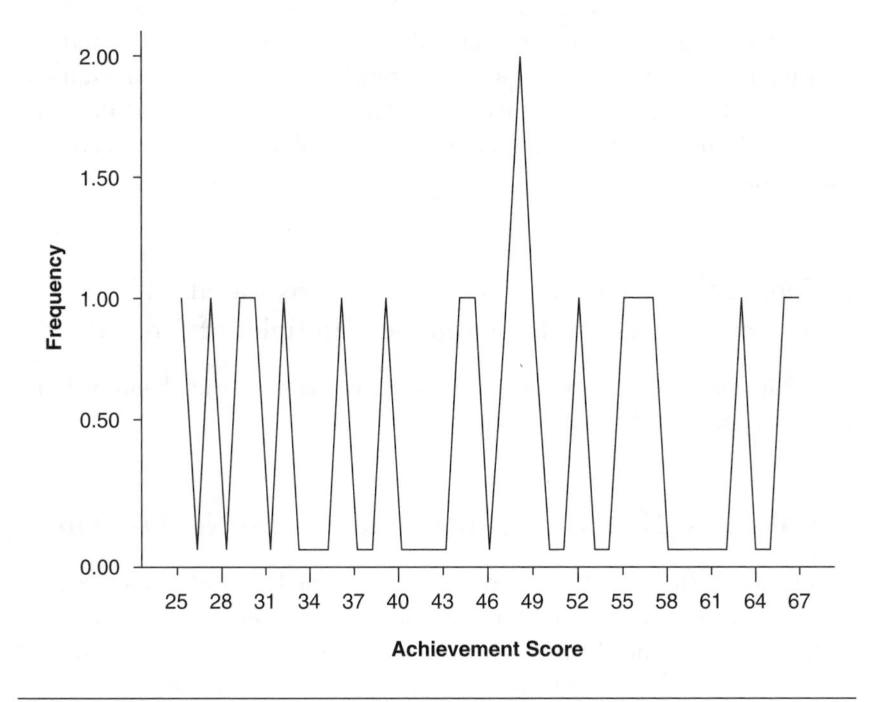

Figure 2.8

Table 2.3

Achievement Score	*Frequency*
20–29	3
30–39	4
40–49	6
50–59	4
60–69	3

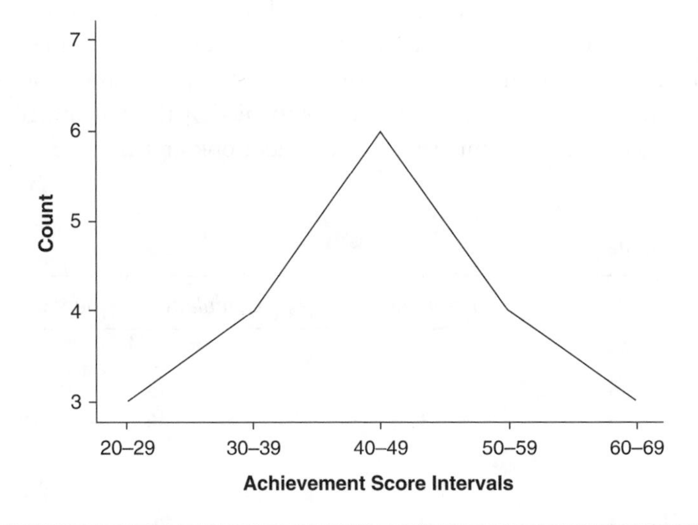

Figure 2.9

spread your data out too far. Generally, somewhere between 5 and 10 total intervals seems to give a good picture of the data. A bad example appears in Table 2.2, where there are 43 intervals, and thus the data are too spread out. A better example appears in Table 2.3, where there are 5 intervals.

2. Choose the Size of Your Interval Widths Based on Understandable Units, for Example, Multiples of 5 or 10

Perhaps because humans generally have five digits on each hand or foot, we intuitively favor base 10 systems.

3. Make Sure That Your Chosen Intervals Do Not Overlap

Look back to the beginning of this chapter at Figure 2.1. Notice that Florence Nightingale violated this rule. She grouped her males into ages 20 to 25, 25 to 30, and 30 to 35. Thus, we cannot know where to place a 25-year-old, as it seems he could fall in either of two categories.

The Cumulative Frequency Distribution

In tabled frequency distributions, statisticians also use the concept of the **cumulative frequency distribution.** This parameter gives a picture of how many cases have been accounted for out of the total number of cases. Look at the example in Table 2.4.

Note that at shoe size 3 and smaller, there is only one pair of shoes. At shoe size 4 and smaller, there are a total of three pairs of shoes (one pair of size 3 plus two pairs of size 4). This cumulative frequency continues until, at size 7 and smaller, all nine pairs of shoes have been accounted for. Tabled frequency distributions are also often accompanied by the percentage of each individual score or the cumulative percentage. Look at Table 2.5.

Table 2.4

Shoe Size	*Frequency*	*Cumulative Frequency*
3	1	1
4	2	3
5	3	6
6	2	8
7	1	9
Total	9	9

Table 2.5

Shoe Size	*Frequency*	*Percentage*	*Cumulative Percentage*
3	1	11.1	11.1 (1/9)
4	2	22.2	33.3 (3/9)
5	3	33.3	66.7 (6/9)
6	2	22.2	88.9 (8/9)
7	1	11.1	100.0 (9/9)
Total	9	99.9[a]	100.0

a. The percentage total in this column may not add up to 100% due to rounding error.

Note that at shoe size 3, there is a total of one pair of shoes out of the total of nine pairs. Therefore, one divided by nine is 11.1%. At shoe size 4, there are two pairs out of a total of nine pairs, and two divided by nine is 22.2%. In the cumulative frequency column, the cumulative percentages are totaled. Shoe size 3 accounts for 11.1% of all the shoes, and thus the total percentage of all shoe sizes at size 3 and below is also 11.1%. Shoe size 4 accounts for 22.2% of all the shoe sizes. The total cumulative percentage of shoe size 4 (and smaller) is 33.3% (obtained by 3/9). At shoe size 7 (and smaller), all nine pairs of the total nine pairs have been accounted for; therefore, the cumulative percentage is 100.0.

Let us return to the raw data presented in Table 2.2 and produce Table 2.6, which includes the percentage and cumulative percentage.

Cumulative Percentages, Percentiles, and Quartiles

Cumulative percentages can be used to identify the position of a score in a distribution. In this previous set of scores measuring achievement, a high score indicated the higher achievement, and a low score meant lower achievement. A raw score of 44 does not have much meaning because its standing relative to the other scores is not known. However, the cumulative percentage shows that a raw score of 44 was in the lower half of all the scores, and 40% of all the subjects scored a 44 or below. *Note:* A raw score of 47 has 50% of all scores at that point or below.

Percentiles are derived from percentages, and they describe the score at or below which a given percentage of the cases falls. The percentile scale is divided up into 100 units. Thus, a raw score of 47 is at the 50th percentile. (The 50th percentile is also called the median of the distribution because it divides the distribution in halves.) A raw score of 44 is at the 40th percentile.

Table 2.6

Achievement Score	Frequency	Percentage	Cumulative Percentage	Achievement Score	Frequency	Percentage	Cumulative Percentage
25	1	5	5	47	1	5	50
26	0	0	5	48	2	10	60
27	1	5	10	49	1	5	65
28	0	0	10	50	0	0	65
29	1	5	15	51	0	0	65
30	1	5	20	52	1	5	70
31	0	0	20	53	0	0	70
32	1	5	25	54	0	0	70
33	0	0	25	55	1	5	75
34	0	0	25	56	1	5	80
35	0	0	25	57	1	5	85
36	1	5	30	58	0	0	85
37	0	0	30	59	0	0	85
38	0	0	30	60	0	0	85
39	1	5	35	61	0	0	85
40	0	0	35	62	0	0	85
41	0	0	35	63	1	5	90
42	0	0	35	64	0	0	90
43	0	0	35	65	0	0	90
44	1	5	40	66	1	5	95
45	1	5	45	67	1	5	100
46	0	0	45				

Quartiles refer to specific points on the percentile scale. The first quartile refers to the 25th percentile, the second quartile refers to the 50th percentile, and the third quartile is the 75th percentile. Percentiles and quartiles are often used in educational assessment.

Stem-and-Leaf Plot

In traditional frequency distributions, particularly when the data are plotted by intervals, each value of an individual score is lost. American statistician John Tukey (1915–2000) created a **stem-and-leaf plot,** which has a number of interesting features: It presents the data horizontally instead of vertically, it preserves each individual score, and extreme scores are readily observed.

To create a stem-and-leaf plot, let's use the data from Table 2.2. With each number, the leftmost digit will become the stem, and the right digit becomes the leaf. The first number in this set is 25, so the left digit 2 will be the stem, and the right digit 5 will be the leaf. The next two numbers in the set, 27 and 29, also share the same stem (2), but they have different leaves

(7 and 9). Thus, a stem-and-leaf plot of the first three numbers in the data would look like this:

stem ⇒ 2 | 579 ⇐ leaves

Thus, attaching the stem (2) with each of its leaves (5, 7, and 9) gives us the original numbers 25, 27, and 29.

The complete stem-and-leaf plot of the data in Table 2.2 would look like this:

2 | 579

3 | 0269

4 | 457889

5 | 2567

6 | 367

Missing interval stems can also be presented. For example, what if the data in Table 2.2 did not have the numbers 63, 66, or 67 but had instead 70, 76, and 77. The stem-and-leaf plot would have looked like this:

2 | 579

3 | 0269

4 | 457889

5 | 2567

6 |

7 | 067

Notice how the interval stem 6 has no leaves. This indicates that there are no numbers in the set in the 60s.

For data with single digits, stem-and-leaf plots are no more useful than a histogram or bar graph.

Nonnormal Frequency Distributions

When frequency distributions are graphically represented, sometimes the resulting line curve has varying symmetrical shapes, and sometimes it has asymmetrical shapes. Often, but not always, a frequency distribution will be mound-shaped. The shape of the mound is referred to as **kurtosis.** In Figure 2.10, there are four types of symmetrical distributions. Example A presents the normal frequency distribution or the bell-shaped curve. Example B has a pointed distribution. This tendency toward pointedness is referred to as **leptokurtosis.** Thus, Example B presents a distribution that is leptokurtic. In Example C, the distribution is flatter than the typical normal distribution. This tendency toward flatness is called **platykurtosis;** thus, the distribution in Example C has platykurtic tendencies. A perfectly normal distribution is said to be mesokurtic.

Many statistical software programs can calculate the kurtosis of a distribution. When you see a value of 0 for kurtosis, it means that the curve is normal or mesokurtic, positive values indicate leptokurtosis, and negative values indicate platykurtosis.

There is one other somewhat mound-shaped curve, called a bimodal distribution. It occurs in situations where there are two most frequently occurring scores but neither of the scores is at the exact center of the distribution. This distribution is presented in Example D. Of course, it is also possible to imagine a trimodal distribution where there are three peaks in the distribution. One study in psychology found a trimodal distribution of the children's ages when they were first admitted to mental health care facilities (about ages 4–5, ages 7–8, and ages 10–12).

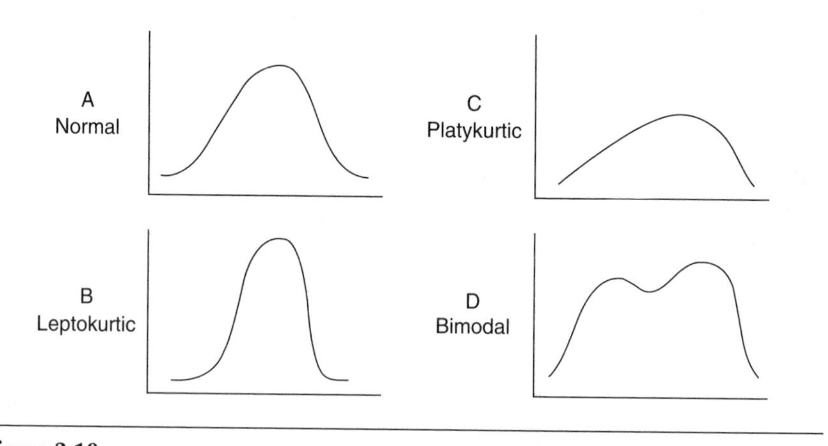

Figure 2.10

On the Importance of the Shapes of Distributions

The labels for the distributions may not have much meaning to you now. However, they are important because they allow statisticians to communicate the shape of distributions quickly even without visual aids such as graphs. In addition, the shapes of distributions of numbers are very important in inferential statistics, where we will make inferences from samples of numbers about the populations from which they were drawn.

Additional Thoughts About Good Graphs Versus Bad Graphs

The purpose of any graph should be to present data both accurately and conceptually clearly. With the proliferation of graphics programs for computers, there has been a proliferation of graphic presentations for data. However, it is already obvious that the most important purpose of statistics, that of conceptual clarity, is occasionally being forgotten in the midst of multicolor graphic options.

Therefore, it will be important for you to keep in mind some common pitfalls of graphic presentations. Watch for them as you prepare your own graphs, and beware of them while trying to interpret others' graphs.

Low-Density Graphs

Tufte (1983) warns of low-density graphs where very few data points are actually presented compared to the number of square inches taken up by the entire graph. A high-density graph is not necessarily good either. Remember, the purpose of a graph is to present the data accurately and clearly. Obviously, if there are very few data points to be presented, the readers may be better served by verbally presenting the data, instead of trying to impress readers with a graph.

Chart Junk

Tufte (1983) also warns of chart junk, which is an attempt to fill up the blank spaces of a graph with trivial or meaningless features. The tendency toward introducing chart junk might be greater in low-density graphs. Chart junk may be simply unnecessary with a good graph. Indeed, it may be a sign of a bad graph. With the advent of computer graphic programs, it has become much easier to create graphs, and it has become much easier to fill graphs with meaningless or unnecessary features such as three-dimensional bars or multidimensional bars and shading. While these additions are impressive, remember that the primary purposes of graphs are clarity of thought and efficiency of presentation. I personally have a great deal of difficulty interpreting the exact height on the *y*-axis of any three-dimensional bar in a histogram.

Changing Scales Midstream (or Mid-Axis)

Examine both axes of a graph. Make sure the scales used for each axis do not suddenly change. For example, if the horizontal axis was plotted by the years 1965, 1975, 1985, 1986, 1987, 1988, 1989, and 1990, the person making the graph would have suddenly changed the measuring scale from 10-year periods to 1-year periods. Perhaps the purpose was to minimize differences in the graph, or perhaps it may have been to emphasize changes in the data. Either way, it may be an unfair distortion of the data.

Labeling the Graph Badly

This can be done in a variety of ways. Frequently, when graphs are reproduced in print, they are reduced in size. Therefore, when you label parts of your graphs, make sure the labels are large enough to survive the reduction. Many advertisements violate this rule intentionally, such that restrictions to their offers appear in print so small that they are overlooked. Graphs should have clear, readable labels. The labels should not be ambiguous, incorrect, or illegible. All too often, business presentations have overhead projections with the labels or explanations produced in standard typewriter font size. I have also witnessed slide presentations of graphs with the same labeling problem. Only those people in the very front row can read such small print, and even to them, the small print is difficult to read.

The Multicolored Graph

The color option for computer graphic programs may introduce confusion into graphs instead of clarity. Although two or more colors may make bar graphs more artistically impressive, additional colors may just confuse the readers. Multicolors may also intentionally fool readers into thinking the graph is more meaningful than a simple black and white graph. Remember, conceptual clarity is a graph's most important purpose.

PowerPoint Graphs and Presentations

In his book, *The Cognitive Style of PowerPoint,* Tufte (2003) warns that the fixed template presentation style of PowerPoint may dumb-down or oversimplify data presentations. His goal is streamlined communication with high impact. He warns that PowerPoint slideshows often have an unending stream of bulleted lists or talking points, which actually reduce the amount of information per slide. He also warns against excessive use of name brands, logos, headers, footers, and titles on slides, which also reduces the amount of important information that can be presented. He notes that cartoon animations in PowerPoint slides often obscure rather than illuminate the subject matter and that associated audio cues often distract an audience's attention. Interestingly, some audiences do prefer PowerPoint presentations to handouts. Nonetheless, it is no excuse for weak or lazy PowerPoint presentations.

History Trivia

De Moivre to Tukey

Who discovered the normal distribution? Some people claim Abraham De Moivre (1667–1754) has the clearest right to the discovery. He was born in France and raised in England. He was a mathematician, and it is known he gave private mathematics lessons in London. It is thought that one of his students may have been an Englishman, Thomas Bayes (1702–1761), who went on to make important theoretical contributions to probability theory. De Moivre published two important works, one in 1711 and another in 1718. It is ironic that their largest appeal was not to mathematicians, but to gamblers, because the works dealt with games of chance. In fact, modern probability theory can trace its roots to letters of correspondence between famous mathematicians of the middle and late 1600s, discussing their attempts to solve and apply rules to gambling games. De Moivre is credited with developing the equation for the normal curve in approximately 1733.

Karl Friedrich Gauss (1777–1855), the German mathematician and astronomer, noticed that whenever large numbers of observations were made regarding the stars and planetoids, large numbers of errors always occurred. Gauss used the mathematical properties of the normal curve to develop a distribution of errors, and it became known as the normal law of error. Quetelet (1796–1874), the Belgian astronomer and mathematician, may have been the first to develop an application of the normal curve other than describing a distribution of errors, instead using it to describe social and biological phenomena. However, it appears Francis Galton received most of the credit for turning the Gaussian law of error into a law of nature, which is applicable to social and biological events.

Francis Galton (1822–1911), an English scientist and cousin of Charles Darwin, argued strongly that Gaussian errors were the exact opposite of what Galton felt should be studied. Gauss had argued that these errors or deviations were to be removed, or allowances were to be made for them. Galton claimed the errors or deviations were the very things he wanted to study or preserve! Galton published many books and articles, primarily on intelligence and inheritance. In 1876, he published a study of twins and the contributions of heredity and the environment, and in it, he coined the famous synonyms *nature* for *heredity* and *nurture* for *environment.* Galton, within the next 10 years, developed the important statistical concept of correlation. It is also interesting to see that the word *error* has persisted in the discipline of statistics when statisticians are now referring to variations or differences between numbers and not actually mistakes or real "errors."

The application of the normal curve to social and biological phenomena is not without its critics. Jum Nunnally (1921–1986), an American professor of psychology, noted that the distribution of psychological and educational test scores is seldom normally distributed, even if there are a large number of scores. He attributed this to the relationship each item on a test has to the others. Because it is expected that the items have varying degrees of relationships to one another, the resulting distribution will be flatter (platykurtic) than the normal distribution. He noted that a perfectly normal distribution would be obtained only with "dead data." More recently, Micceri (1989) surveyed 440 large-sample distributions of measures of achievement and psychological characteristics. He found that all 440 samples significantly deviated from the normal distribution. He likened the finding of a normal distribution of data to the probability of finding a unicorn.

As you will find later in the book, some statistical tests assume that the sample of data to be analyzed comes from a population that is normally distributed. However, many of these statistical tests possess a characteristic known as robustness; that is, correct statistical decisions based on the tests will still be correct despite violations of the assumption of normality. Thus, while Micceri (1989) clearly demonstrated many deviations from normality in his 440 samples, he did not clearly demonstrate any repercussions of the violations.

It is interesting that Galton, even in his own time, recognized the potential limitations of the normal curve. In his biography, he states that he may have "pushed the application of the Law of Frequency of Error somewhat too far." However, consistent with modern thought, Galton believed that "the applicability of that law is more than justified within . . . reasonable limits."

Edward Tufte (born 1940), a contemporary Yale Statistics Professor Emeritus, has self-published seven books on the effective presentation of data. He has been a vocal critic of the statistical and presentation foibles of NASA management and engineers that especially contributed to the *Challenger* and *Columbia* space shuttle disasters. In one of his books, he also criticizes ready-made software presentation packages that he feels mislead, misdirect, and water down more effective methods for presenting data. More important, he is not just a statistical critic; he also offers many workshops on constructive and creative presentations of data.

The stem-and-leaf creator, John Wilder Tukey (1915–2000), was a Professor of Statistics at Princeton for his entire academic career (although he also worked for Bell Labs, Inc.). Like his contemporary Tufte, Tukey was also concerned with discovering ways to present data effectively. Tukey encouraged statisticians to reject the role of "guardians of proven truth" and to resist providing all-encompassing single solutions to real-world problems. He also made important and long-lasting contributions to the important inferential statistical test, analysis of variance (which will be discussed in Chapter 9).

Key Terms and Definitions

Bar graph—A graph that often represents nominal data in rectangular columns.

Bimodal distribution—A frequency distribution that has twin peaks.

Cumulative frequency distribution—A frequency distribution where the distribution of scores is progressively represented by the total frequency.

Frequency distribution—A set of scores arranged in order of magnitude along the *x*-axis, and the frequency of each score is represented along the *y*-axis.

Frequency histogram—A graphic representation of a set of scores where the individual scores are represented by a bar whose height corresponds to the frequency of the score.

Frequency polygon—A graphic representation of a frequency distribution where the individual scores are grouped into class intervals.

Kurtosis—Refers to the peakedness or flatness of the overall shape of a frequency distribution.

Leptokurtosis—A frequency distribution that has a tendency toward peakedness.

Multivariate—A term that usually indicates multiple variables, including multiple independent variables and/or multiple dependent variables.

Negative skew—An asymmetrical frequency distribution whose left tail is longer than the right.

Normal curve or bell-shaped curve—The most frequently occurring distribution whose shape resembles a bell.

Percentiles—A distribution that is divided into hundredths.

Platykurtosis—A frequency distribution that has a tendency toward flatness.

Positive skew—An asymmetrical frequency distribution whose right tail is longer than the left.

Quartiles—A frequency distribution that is divided into fourths.

Skewed distributions—An asymmetrical frequency distribution where either the right tail or the left tail is much longer than the other.

Stem-and-leaf plot—A representation of a distribution where the individual scores are preserved.

Univariate—A term that usually indicates a single dependent variable, although multiple independent variables may be involved.

Chapter 2 Practice Problems

1. Name the four types of distributions.
2. Describe the qualities of kurtosis and skewness.
3. Using the following information, construct a bar graph, and remember to label both axes of the graph. A clinical psychologist is comparing her net income for the first 6 months of the year. In January, she made $8500; February, $5000; March, $2500; April, $3750; May, $4500; and June, $4900.
4. A psychologist studying intelligence tested the intelligence of 30 college psychology students using the Wechsler Adult Intelligence Scale-Third Edition (WAIS-III). Following is a table of the full-scale intelligence scores the psychologist obtained:

103	92	113	110	122	122
115	100	133	111	131	108
108	121	110	124	100	107
98	110	109	127	99	111
122	109	103	97	113	101

For the data presented above,

a. Create a table showing the cumulative frequency distribution of the individual scores.
b. Create a stem-and-leaf plot of the data.
c. Create a table using intervals to summarize the data.

5. Based on the table you created for response 4c above, create a frequency distribution graph. Describe the graph's shape and skewness, if any.

Chapter 2 Test Questions

1. Perhaps the oldest presentation in history of descriptive statistics was
   a. a frequency distribution
   b. graphs and tables
   c. a frequency polygon
   d. a pie chart

2. In her bar graph presentation to the queen of England, Florence Nightingale controlled for which two rival hypotheses?
   a. age and gender
   b. age and ethnicity
   c. age and war
   d. health and conditions
   e. war and gender

3. What did Dr. Snow do to show very effectively that his hypothesis was correct and thereby demonstrated the cause-effect relationship he advocated?
   a. drank from the Broad Street well
   b. removed the pump handle from the Broad Street well
   c. isolated the bacteria from the Broad Street well
   d. performed an autopsy on one of the victims before the city council
   e. all of the above

4. In Figure 2.3 (truck brands), the major problem of the presentation was
   a. cutting off the bottom 95% of the bars' heights
   b. hiding the brand names of the competing trucks
   c. too many useless colors
   d. meaningless three-dimensional bars
   e. all of the above

5. The main difference between a bar graph and a histogram is
   a. two-dimensional bars instead of three-dimensional bars
   b. vertical instead of a horizontal presentation
   c. bar graphs have spaces between bars and histograms don't
   d. they are identical in every aspect

6. The normal curve is also called the
   a. vertical frequency distribution
   b. bimodal distribution

c. kurtotic curve
d. bell-shaped curve

7. A positively skewed distribution is also said to be
   a. skewed normally
   b. skewed left
   c. skewed down
   d. skewed up
   e. none of the above

8. In a frequency distribution table where a majority of the frequencies are 1, the best solution would be to
   a. group the data into intervals
   b. use cumulative frequency as the dependent variable
   c. use percentages instead of frequencies
   d. use a frequency polygon
   e. all of the above would be acceptable solutions

9. Which of the following is NOT true about Tukey's stem-and-leaf plots?
   a. presents the data horizontally instead of vertically
   b. presents all of the numbers in the set
   c. can result in a single stem for some data sets
   d. represents missing values with 0 or 99

10. A very pointed narrow frequency distribution graph is said to be
    a. positively skewed
    b. negatively skewed
    c. leptokurtic
    d. mesokurtic
    e. platykurtic

11. A very flat frequency distribution graph is said to be
    a. positively skewed
    b. negatively skewed
    c. leptokurtic
    d. mesokurtic
    e. platykurtic

12. A perfectly normal frequency distribution graph is said to be
    a. positively skewed
    b. negatively skewed
    c. leptokurtic
    d. mesokurtic
    e. platykurtic

13. Which of the following was NOT a characteristic of bad graphs according to Tufte?
    a. low density
    b. chart junk
    c. changing scales

d. labeling badly
e. the use of only black and white in a graph

14. Who used the word *statistics* in the English language as a method of determining a country's inhabitants' happiness?
    a. De Moivre
    b. Sinclair
    c. Gauss
    d. Galton

15. Currently, the best substitute for the word *errors* in statistics is
    a. problems
    b. klinkers
    c. variations
    d. mistakes

16. In a study by Micceri of 440 large-sample distributions, what percentages were significantly different from the true normal distribution?
    a. 54%
    b. 79%
    c. 92%
    d. 100%

17. How many graphs and tables did the rocket manufacturers prepare to convince NASA that the seals in the Space Shuttle *Challenger* might fail in cold weather?
    a. 1
    b. 2
    c. 13
    d. 25

18. How many graphs and tables in the previous question showed the direct comparison between the number of seal failures and temperature?
    a. 0
    b. 1
    c. 2
    d. 13
    e. all 25 but NASA had already made up its mind

19. What error did Florence Nightingale commit in her bar graph presentation to the queen of England?
    a. interval widths were too large
    b. interval widths were too small
    c. overlapping interval widths
    d. too many colors

20. To display vividly the cause of some types of ulcers, a researcher
    a. cured the patient with milk and a bland diet
    b. injected himself with the bacteria from a patient's ulcer
    c. injected a patient with ulcers with an antibiotic
    d. injected a patient without ulcers with bacteria from a patient's ulcer

21. For the following set of numbers 21, 22, 25, 25, 27, 29, 30, 32, 33, 34, 35, 36, 36, 38, 39, 39, 39, 40, 42, 44, 45, 56, 57, 59, 60, 60, 60, set up a frequency distribution table with interval widths of 10 (starting at 20–29). What is the frequency of the second interval?
    a. 6
    b. 8
    c. 11
    d. 4

22. For the previous set of numbers, what is the percentage of frequency of the second interval rounded to one decimal place?
    a. 20.7%
    b. 40.7%
    c. 40.74%
    d. 40.8%

23. For the previous set of numbers, what is the cumulative percentage, including the second interval rounded to one decimal place?
    a. 62.96%
    b. 62.9%
    c. 63.0%
    d. 62.9629%

24. In a stem-and-leaf plot of the previous data, what would the second stem look like?
    a. 3 | 12345668999
    b. 3 | 0123456678999
    c. 3 | 02345668999
    d. 3032333435(1)36(2)38(1)39(3)

25. In a stem-and-leaf plot of the previous data, what would the third stem look like?
    a. 4 | 0225
    b. 40 | 1, 42 | 1, 44 | 1, 45 | 1
    c. 5 | 679
    d. 6 | 000

3 Statistical Parameters

Measures of Central Tendency and Variation

Chapter 3 Goals

- Understand and compute measures of central tendency
- Understand and compute measures of variation
- Learn the differences between klinkers and outliers and how to deal with them
- Learn how Tchebysheff's theorem led to the development of the empirical rule
- Learn about the sampling distribution of means, the central limit theorem, and the standard error of the mean
- Use the empirical rule for prediction

Measures of Central Tendency

In addition to graphs and tables of numbers, statisticians often use common parameters to describe sets of numbers. There are two major categories of these parameters. One group of parameters measures how a set of numbers is centered around a particular point on a line scale or, in other words, where (around what value) the numbers bunch together. This category of parameters is called **measures of central tendency.** You already know and have used the most famous statistical parameter from this category, which is the **mean** or average.

The Mean

The mean is the arithmetic average of a set of scores. There are actually different kinds of means, such as the harmonic mean (which will be discussed later in the book) and the geometric mean. We will first deal with the arithmetic mean. The mean gives someone an idea where the center lies for a set of scores. The arithmetic mean is obtained by taking the sum of all the numbers in the set and dividing by the total number of scores in the set.

You already learned how to derive the mean in Chapter 1. Here again is the official formula in proper statistical notation:

$$\bar{x} = \frac{\sum_{i=1}^{k} x_i}{N},$$

where

$\bar{x}$ = the mean,

$\sum_{i=1}^{k} x_i$ = the sum of all the scores in the set,

N = the number of scores or observations in the set.

Note that from this point on in the text, we will use the shorthand symbol Σx instead of the proper $\sum_{i=1}^{k} x_i$.

The mean has many important properties that make it useful. Probably its most attractive quality is that it has a clear conceptual meaning. People almost automatically understand and easily form a picture of an unseen set of numbers when the mean of that set is presented alone. Another attractive quality is that the mathematical formula is simple and easy. It involves only adding, counting, and dividing. The mean also has some more complicated mathematical properties that also make it highly useful in more advanced statistical settings such as inferential statistics. One of these properties is that the mean of a sample is said to be an **unbiased estimator** of the population mean. But first, let us back up a bit.

The branch of statistics known as inferential statistics involves making inferences or guesses from a sample about a population. As members of society, we continually make decisions, some big (such as what school to attend, who to marry, or whether to have an operation) and some little (such as what clothes to wear or what brand of soda to buy). We hope that our big decisions are based on sound research. For example, if we decide to take a drug to lower high blood pressure, we hope that the mean response of the participants to the drug is not just true of the sample but also of the population, that is, all people who could take the drug for high blood pressure. So, if we take the mean blood pressure of a sample of patients after taking the drug, we hope that it will serve as an unbiased estimator of the population mean; that is, the mean of a sample should have no tendency to overestimate or underestimate the population mean, μ (which is also written *mu* and pronounced mew). Thus, if consecutive random samples are drawn from a larger population of numbers, each sample mean is just as likely to be above μ as it is to be below μ. This property is also useful because it means that the population formula for μ is the same as the sample formula for $\bar{x}$. These formulas are as follows:

	Sample	*Population*
Mean	$\bar{x} = \frac{\sum x}{N}$	$\mu = \frac{\sum x}{N}$

The Median

Please think back to when you heard income or wealth reports in the United States. Can you recall if they reported the mean or the median income (or wealth)? More than likely, you heard reports about the median and not the mean. Although the mean is the mostly widely used measure of central tendency, it is not always appropriate to use it. There may be many situations where the **median** may be a better measure of central tendency. The median value in a set of numbers is that value that divides the set into equal halves when all the numbers have been ordered from lowest to highest. Thus, when the median value has been derived, half of all the numbers in the set should be above that score, and half should be below that score. The reason that the median is used in reports about income or wealth in the United States is that income and wealth are unevenly distributed (i.e., they are not normally distributed). A plot of the frequency distribution of income or wealth would reveal a skewed distribution. Now, would the resulting distribution be positively skewed or negatively skewed? It is easier to answer that question if you think in terms of wealth. Are *most* people in the United States wealthy and there's just a few outlying poor people, or do most people have modest wealth and a few people own a huge amount? Of course, the latter is true. It is often claimed that the wealthiest 3% of U.S. citizens own about 40% of the total wealth. Thus, wealth is positively skewed. A mean value of wealth would be skewed or drawn to the right by a few higher values, giving the appearance that the average person in the United States is far wealthier than he or she really is. In all skewed distributions, the mean is strongly influenced by the few high or low scores. Thus, the mean may not truthfully represent the central tendency of the set of scores because it has been raised (or lowered) by a few outlying scores. In these cases, the median would be a better measure of central tendency. The formula for obtaining the median for a set of scores will vary depending on the nature of the ordered set of scores. The following two methods can be used in many situations.

Method 1

When the scores are ordered from lowest to highest and there are an odd number of scores, the middle value will be the median score. For example, examine the following set of scores:

7, 9, 12, 13, 17, 20, 22

Because there are seven scores and 7 is an odd number, then the middle score will be the median value. Thus, 13 is the median score. To check whether this is true, look to see whether there is exactly the same number of scores above and below 13. In this case, there are three scores above 13 (17, 20, and 22) and three scores below 13 (7, 9, and 12).

Method 2

When the scores are ordered from lowest to highest and there are an even number of scores, the point midway between the two middle values will be the median score. For example, examine the following set of scores:

2, 3, 5, 6, 8, 10

There are six scores, and 6 is an even number; therefore, take the average of 5 and 6, which is 5.5, and that will be the median value. Notice that in this case, the median value is a hypothetical number that is not in the set of numbers. Let us change the previous set of numbers slightly and find the median:

2, 3, 5, 7, 8, 10

In this case, 5 and 7 are the two middle values, and their average is 6; therefore, the median in this set of scores is 6. Let us obtain the mean for this last set, and that is 5.8. In this set, the median is actually slightly higher than the mean. Overall, however, there is not much of a difference between these two measures of central tendency. The reason for this is that the numbers are relatively evenly distributed in the set. If the population from which this sample was drawn is normally distributed (and not skewed), then the mean and the median of the sample will be about the same value. In a perfectly normally distributed sample, the mean and the median will be exactly the same value.

Now, let us change the last set of numbers once again:

2, 3, 5, 7, 8, 29

Now the mean for this set of numbers is 9.0, and the median remains 6. Notice that the mean value was skewed toward the single highest value (29), while the median value was not affected at all by the skewed value. The mean in this case is not a good measure of central tendency because five of the six numbers in the set fall below the mean of 9.0. Thus, the median may be a better measure of central tendency if the set of numbers has a skewed distribution. See Figure 3.1 for graphic examples.

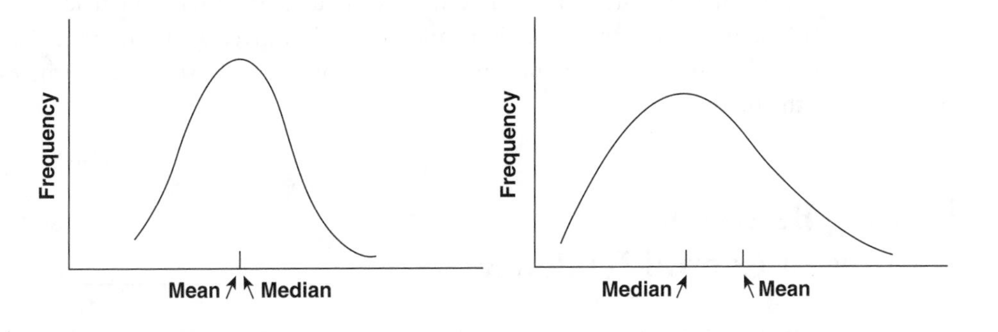

Figure 3.1

If there are ties at the median value when you use either of the two previous methods, then you should consult an advanced statistics text for a third median formula, which is much more complicated than the previous two methods. For example, examine the following set:

2, 3, 5, 5, 5, 10

There are an even number of scores in this set, and normally we would take the average of the two middle values. However, there are three 5s, and that constitutes a tie at the median value. Notice that if we used 5 as the value of the median, there is one score above the value 5, and there are two scores below 5. Therefore, 5 is not the correct median value.

The Mode

The **mode** is a third measure of central tendency. The mode score is the most frequently occurring number in a set of scores. In the previous set of numbers, 5 would be the mode score because it occurs at a greater frequency than any other number in that set. Notice that the mode score in that set is 5, and the frequency (how many are there) of the mode score is 3 because there are three 5s in the set.

It is also possible to have two or more mode scores in a set of numbers. For example, examine this set:

2, 3, 3, 3, 4, 5, 6, 6, 6, 8

In this set, there are two modes: One mode score is 3, and the other mode score is 6. The frequency of both mode scores is 3. A distribution that has two different modes is said to be bimodally distributed. The mode score can change drastically across different samples, and thus it is not a particularly good overall measure of central tendency. The mode probably has its greatest value as a measure with nominal or categorical scales. For example, we might report in a study that there were 18 male and 14 female participants. Although it may appear obvious and highly intuitive, we know that there were more male than female participants because we can see that 18 (males) is the mode score.

Choosing Between Measures of Central Tendency

In one of my journal articles, "Dreams of the Dying" (Coolidge & Fish, 1983), an undergraduate student and I obtained dream reports from 14 dying cancer patients. In the method section of the article, we reported the subjects' ages. Rather than list the ages for all 14 subjects, what category of statistical parameters would be appropriate? Of course, it would be the category of measures of central tendency. The following numbers represent the subjects' ages at the time of their deaths:

28, 34, 40, 40, 42, 43, 45, 48, 59, 59, 63, 63, 81, 88

The mean for this set of scores is 52.4, and the median is 46.5 [(45 + 48)/2]. Typically, a researcher would not report both the mean and the median, so which of the two measures would be reported? A graph of the frequency distribution (by intervals of 10 years) shows that the distribution appears to be skewed right. See Figure 3.2 for two versions of the frequency distribution.

Because of this obvious skew, the mean is being pulled by the extreme scores of 81 and 88. Thus, in this case, we reported the median age of the subjects instead of the mean.

In most statistical situations, the mean is the most commonly used measure of central tendency. Besides its ability to be algebraically manipulated,

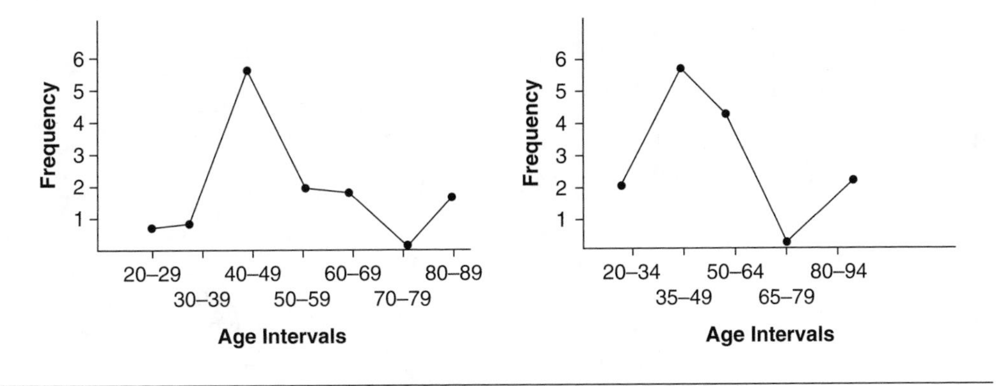

Figure 3.2

which allows it to be used in conjunction with other statistical formulas and procedures, the mean also is resistant to variations across different samples.

Klinkers and Outliers

Sometimes when we are gathering data, we may have equipment failure, or in a consumer preference study, we may have a subject who does not speak English. A datum from these situations may be simply wrong or clearly inappropriate. Abelson (1995) labels these numbers **klinkers.** Abelson argues that when it is clearly inappropriate to keep klinkers, they should be thrown out, and additional data should be gathered in their place. Can I do this? Is this ethical? You may legitimately ask yourself these questions. Tukey (1969) and Abelson (1995) both warn against becoming too "stuffy" and against the "sanctification" of statistical rules. If equipment failure has clearly led to an aberrant score, or if some participants in a survey failed to answer some questions because they did not speak English, then our course is clear. Keeping these data would ruin the true spirit of the statistical investigative process. We are not being statistically conservative but foolish if we kept such data.

The other type of aberrant score in our data is called an **outlier.** Outliers are deviant scores that have been legitimately gathered and are not due to equipment failures. While we may legitimately throw out klinkers, outliers are a much murkier issue. For example, the sports section of a newspaper reported that the Chicago Bulls were the highest paid team (1996–1997) in the National Basketball Association, with an average salary of $4.48 million. However, let's examine the salaries and see if the mean is an accurate measure of the 13 Bulls players.

As a measure of central tendency, the mean salary is misleading. Only two players have salaries above the mean, Jordan and Rodman, while 11 players are below the mean. A better measure of central tendency for these data would be the median. Randy Brown's salary of $1,300,000 would be the median salary since 6 players have salaries above that number and 6 players

Table 3.1

Player	*Salary*
Jordan	$30,140,000
Rodman	$9,000,000
Kukoc	$3,960,000
Harper	$3,840,000
Longley	$2,790,000
Pippen	$2,250,000
Brown	$1,300,000
Simpkins	$1,040,000
Parish	$1,000,000
Wennington	$1,000,000
Kerr	$750,000
Caffey	$700,000
Buechler	$500,000

are below that number. It is clear that Michael Jordan's salary is skewing the mean, and his datum would probably be considered an outlier. Without Jordan's salary, the Bulls's average pay would be about $2,300,000, which is less than half of what the Bulls's average salary would be with his salary. Furthermore, to show how Jordan's salary skews the mean, Michael Jordan made about $3,000,000 more than the whole rest of the team's salaries combined!

We will revisit the issue of outliers later in this chapter. Notice in Jordan's case, we did not eliminate his datum as an outlier. First, we identified it as an outlier (because of its effect of skewing the mean of the distribution) and reported the median salary instead. Second, we analyzed the data (team salaries) both ways, with Jordan's salary and without, and we reported both statistical analyses. There is no pat answer to dealing with outliers. However, the latter approach (analyzing the data both ways and reporting both of the analyses) may be considered semi-conservative. The more conservative position would be to analyze the data and report them with the outlier and have no other alternative analyses without the outlier.

Uncertain or Equivocal Results

One of my colleagues sums up statistics to a single phrase, "it's about measurement." Often, as in the case of klinkers and outliers, measurement is not as straightforward as it would appear. Sometimes measurement issues arise because there is a problem with the measurement procedures or measurement. When the results of the measurement are unclear, uncertain, or unreadable, they are said to be **equivocal.**

Intermediate results occur when a test outcome is neither positive nor negative but falls between these two conditions. An example might be a home pregnancy kit that turns red if a woman is pregnant and stays white if she is not. If it turns pink, how are the results to be interpreted? **Indeterminate results** may occur when a test outcome is neither positive nor negative and does not fall between these two conditions. Perhaps, in the pregnancy kit example, this might occur if the kit turns green. Finally, it may be said that **uninterpretable results** have occurred when a test has not been given according to its correct instructions or directions, or when it yields values that are completely out of range.

Measures of Variation

The second major category of statistical parameters is **measures of variation.** Measures of variation tell us how far the numbers are scattered about the center value of the set. They are also called measures of dispersion. There are three common parameters of variation: the **range, standard deviation,** and **variance.** While measures of central tendency are indispensable in statistics, measures of variation provide another important yet different picture of a distribution of numbers. For example, have you heard of the warning against trying to swim across a lake that averages only 3 feet deep? While the mean does give a picture that the lake on the whole is shallow, we intuitively know that there is danger because while the lake may average 3 feet in depth, there may be much deeper places as well as much shallower places. Thus, measures of central tendency are useful in understanding how scores cluster about a center value, and measures of variation are useful in understanding how far, wide, or deep the high scores are scattered about the center value.

The Range

The range is the simplest of the measures of variation. The range describes the difference between the lowest score and the highest score in a set of numbers. Typically, statisticians do not actually report the range value, but they do state the lowest and highest scores. For example, given the set of scores 85, 90, 92, 98, 100, 110, 122, the range value is $122 - 85 = 37$. Therefore, the range value for this set of scores is 37, the lowest score is 85, and the highest score is 122.

It is also important to note that the mean for this set is 99.6, and the median score is 98. However, neither of these measures of central tendency tells us how far the numbers are from this center value. The set of numbers 95, 96, 97, 98, 99, 103, 109 would also have a mean of 99.6 and a median of 98. However, notice how the range, as a measure of variation, tells us that in the first set of numbers, they are widely distributed about the center (range = 37), while in the second set, the range is only 14 ($109 - 95 = 14$). Thus, the second set varies less about its center than the first set.

Let us refer back to the ages of the dream subjects in the previously mentioned study. Although the mean of the subjects' ages was 52.4, we have no idea how the ages are distributed about the mean. In fact, because the median was reported, the reader may even suspect that the ages are not evenly distributed about the center. The reader might correctly guess that the ages might be skewed, but the reader would not know whether the ages were positively or negatively skewed. In this case, the range might be useful. For example, it might be reported that the scores ranged from a low of 28 to a high of 88 (although the range value itself, which is 60, might be of little conceptual use). Although the range is useful, the other two measures of variation, standard deviation and variance, are used far more frequently. The range is useful as a preliminary descriptive statistic, but it is not useful in more complicated statistical procedures, and it varies too much as a function of sample size (the range goes up when the sample size goes up). The range also depends only on the highest and lowest scores (all of the other scores in the set do not matter), and a single aberrant score can affect the range dramatically.

The Standard Deviation

The standard deviation is a veritable bulwark in the sea of statistics. Along with the mean, the standard deviation is a theoretical cornerstone in inferential statistics. The standard deviation gives an approximate picture of the average amount each number in a set varies from the center value. To appreciate the standard deviation, let us work with the idea of the average deviation. Let us work with a small subsample of the ages of the dream subjects:

28, 42, 48, 59, 63

Their mean is 48.0. Let us see how far each number is from the mean.

Note that the positive and negative signs tell us whether an individual number is above or below the mean. The size or magnitude of the distance score tells us how far that number is from the mean.

Each Number (x_i)	*Mean* ($\bar{x}$)	*Distance from Mean* ($x_i - \bar{x}$)
28	48	−20
42	48	−6
48	48	0
59	48	+11
63	48	+15

To get the average deviation for this set of scores, we would normally sum the five distance values and divide by 5 because there were five scores. In this case, however, if we sum −20, −6, 0, +11, and +15, we would get 0 (zero), and 0 divided by 5 is 0.

One solution to this dilemma would be to take the absolute value of each distance. This means that we would ignore the negative signs. If we now try to average the absolute values of the distances, we would obtain the sum of 20, 6, 0, 11, and 15, which is 52, and 52 divided by 5 is 10.4. Now, we have a picture of the average amount each number varies from the mean, and that number is 10.4.

The average of the absolute values of the deviations has been used as a measure of variation, but statisticians have preferred the standard deviation as a better measure of variation, particularly in inferential statistics. One reason for this preference, especially among mathematicians, is that the absolute value formula cannot be manipulated algebraically.

The formula for the standard deviation for a population is

$$\sigma = \sqrt{\frac{\sum(x_i - \bar{x})^2}{N}}$$

Note that σ or sigma represents the population value of the standard deviation. You previously learned Σ as the command to sum numbers together. Σ is the capital Greek letter, and σ is the lowercase Greek letter. Also note that although they are pronounced the same, they have radically different meanings. The sign Σ is actually in the imperative mode; that is, it states a command (to sum a group of numbers). The other value σ is in the declarative mode; that is, it states a fact (it represents the population value of the standard deviation).

The sample standard deviation has been shown to be a biased estimator of the population value, and consequently, there is bad news and good news. The bad news is that there are two different formulas, one for the sample standard deviation and one for the population standard deviation. The good news is that statisticians do not often work with a population of numbers. They typically only work with samples and make inferences about the populations from which they were drawn. Therefore, we will only use the sample formula. The two formulas are presented as follows:

	Sample	*Population*
Standard Deviation	$S = \sqrt{\frac{\sum(x_i - \bar{x})^2}{N - 1}}$	$\sigma = \sqrt{\frac{\sum(x_i - \bar{x})^2}{N}}$

where S (capital English letter S) stands for the sample standard deviation.

Correcting for Bias in the Sample Standard Deviation

Notice that the two formulas only differ in their denominators. The sample formula has $N - 1$ in the denominator, and the population has only

N. When it was determined that the original formula (containing only N in the denominator) consistently underestimated the population value when applied to samples, the correction -1 was added to correct for the bias. Note that the correction makes the numerator larger, and that makes the value of the sample standard deviation larger (if we divide the numerator by a large number, then it makes the numerator smaller; if we divide the numerator by a smaller number, then that makes the numerator larger). The correction for bias has its greatest effect in smaller samples, for example, dividing by 9 instead of 10. In larger samples, the power of the correction is diminished, yet statisticians still leave the correction in the formula, even in the largest samples.

How the Square Root of x^2 Is Almost Equivalent to Taking the Absolute Value of x

As mentioned earlier, statisticians first used the absolute value method of obtaining the average deviation as a measure of variation. However, squaring a number and then taking the square root of that number also removes negative signs while maintaining the value of the distance from the mean for that number. For example, if we have a set of numbers with a mean of 4 and our lowest number in the set is 2, then $2 - 4 = -2$. If we square -2, we get 4, and the square root of $4 = 2$. Therefore, we have removed the negative sign and preserved the original value of the distance from the mean. Thus, when we observe the standard deviation formula, we see that the numerator is squared, and we take the square root of the final value. However, if we take a set of numbers, the absolute value method for obtaining the standard deviation and the square root of the squares method will yield similar but not identical results. The square root of the square method has the mathematical property of weighting numbers that are farther from the mean more heavily. Thus, given the previous subset of numbers

28, 42, 48, 59, 63

the absolute value method for standard deviation (without the correction for bias) yielded a value of 10.4, while the value of the square root of the squares method is 12.5.

The Computational Formula for Standard Deviation

One other refinement of the standard deviation formula has also been made, and this change makes the standard deviation easier to compute. The following is the computational formula for the sample standard deviation:

$$S = \sqrt{\frac{\sum x^2 - \frac{(\sum x)^2}{N}}{N-1}}$$

Remember that this computational formula is exactly equal to the theoretical formula presented earlier (the proof of their equality is presented in the appendix). The computational formula is simply easier to compute. The theoretical formula requires going through the entire data three times: once to obtain the mean, once again to subtract the mean from each number in the set, and a third time to square and add the numbers together. Note that on most calculators, Σx and Σx^2 can be performed at the same time; thus, the set of numbers will only have to be entered in once. Of course, many calculators can obtain the sample standard deviation or the population value with just a single button (after entering all of the data). You may wish to practice your algebra, nonetheless, with the computational formula and check your final answer with the automatic buttons on your calculator afterwards. Later in the course, you will be required to pool standard deviations, and the automatic standard deviation buttons of your calculator will not be of use. Your algebraic skills *will* be required, so it would be good to practice them now.

The Variance

The variance is a third measure of variation. It has an intimate mathematical relationship with standard deviation. Variance is defined as the average of the square of the deviations of a set of scores from their mean. In other words, we use the same formula as we did for the standard deviation, except that we do not take the square root of the final value. The formulas are presented as follows:

	Sample	*Population*
Variance	$s^2 = \frac{\sum(x_i - \bar{x})^2}{N-1}$	$\sigma^2 = \frac{\sum(x_i - \bar{x})^2}{N}$

Statisticians frequently talk about the variance of a set of data, and it is an often-used parameter in inferential statistics. However, it has some conceptual drawbacks. One of them is that the formula for variance leaves the units of measurement squared. For example, if we said that the standard deviation for shoe sizes is 2 inches, it would have a clear conceptual meaning. However, imagine if we said the variance for shoe sizes is 4 inches squared. What in the world does "inches squared" mean for a shoe size?

This conceptual drawback is one of the reasons that the concept of standard deviation is more popular in descriptive and inferential statistics.

The Sampling Distribution of Means, the Central Limit Theorem, and the Standard Error of the Mean

A **sampling distribution** is a theoretical frequency distribution that is based on repeated (a large number of times) sampling of n-sized samples from a given population. If the means for each of these many samples are formed into a frequency distribution, the result is a sampling distribution. In 1810, French mathematician Pierre LaPlace (1749–1827) formulated the **central limit theorem**, which postulated that if a population is normally distributed, the sampling distribution of the means will also be normally distributed. However, more important, even if the scores in the population are not normally distributed, the sampling distribution of means will still be normally distributed. The standard deviation of a sampling distribution of means is called the **standard error of the mean.** As the sample size of the repeated samples becomes larger, the standard error of the mean becomes smaller. Also, as the sample size increases, the sampling distribution approaches a normal distribution. At what size a sample does the sampling distribution approach a normal distribution? Most statisticians have decided upon $n = 30$. It is important to note, however, if the population from which the sample is drawn is normal, then any size sample will lead to a normally distributed sampling distribution. The characteristics of the sampling distribution serve as an important foundation for hypothesis testing and the statistical tests presented later in this book. However, at this point in a statistics course, it might be important to remember at least this: In inferential statistics, a sample size becomes arbitrarily large at $n \geq 30$.

The Use of the Standard Deviation for Prediction

Pafrutti Tchebysheff (1821–1894), a Russian mathematician, developed a theorem, which ultimately led to many practical applications of the standard deviation. Tchebysheff's theorem could be applied to samples or populations, and it stated that specific predictions could be made about how many numbers in a set would fall within a standard deviation or standard deviations from the mean. However, the theorem was found to be conservative, and statisticians developed the notion of the **empirical rule.** The empirical rule holds only for normal distribution or relatively mound-shaped distributions.

The empirical rule predicts the following:

1. Approximately 68% of all numbers in a set will fall within ±1 standard deviation of the mean.
2. Approximately 95% of all numbers in a set will fall within ±2 standard deviations of the mean.
3. Approximately 99% of all numbers in a set will fall within ±3 standard deviations of the mean.

For example, let us return to the ages of the subjects in the dream study previously mentioned:

28, 34, 40, 40, 42, 43, 45, 48, 59, 59, 63, 63, 81, 88

The mean is 52.4. The sample standard deviation computational formula is as follows:

$$S = \sqrt{\frac{\sum x^2 - \frac{(\sum x)^2}{N}}{N-1}} = \sqrt{\frac{42{,}287 - \frac{(733)^2}{14}}{13}} = \sqrt{\frac{42{,}287 - \frac{537{,}289}{14}}{13}}$$

$$S = \sqrt{\frac{42{,}287 - 38{,}377.7857}{13}} = \sqrt{\frac{3{,}909.2143}{13}} = \sqrt{300.7088} = 17.34$$

Thus, $S = 17.34$.

Now, let us see what predictions the empirical rule will make regarding this mean and standard deviation.

$$1.\ \bar{x} + 1\,S = 52.4 + 17.3 = 69.7$$
$$\bar{x} - 1\,S = 52.4 - 17.3 = 35.1$$

Thus, the empirical rule predicts that approximately 68% of all the numbers will fall within this range of 35.1 years old to 69.7 years old.

If we examine the data, we find that 10 of the 14 numbers are within that range, and 10/14 is about 70%. We find, therefore, that the empirical rule was relatively accurate but conservative.

$$2.\ \bar{x} + 2\,S = 52.4 + 2\,(17.3) = 52.4 + 34.6 = 87.0$$
$$\bar{x} - 2\,S = 52.4 - 2\,(17.3) = 52.4 - 34.6 = 17.8$$

Inspection of the data reveals that 13 of the 14 numbers in the set fall within two standard deviations of the mean, or approximately 93%. The empirical rule predicted about 95%; thus, it was relatively accurate for these data, but this time it was too liberal a prediction.

$$3.\ \bar{x} + 3\ S = 52.4 + 3\ (17.3) = 52.4 + 51.9 = 104.3$$

$$\bar{x} - 3\ S = 52.4 - 3\ (17.3) = 52.4 - 51.9 = 0.5$$

All 14 of the 14 total numbers fall within three standard deviations of the mean. The empirical rule predicted 99%, and again we see the rule was relatively accurate and conservative.

Practical Uses of the Empirical Rule: As a Definition of an Outlier

Some statisticians use the empirical rule to define outliers. For example, a few statisticians define an outlier as a score in the set that falls outside of ±3 standard deviations of the mean. However, caution is still urged before deciding to eliminate even these scores from an analysis because although they may be improbable, they still occur.

Practical Uses of the Empirical Rule: Prediction and IQ Tests

The empirical rule has great practical significance in the social sciences and other areas. For example, IQ scores (on Wechsler's IQ tests) have a theoretical mean of 100 and a standard deviation of 15. Therefore, we can predict with a reasonable degree of accuracy that 68% of a random sample of normal people taking the test should have IQs between 85 and 115.

Furthermore, only 5% of this sample should have an IQ below 70 or above 130 because the empirical rule predicted that 95% would fall within two standard deviations of the mean. Because IQ scores are assumed to be normally distributed and both tails of the distribution are symmetrical, we can predict that 2.5% of people will have an IQ less than 70 and 2.5% will have IQs greater than 130.

What percentage of people will have IQs greater than 145? An IQ of 145 is exactly three standard deviations above the mean. The empirical rule predicts that 99% should fall within ±3 standard deviations of the mean. Therefore, of the 1.0% who fall above 145 or below 55, 0.5% will have IQs above 145.

Some Further Comments

The two categories of parameters, measures of central tendency and measures of variation, are important in both simple descriptive statistics and in inferential statistics. As presented, you have seen how parameters from both categories are necessary to describe data. Remember that the purpose of statistics is to summarize numbers clearly and concisely. The parameters, mean and standard deviation, frequently accomplish these two goals, and a parameter from each category is necessary to describe data. However, being able to understand the data clearly is the most important goal of statistics. Thus, not always will the mean and standard deviation be the appropriate parameters to describe data. Sometimes, the median will make better sense of the data, and most measures of variability do not make sense for nominal or categorical data.

History Trivia

Fisher to Eels

Ronald A. Fisher (1890–1962) received an undergraduate degree in astronomy in England. After graduation, he worked as a statistician and taught mathematics. At the age of 29, he was hired at an agricultural experimental station. Part of the lure of the position was that they had gathered approximately 70 years of data on wheat crop yields and weather conditions. The director of the station wanted to see if Fisher could statistically analyze the data and make some conclusions. Fisher kept the position for 14 years. Consequently, modern statistics came to develop some strong theoretical "roots" in the science of agriculture.

Fisher wrote two classic books on statistics, published in 1925 and 1935. He also gave modern statistics two of its three most frequently used statistical tests, *t* tests and analysis of variance. Later in his career, in 1954, he published an interesting story of a scientific discovery about eels and the standard deviation. The story is as follows:

Johannes Schmidt was an ichthyologist (one who studies fish) and biometrician (one who applies mathematical and statistical theory to biology). One of his topics of interest was the number of vertebrae in various species of fish. By establishing means and standard deviations for the number of vertebrae, he was able to differentiate between samples of the same species depending on where they were spawned. In some cases, he could even differentiate between two samples from different parts of a fjord or bay.

However, with eels, he found approximately the same mean and same large standard deviation from samples from all over Europe, Iceland, and Egypt. Therefore, he inferred that eels from all these different places had the same breeding ground in the ocean. A research expedition in the Western Atlantic Ocean subsequently confirmed his speculation. In fact, Fisher notes, the expedition found a different species of eel larvae for eels of the Eastern rivers of North America and the Gulf of Mexico.

Key Terms, Symbols, and Definitions

Central limit theorem—A mathematical proposition that states that if a population's scores are normally distributed, the sampling distribution of means will also be normally distributed, and if the scores in the population are not normally distributed, the sampling distribution of means will still be normally distributed.

Empirical rule—A rule of thumb that dictates that approximately 68% of all scores will fall within plus or minus one standard deviation of the mean in a normal distribution, 95% will fall within plus or minus two standard deviations, and 99% will fall within plus or minus three standard deviations of the mean.

Equivocal results—Test results that are unclear, uncertain, or unreadable.

Indeterminate results—A condition when a test's outcome is neither positive nor negative and does not fall between these two conditions.

Intermediate results—A condition when a test's outcome is neither positive nor negative but falls between these two conditions.

Klinkers—A term used for a datum that is clearly wrong or inappropriate (e.g., a number not on the measuring scale).

Mean ($\bar{x}$)—The arithmetical average of a group of scores.

Measures of central tendency—Parameters that measure the center of a frequency distribution.

Measures of variation—Parameters that measure the tendency of scores to be close or far away from the center point.

Median—The center of a distribution of scores, such that half of the scores are above that number, and half of the scores in the distribution are below that number.

Mode—The most frequently occurring score.

Outliers—These are deviant scores that have been legitimately gathered and are not due to equipment failures, yet they are uncommon scores or they range far from the central tendency of the data set.

Range—The lowest score and the highest score in a distribution.

Sampling distribution—A theoretical frequency distribution based on repeated sampling of *n*-sized samples from a given population. If the means of these samples are formed into a frequency distribution, the result is a sampling distribution.

Standard deviation (σ ***or SD***)—A parameter of variability of data about the mean score.

Standard error of the mean—The standard deviation of a sampling distribution of means.

Unbiased estimator—A sample parameter that neither overestimates nor underestimates the population value.

Uninterpretable results—A condition when a test's outcome is invalid because it has not been given according to its correct instructions or directions.

Variance (σ^2)—A parameter of variability of data about the mean score, which is the square of the standard deviation.

Chapter 3 Practice Problems

1. Find the mean, median, and mode for each of the following sets of scores:
   a. 35, 55, 80, 72, 55, 66, 74
   b. 110, 115, 102, 102, 107, 102, 108, 110
   c. 21, 19, 18, 30, 16, 30
   d. 24, 26, 27, 22, 23, 22, 26

2. To familiarize yourself with the measures of variation, compute the range, standard deviation, and variance for each set of scores.
   a. 6, .5, .8, .6, .2, .3, .9, .9, .7
   b. 9.62, 9.31, 9.15, 10.11, 9.84, 10.78, 9.08, 10.17, 11.23, 12.45
   c. 1001, 1253, 1234, 1171, 1125, 1099, 1040, 999, 1090, 1066, 1201, 1356
3. State the differences between klinkers and outliers. How do statisticians deal with each?
4. State the three predictions made by the empirical rule.

Chapter 3 Test Questions

1. Which of the following is NOT true of the mean?
   a. it is appropriate for normal and skewed distributions
   b. it is the same as the arithmetic average
   c. it belongs in the category of parameters of central tendency
   d. it is considered an unbiased estimator
2. The "unbiased" aspect of an unbiased estimator indicates that
   a. it holds for all ethnic groups
   b. it underestimates the population value with the same tendency as it overestimates the population value
   c. it works in skewed as well as nonskewed distributions
   d. all of the above
3. The most important reason that the median is preferred to the mean is
   a. it is more useful than the mean in inferential statistics
   b. it is an unbiased estimator of the population median value
   c. it is a more accurate measure of central tendency in highly skewed distributions
   d. it is a better measure of variation
4. What is the mean (rounded to two decimal places) of the following set of first-time statistics students' ages: 18, 18, 18, 19, 19, 19, 19, 20, 22, 22, 23, 23, 23, 24, 25, 26, 26, 29, 33, 64?
   a. 24.5
   b. 25
   c. 24.50
5. What is the median for the previous set of scores?
   a. 22.5
   b. 22
   c. 23
   d. 23.5
6. What is the mode score, and what is its frequency?
   a. 18 and 3
   b. 23 and 3
   c. 19 and 4
   d. 4 and 19

7. What is the standard deviation (rounded to one decimal place) for the previous set of scores?
   a. 10.12
   b. 10.11
   c. 10.1
   d. 10.2

8. What is the variance (rounded to one decimal place) of the previous set of scores?
   a. 102.4
   b. 102.3
   c. 102.368
   d. 102

9. What is the range of the previous set of scores?
   a. 18
   b. 64
   c. 46
   d. 490

10. With respect to the mean and the median values of the previous set of scores, which of the following is true?
    a. the median is lower than the mean, suggesting a negative skew
    b. the median is lower than the mean, suggesting a positive skew
    c. the standard deviation is lower than the variance, suggesting a positive skew
    d. both b and c are correct

Problems 11–15. The following is a list of annual salaries (in dollars) for a group of mid-level managers in a large company.

98,000	115,000	75,000	88,000
65,000	107,000	72,000	71,000
88,000	57,000	88,000	73,000
97,000	85,000	87,000	73,000
93,000	88,000	71,000	65,000
87,000	83,000	75,000	44,000
81,000	76,000	81,000	89,000

11. What is the mean of the salaries (rounded to a whole number)?
    a. 81,000
    b. 81,143
    c. 88,000
    d. 14,864

12. What is the median of the salaries (rounded to a whole number)?
    a. 81,000
    b. 81,111
    c. 88,000
    d. 14,864

13. What is the mode salary (rounded to a whole number)?
    a. 81,000
    b. 81,111
    c. 88,000
    d. 14,864
14. What is the standard deviation of the salaries (rounded to a whole number)?
    a. 81,000
    b. 81,111
    c. 88,000
    d. 14,701
15. The top salary in the previous set was $115,000. Would you consider that salary
    a. a klinker
    b. an outlier
    c. neither because it falls well within three standard deviations of the mean

Problems 16–21. The following is a list of annual advertising expenses (in dollars) for a group of states' mental health centers.

1500	11,000	7500	800
600	2350	3000	3800
500	6000	8000	500
1750	1250	50	3400
9000	890	1600	2200

16. What is the mean of the advertising expenses (rounded to a whole number)?
    a. $3284
    b. $3285
    c. $1975
    d. $3241
17. What is the median of the advertising expenses (rounded to a whole number)?
    a. $3284
    b. $3285
    c. $1975
    d. $3241
18. What is the mode advertising expense (rounded to a whole number)?
    a. $500
    b. $3285
    c. $1975
    d. $3241
19. What is the standard deviation of the advertising expenses (rounded to a whole number)?
    a. $3284
    b. $3285
    c. $1975
    d. $3241

20. The lowest advertising expense in the previous set was $50. Would you consider that expense
    a. a klinker
    b. an outlier
    c. neither because it falls well within three standard deviations of the mean

21. What is the range of the previous advertising expenses?
    a. $11,050
    b. $10,050
    c. $10,950
    d. none of the above

22. The most common IQ tests have a standardized mean of 100 and a standard deviation of 15. Based on this information, use the empirical rule to answer the following question: If approximately 348 prospective college students are tested for their IQ, approximately how many will fall within one standard deviation of the mean?
    a. 345
    b. 338
    c. 313
    d. 237

23. Approximately how many students will have at least an IQ of 100 but not greater than 115?
    a. 68%
    b. 169
    c. 174
    d. 118

24. Approximately how many students will have IQs greater than 100?
    a. 118
    b. 174
    c. 169
    d. 237

25. Approximately how many students should have IQs greater than 130?
    a. 9
    b. 18
    c. 27
    d. 36

4 Standard Scores, the *z* Distribution, and Hypothesis Testing

Chapter 4 Goals

- Understand standard scores, such as *z* scores and *T* scores
- Learn about the classic *z* distribution
- Learn how to calculate *z* scores
- Learn how to convert *z* scores to other types of standard scores
- Learn how to test predictions of the empirical rule with the *z* distribution
- Learn how the *z* distribution is used to test experimental hypotheses
- Become adept with the use of the *z* distribution and *T* scores
- Learn how to convert *z* scores to percentiles

Imagine if a child comes home from school and states that he or she received a 10 on a spelling test and a 20 on an arithmetic test. Did the child do twice as well on the arithmetic test? You can perceive intuitively that there is no way to compare the two scores because we do not know what the maximum score was on either test. This example serves as an introduction to the idea of standard scores and their distribution. Standard scores are values where the mean and the standard deviation are already established, known, or given. Thus, standard scores inherently already have meaning, and a new score can be judged as good, fair, or poor depending on the given mean and standard deviation.

Standard Scores

As noted previously, a **standard score** is a number in a set where the mean and standard deviation of the set are already known or given. One of the most famous standard scores is the intelligence quotient or IQ. For example, IQ scores are a type of standard score, and the mean for a population of participants is 100, and the standard deviation is 15. The Scholastic Aptitude Test (SAT) score is also a type of standard score. Its mean is 500, and the standard deviation is 100. The Graduate Record Examination (GRE) and the Graduate Management Achievement Test (GMAT) are also standard scores, also with a mean of 500 and a standard deviation of 100. Interestingly, both the SAT and the GRE scores are based on a much older standard score known as the ***T* score.** *T* scores have a mean of 50 and a standard deviation of 10. You can see quickly that SATs, GREs, and GMATs are multiples of 10 of *T* scores.

Standard scores are a useful way of summarizing and comparing scores across different kinds of rating scales or tests. For example, if a high school student's raw score on the quantitative portion of the SAT is 67, has the student done well or poorly? Because we do not know the mean and standard deviation of the standardization sample for the quantitative portion of the SAT, we cannot ascertain whether the student has done well. Furthermore, with only raw scores, we cannot compare scores across different scales. For example, is this same student's score of 67 on the quantitative portion of the SAT better than a 55 raw score on the verbal portion? Without knowing the number of items and the means and standard deviations of each scale, we cannot make any appropriate comparisons.

Imagine the advantages of reporting scores as standard scores. As soon as we know a student's standard score, we would know immediately where he or she stands relative to a standardization sample measured on that same scale. If the student's SAT verbal score was 650 and his or her quantitative score was 450, we would immediately know that the student did better on the verbal portion than the quantitative portion.

On IQ tests, the use of two standard deviations above or below the mean is often used to establish criteria for gifted children programs and for eligibility for programs benefiting mentally retarded children. For example, among other criteria, some gifted children programs require an IQ of 130 or greater. This criterion is obviously two standard deviations (2 × 15) above the mean IQ (100). Similarly, many programs for mentally retarded persons use the criterion of an IQ of 70 or below, which is two standard deviations below the mean IQ. Because the empirical rule states that approximately 95% of all scores will fall within plus or minus two standard deviations of the mean, then approximately 2.5% of IQ scores would be above 130, and 2.5% of the IQ scores would fall below a score of 70. On this statistical basis, 2.5% of children would be considered "gifted" and 2.5% would be considered "mentally retarded." Would the resulting children actually or behaviorally be gifted or retarded? Of course not, but these statistical criteria serve as the basis of a beginning for the establishment of more relevant and associated criteria.

The Classic Standard Score: The *z* score and the *z* Distribution

One important and the oldest type of standard score is the ***z* score.** The mean of a z distribution is 0.00, and the value of one standard deviation is 1.00. z scores are the initial step in converting any raw score into any other kind of standard score. Most statisticians convert their raw scores into z scores and then into some other standard scores because z scores have two unfortunate characteristics. One is that a z score of 0 has bad connotations. For example, imagine if a person takes a major test at school and he or she comes home and reports a zero score. People might automatically assume that the person did not do very well at all, yet if the person is reporting a z score, that person scored right at the mean. z scores thus have a strong chance of being misinterpreted by people without statistical training. Indeed, even trained statisticians have unconscious difficulty with family members coming home and reporting zeros on tests, even if they know that the scores are z scores.

A second problem with z scores is that they involve negative numbers. Statisticians and nonstatisticians have a natural aversion to negative numbers. We do not like them in our checking accounts, on final exams, or to identify football players ("Hey, look, there goes player number minus 7"). In the ***z* distribution,** any z score less than the mean of the z distribution will be a negative number, and in most samples, that will be approximately 50% of all of the scores. As shown previously when we derived the standard deviation formula, negative numbers are more difficult to deal with mathematically than positive numbers.

Despite the mathematical and psychological problems presented by negative z scores, the z distribution lies at the heart of inferential statistics. This is because statisticians use the z distribution to test experimental hypotheses, such as whether monetary bonuses improve worker performance, whether a new cancer drug is effective, or whether brain-injured patients will score higher on some psychological test than a non-brain-injured group.

Let us briefly practice converting from SAT scores back into z scores without using a formula. What would the earlier student's verbal score of 650 be if it were reported as a z score? Because the value of one standard deviation of an SAT score is 100, and the student scored 150 points above the SAT mean of 500, then +150 divided by 100 is 1.5. Since the value of one standard deviation of a z score is 1.00, then we know the student has a z score of 1.50 (or 1.50 standard deviations above the mean z score of 0.00). What about the student's quantitative z score? Because his or her SAT score for the quantitative portion was 450 and the mean SAT score is 500, the student was −50. Dividing this value by the value of the SAT's standard deviation of 100 yields a z score of −0.50 (remember z scores are typically reported to two decimal places). Thus, an SAT score of 600 is the equivalent of a z score of 1.00 and a *T* score of 60. An SAT score of 450 is the equivalent of a *z* score of −0.50 and a *T* score of 45.

Calculating *z* Scores

To change scores into *z* scores, the mean and the standard deviation of the scores must be known (or at least derivable, meaning that we assume that we would have access to all of the scores in the set). If you have the mean and standard deviation for a set of scores, then any individual number can be converted to a *z* score by the following formula:

z Score Formula

$$z \text{ score} = \frac{x_i - \bar{x}}{S}$$

where

x_i = any individual score in the set of original numbers,

$\bar{x}$ = the mean of the original set of numbers,

S = the standard deviation for the original set of numbers.

Other standard scores can be converted into *z* scores or vice versa. In a previous example, a student had a *T* score of 60 on the verbal portion of the SAT and a *T* score of 45 on the quantitative part. Now, let us use the formula for obtaining *z* scores to convert between the *T* scores and *z* scores. For the verbal score:

$$z \text{ score} = \frac{x_i - \bar{x}}{S} = \frac{600 - 500}{100} = \frac{100}{100} = 1.00$$

Thus, we see formulaically how the *T* score of 60 converts to a *z* score of +1.00.

To convert the quantitative score of 450 into a *z* score, we follow the same procedure:

$$z \text{ score} = \frac{x_i - \bar{x}}{S} = \frac{450 - 500}{100} = \frac{-50}{100} = -0.50$$

More Practice on Converting Raw Data Into *z* Scores

Let us convert the following raw scores, obtained by 10 nursing students on their first psychopharmacology exam, into *z* scores:

67, 74, 77, 81, 85, 89, 92, 93, 94, 99

First, obtain the mean and standard deviation. Note that these parameters should be accurate to at least three decimal places, so that the resulting z score is accurate to at least two decimal places.

$$\bar{x} = \frac{\sum x}{N}$$

$$\bar{x} = \frac{851}{10}$$

$$\bar{x} = 85.100$$

$$S = \sqrt{\frac{\sum x^2 - \frac{\left(\sum x\right)^2}{N}}{N-1}}$$

$$S = \sqrt{\frac{73351 - \frac{724201}{10}}{9}}$$

$$S = \sqrt{\frac{73351 - 72420.1}{9}}$$

$$S = \sqrt{\frac{930.9}{9}}$$

$$S = \sqrt{103.433}$$

$$S = 10.170$$

$$\text{Formula for } z\text{: } z = \frac{x_i - \bar{x}}{S}$$

Thus, for the raw score of 67:

$$z = \frac{67 - 85.100}{10.170}$$

Raw Score	*z Score*
67	−1.78
74	−1.09
77	−0.80
81	−0.40
85	−0.01
89	0.38
92	0.68
93	0.78
94	0.88
99	1.37

It is important to remember that the z is a standard score whose distribution is assumed to be normally distributed. When the raw scores from any distribution are converted to z scores, the resulting distribution will approximate the normal distribution. This is because the z distribution is a perfectly normal distribution based on an infinite number of scores. By converting from raw scores to z scores, interpretations can then be made as if the scores are normally distributed. However, that is not to say that this assumption is correct. Perhaps the raw scores come from a nonnormal distribution. If raw scores were converted into z scores from a distribution that deviates strongly from normality, the resulting inferences might be inappropriate. Fortunately, many raw score distributions in nature approximate the normal distribution. Therefore, the transformations from raw scores to z scores, in most cases, will be appropriate. How can you tell whether your data are normally distributed? Simply graph your data in a frequency distribution!

Converting From *z* Scores to Other Types of Standard Scores

Suppose that you wanted to convert the previous z scores into some other type of standard score. Once the z scores have been obtained, transforming them into any other type of standard score is easy and can be performed with the following formula:

$$z' = \bar{x}' + (S')(z \text{ score})$$

where

z' (z prime) = the new standard score that you are trying to obtain (this is unknown in this equation; it is what you are trying to find),

$\bar{x}'$ ($\bar{x}$ prime) = the mean of the new standard score (this is a known quantity),

S' (S prime) = the standard deviation of the new standard score (also a known quantity).

Thus, if the teacher of the nursing students wanted to convert the z scores into T scores ($\bar{x} = 50$, $S = 10$), to eliminate the bad connotations associated with the minus signs of the z scores, he or she would do the following:

Formula to convert from a z score to another type of standard score:

$$z' = \bar{x}' + (S')(z \text{ score})$$

Thus, for the first score in the set, 67,

$$T = 50 + (10)(-1.78)$$
$$T = 50 + (-17.8)$$
$$T = 50 - 17.8$$
$$T = 32.2$$

$T = 32$ (T scores are typically rounded to whole numbers)

Raw Score	*z Score*	T *Score*
67	−1.78	32
74	−1.09	39
77	−0.80	42
81	−0.40	46
85	−0.01	50
89	0.38	54
92	0.68	57
93	0.78	58
94	0.88	59
99	1.37	64

Note that the original raw scores give the students no indication of their standing relative to the class. With performance scores converted to z scores or T scores, the students can tell where they stand versus the other students' scores.

The *z* Distribution

See Appendix A, page 375 of this book. It contains the z distribution. Table 4.1 has selected values of the z distribution, so that you can become comfortable reading the complete distribution.

Table 4.1 Selected Values of the *z* Distribution

z	*B*	*C*
0.00	.0000	.5000
0.50	.1915	.3085
1.00	.3413	.1587
1.64	.4495	.0505
1.96	.4750	.0250
2.00	.4772	.0228
2.57	.4949	.0051
3.00	.4987	.0013
4.00	.49997	.00003

Appendix A shows the proportion of scores that can be expected from the mean of the z score (0.00) to a given z score. In fact, the predictions that were made for the empirical rule were obtained from this z distribution. It can also be observed that the predictions of the empirical rule had rounding errors in them.

The values under B in Table 4.1 give the proportion of scores from positive z scores of 0.00 (the mean of all z scores) to a target z score. The values under C in Table 4.1 give the proportion of scores from the target z score to an infinitely high z score. Remember, a proportion can be read and interpreted just like decimals. In Table 4.1, look at a z score of 1.00; what is the B area proportion? It is .3413, and this means that .3413 of all z scores can be accounted for from a z score of 0.00 to a z score of 1.00. This proportion may also be converted to a percentage by moving the decimal place two places over to the right. Thus, 34.13% of all z scores occur from a z score of 0.00 to a z score of 1.00.

How many z scores can be accounted for from a z score of 0.00 to infinity? The total proportion of the z distribution is 1.00. A z score of 0.00 divides the z distribution exactly in half; thus, the answer to the previous question is .5000. Another way to solve this problem is to use the C area. Remember the C area gives the proportion of scores from the target z score to infinity. Thus, if our target z is 0.00 (because the question asked *from* a z score of 0.00 *to* infinity), then the C area answers the question because the C area reports *from* the target z score (in this case 0.00) *to* infinity. The C area in this case is .5000.

Notice in Table 4.1 that the B area and the C area always add up to .5000. This makes sense because the B area gives the proportion from a z score of 0.00 to a target z, and the C area gives the proportion from the target z to infinity. Thus, one half of the z distribution can be accounted for by the B and C areas combined.

Interpreting Negative *z* Scores

To interpret negative z scores, just imagine the positive z scores as negative. For example, imagining the z score of 1.00 as a $z = -1.00$ still yields a B proportion of .3413 and a C proportion of .1587. Thus, from a z score of 0.00 to a z score of −1.00 still accounts for .3413 or 34.13% of all the total scores. The C area shows what proportion of the scores is from the target z score (in this case −1.00) to an infinite negative z score, and that is .1587 or 15.87% of the total z scores.

Testing the Predictions of the Empirical Rule With the *z* Distribution

The empirical rule states that approximately 68% of all scores will fall within ±1 standard deviation of the mean. If the proportion of scores from a z of 0.00 to a z score of 1.00 is .3413, then just double this proportional value to obtain the *exact* proportion of scores that fall within ±1 standard deviation of the mean. The reason that this value is doubled is that we want to find the total proportion of scores from the mean z score of 0.00 to the value of one standard deviation on *each* side of the mean. This means from a z of 0.00 to a z score of 1.00 and from a z score of 0.00 to a z score of −1.00. It is important to remember that the z scores can be positive or negative depending on whether the target scores are less than or greater than the mean. Thus, the answer to the previous question would be .6826 or 68.26% of the total scores. The proportional answers must always be positive. Remember, the question in most cases will be what proportion of all cases can be accounted for by a z score. It would make no sense whatsoever to have a negative proportion or negative percentage of cases. The empirical rule gives only a rough or approximate estimate of the real proportion of scores. Remember the empirical rule states that ±2 standard deviations from the mean would account for 95% of all the z scores? Let us check the accuracy of that statement.

A z score of 2.00 would be two standard deviations from the mean. The B area would have the proportion of scores from a z of 0.00 to a z of 2.00, and that is .4772. Doubling this value yields the value .9544 or 95.44%. Again, the proportion was doubled because we wanted to find the proportion of scores from a z score mean of 0.00 to one standard deviation on each side of the mean. Now we know that 95.44% of all z scores fall between a z score of −2.00 and +2.00, and the empirical rule stated 95%. Therefore, the empirical rule was incorrect by .0044 or .44%.

How accurate is the empirical rule at three standard deviations? It stated that 99% of all scores would fall within three standard deviations of the mean. A z score of 3.00 yields a B proportion of .4987. If we add the proportions of z scores from a z score of 0.00 to a z of 3.00 and from a z of 0.00 to a z of −3.00, we obtain .9974 or 99.74% of all scores fall within ±3 standard deviations of the mean on the z distribution, and the empirical rule would be incorrect by .0074 or .74%.

Why Is the *z* Distribution So Important?

Standard scores and the z distribution may not appear overly important to you at this point. However, the z distribution and distributions like it have more meaningful purposes. The z distribution is a perfectly normal distribution based on an infinite number of scores. Approximately normal distributions occur more frequently than any other kind of distribution in nature. Throughout our exploration of inferential statistics, we will assume that our data come from a normally distributed population or, at the very least, that it is relatively mound-shaped. If this assumption is true or mostly true, then we can use the z distribution to test experimental hypotheses noted earlier such as whether monetary bonuses improve worker performance, whether a new cancer drug is effective, or whether brain-injured patients will score higher on some neuropsychological test than a non-brain-injured group.

How We Use the *z* Distribution to Test Experimental Hypotheses

z scores are used as the foundation for the testing of experimental hypotheses. The following is a rough idea of how this process works. Let us imagine that we are trying to determine whether a new teaching technique is more effective in improving reading scores in children with dyslexia than the traditional method. The independent variable would be the two types of treatment (new vs. traditional teaching technique). The dependent variable might be the reading scores on a standardized test after treatment. We will take all of the students' scores in both groups and convert them by a complicated formula into a single *z*-like score. To the extent that the new single *z*-like score is different from zero (greater than or less than $z = 0$), we will conclude that the teaching methods produce reading score means that are different from each other. To the extent that the new single *z*-like score is close to zero, we will conclude that the teaching methods produce the same results. Because 95.44% of all *z* scores should fall within plus or minus two standard deviations of the mean (−2.00 to +2.00) by chance alone, we will assume that only chance is operating if we obtain a *z*-like score within this range; that is, the two treatments do not produce different reading score results. If all of the scores in the experiment result in a *z*-like score greater than +2.00 or less than a *z* score of −2.00, then it will be concluded that the experimental treatments are different (because this would not likely be due to chance, or there would be about a 4.56% or less probability that our results occurred by chance). See Figure 4.1 for a picture of these decisions based on a *z* distribution.

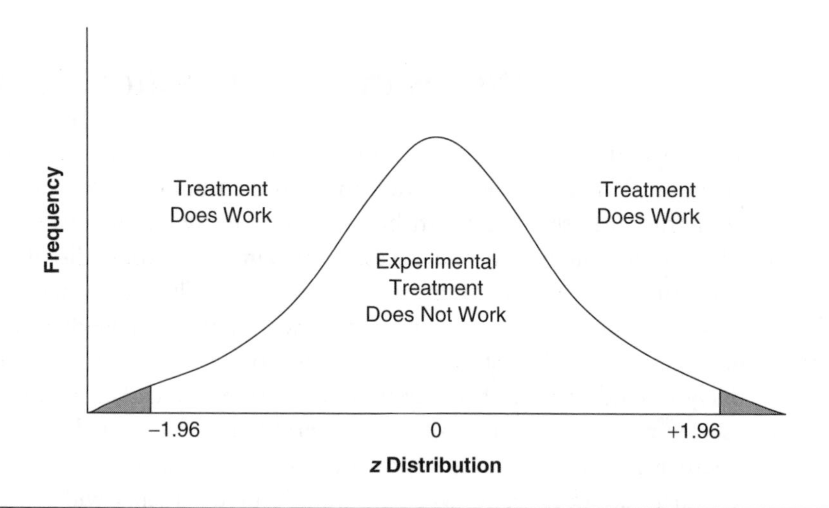

Figure 4.1

For example, with regard to the previous question, "Are the two worker treatments different from each other?" it is automatically assumed in statistics that the treatments do not differ. As you will learn later in this course, this starting position in all experiments is referred to as the null hypothesis.

More Practice With the *z* Distribution and *T* Scores

Example 1: Finding the area in a *z* distribution that falls above a known score where the known score is above the mean

Question: What *percentage* of all *T* scores will be greater than a *T* score of 65?

Answer: First, you may find it highly useful to draw a picture of the question. See Figure 4.2.

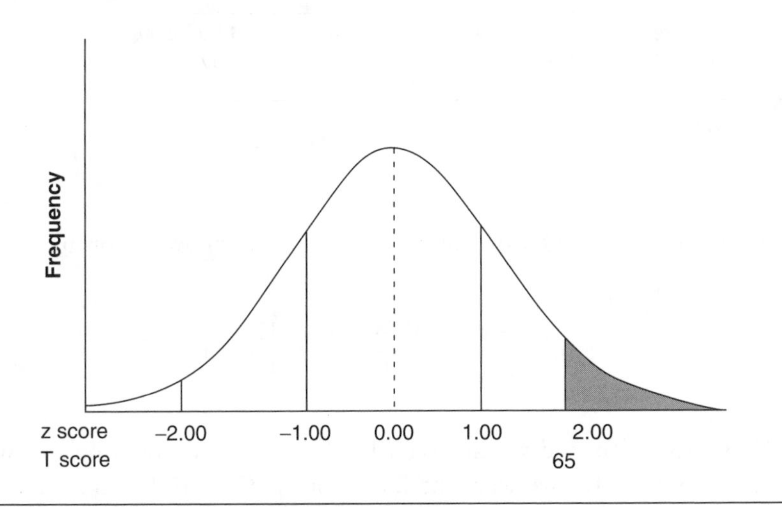

Figure 4.2

Next, convert the *T* score into a *z* score by the following formula:

$$z = \frac{x_i - \bar{x}}{S} = \frac{65 - 50}{10} = \frac{15}{10} = 1.50$$

Now look in Appendix A and see what B and C areas are associated with a z score of 1.50. Do we want the B area or the C area? The B area represents the proportion of scores from a z of 0.00 to a z of 1.50. The C area represents the proportion of scores from a z of 1.50 to an infinite z. We want the C area that will give the proportion of scores greater than a

z of 1.50, which is equivalent to a T score of 65. The proportion in the C area is .0668, or 6.68% of all T scores will be greater than a T score of 65.

Question: What percentage of all IQ scores will be greater than an IQ score of 117?

Answer: Draw a picture of the question. See Figure 4.3.

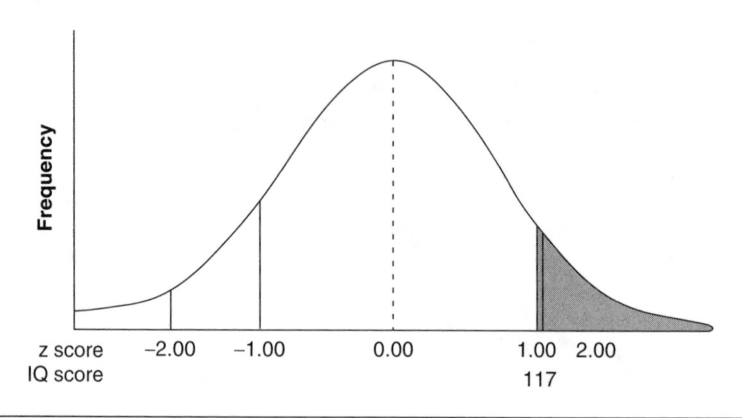

Figure 4.3

Next, convert the IQ score into a z score by the following formula:

$$z = \frac{x_i - \bar{x}}{S} = \frac{117 - 100}{15} = \frac{17}{15} = 1.13$$

Now look in Appendix A and see what B and C areas are associated with a z score of 1.13. Do we want the B area or the C area? The B area represents the proportion of scores from a z of 0.00 to a z of 1.130. The C area represents the proportion of scores from a z of 1.13 to an infinite z. We want the C area that will give the proportion of scores greater than a z of 1.13, which is equivalent to an IQ score of 117. The proportion in the C area is .1292, or 12.92% of all IQ scores will be greater than an IQ score of 117.

Example 2: Finding the area in a z distribution that falls below a known score where the known score is above the mean

Question: What percentage of all T scores will be less than a T score of 61?

Answer: It is highly useful to draw a picture of the question. See Figure 4.4.

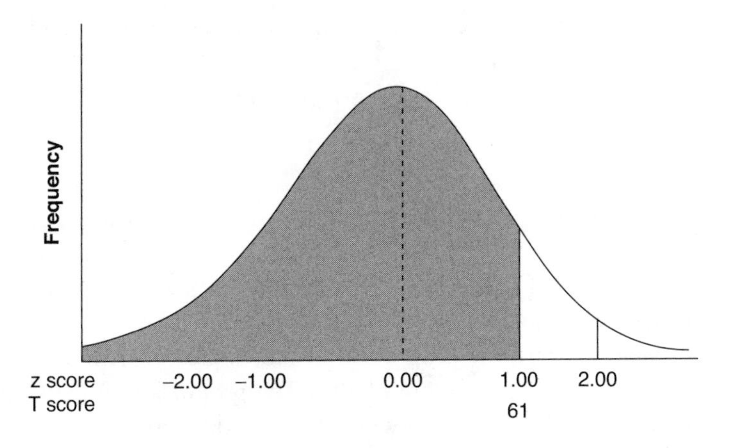

Figure 4.4

Next, convert the T score into a z score by the following formula:

$$z = \frac{x_i - \bar{x}}{S} = \frac{61 - 50}{10} = \frac{11}{10} = 1.10$$

Now look in Appendix A and see what B and C areas are associated with a z score of 1.10. Do we want the B area or the C area? The B area represents the proportion of scores from a z of 0.00 to a z of 1.10. The C area represents the proportion of scores from a z of 1.10 to an infinite z. We want the B area that will give the proportion of scores greater than a z of 0.00 to a z of 1.10, which is .3643. Now all of the area under the z distribution less than a z of 0.00 is .5000. Thus, we add .5000 to .3643, which gives us a proportion of .8643, or 86.43% of all scores will be less than a T score of 61.

Question: What percentage of all IQ scores will be less than an IQ score of 117?

Answer: Again, you may find it useful to draw a picture of the question. See Figure 4.5.

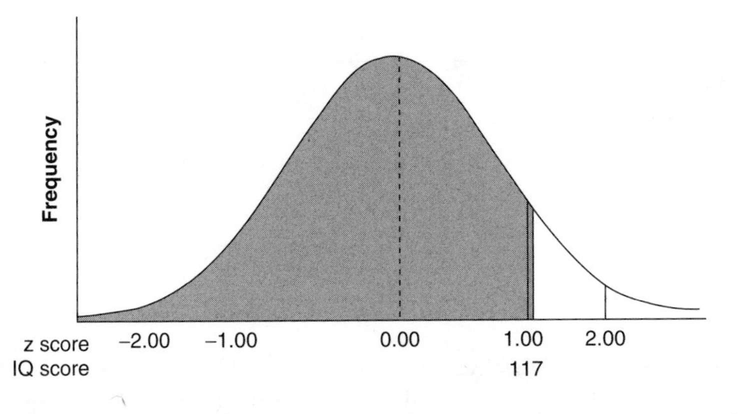

Figure 4.5

Next, convert the IQ score into a z score by the following formula:

$$z = \frac{x_i - \bar{x}}{S} = \frac{117 - 100}{15} = \frac{17}{15} = 1.13$$

Now look in Appendix A and see what B and C areas are associated with a z score of 1.13. Do we want the B area or the C area? The B area represents the proportion of scores from a z of 0.00 to a z of 1.130. The C area represents the proportion of scores from a z of 1.13 to an infinite z. We want the C area that will give the proportion of scores greater than a z of 1.13, which is equivalent to an IQ score of 117. The proportion in the C area is .1292, or 12.92% of all IQ scores will be greater than an IQ score of 117.

Example 3: Finding the area in a *z* distribution that falls below a known score where the known score is below the mean

Question: What percentage of *T* scores is less than a *T* score of 31?

Answer: First, draw a picture of the problem.

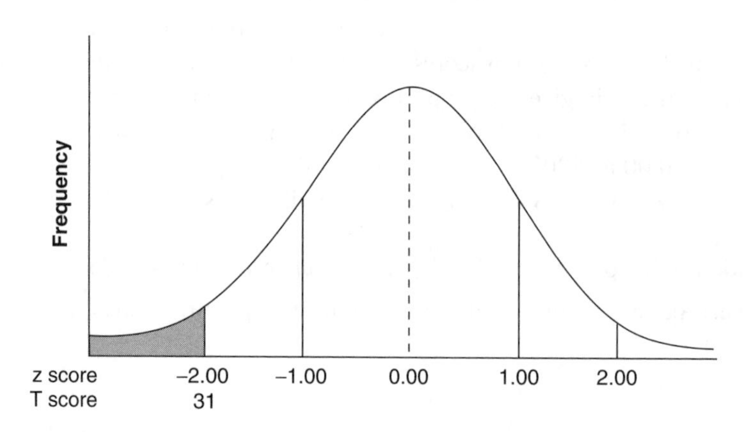

Figure 4.6

Next, convert the *T* score into a z score by the following formula:

$$z = \frac{x_i - \bar{x}}{S} = \frac{31 - 50}{10} = \frac{-19}{10} = -1.90$$

Now look in Appendix A and see what B and C areas are associated with a z score of −1.90. Do we want the B area or the C area? The B area represents the proportion of scores from a z of 0.00 to a z of −1.90. The C area represents the proportion of scores from a z of -1.90 to an infinite negative z. We want the C area that will give the proportion of scores less than a z of -1.90. The proportion in the C area is .0287, or 2.87% of all T scores will be less than a T score of 31.

Question: What percentage of IQ scores is less than an IQ score of 70?

Answer: First, draw a picture of the problem. See Figure 4.7.

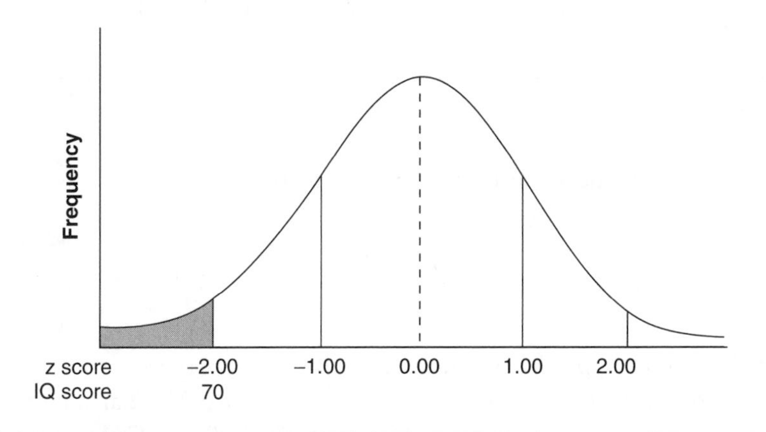

Figure 4.7

Next, convert the IQ score into a z score by the following formula:

$$z = \frac{x_i - \bar{x}}{S} = \frac{70 - 100}{15} = \frac{-30}{15} = -2.00$$

Now look in Appendix A and see what B and C areas are associated with a z score of -2.00. Do we want the B area or the C area? The B area represents the proportion of scores from a z of 0.00 to a z of -2.00. The C area represents the proportion of scores from a z of -2.00 to an infinite negative z. We want the C area that will give the proportion of scores less than a z of -2.00. The proportion in the C area is .0228, or 2.28% of all IQ scores will be less than an IQ score of 70.

Example 4: Finding the area in a z distribution that falls above a known score where the known score is below the mean

Question: What percentage of T scores will be greater than a T score of 31?

Answer: A picture of this problem is almost mandatory. It will help you to think clearly about the problem. See Figure 4.8.

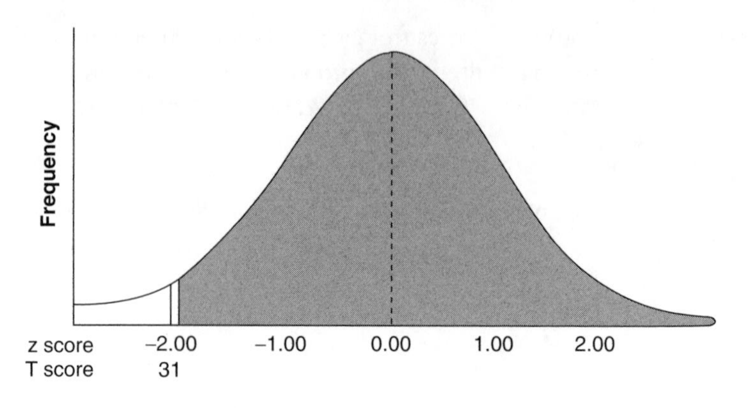

Figure 4.8

Next, convert the *T* score into a *z* score by the following formula:

$$z = \frac{x_i - \bar{x}}{S} = \frac{31 - 50}{10} = \frac{-19}{10} = -1.90$$

Now look in Appendix A and see what B and C areas are associated with a *z* score of −1.90. Do we want the B area or the C area? The B area represents the proportion of scores from a *z* of 0.00 to a *z* of −1.90. The C area represents the proportion of scores from a *z* of −1.90 to an infinite negative *z*. This time, we want the B area that will give the proportion of scores greater than a *z* of −1.90 to the mean *z* score of 0.00. This proportion in the B area is .4713. However, the question asked for the percentage of ALL *T* scores above 31 (not just to the mean *z* of 0.00). What proportion of all *z* scores is above a *z* of 0.00? The answer is .5000. Therefore, to obtain our final answer, we have to add the B area proportion and .5000 together. Thus, we obtain the proportion of .9713, or 97.13% of all the scores are above a *T* score of 31.

Question: What percentage of IQ scores will be greater than an IQ score of 80?

Answer: A picture, please! See Figure 4.9.

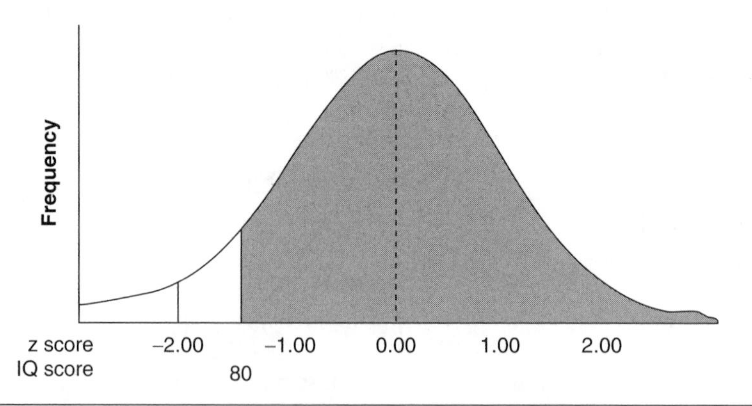

Figure 4.9

Next, convert the IQ score into a z score by the following formula:

$$z = \frac{x_i - \bar{x}}{S} = \frac{80 - 100}{15} = \frac{-20}{15} = -1.33$$

Now look in Appendix A and see what B and C areas are associated with a z score of -1.33. Do we want the B area or the C area? The B area represents the proportion of scores from a z of 0.00 to a z of -1.33. The C area represents the proportion of scores from a z of -1.33 to an infinite negative z. This time, we want the B area that will give the proportion of scores greater than a z of -1.33 to the mean z score of 0.00. This proportion in the B area is .4082. However, the question asked for the percentage of ALL IQ scores above 80 (not just to the mean z of 0.00). What proportion of all z scores is above a z of 0.00? Again, the answer is .5000. Therefore, to obtain our final answer,we have to add the B area proportion and .5000 together. Thus, .4082 + .5000 = .9082, or 90.82% of all the scores are above an IQ score of 80.

Example 5: Finding the area in a z distribution that falls between two known scores where both known scores are above the mean

Question: What percentage of T scores will be between a T score of 55 and a T score of 69?

Answer: First, draw a picture of the problem. See Figure 4.10.

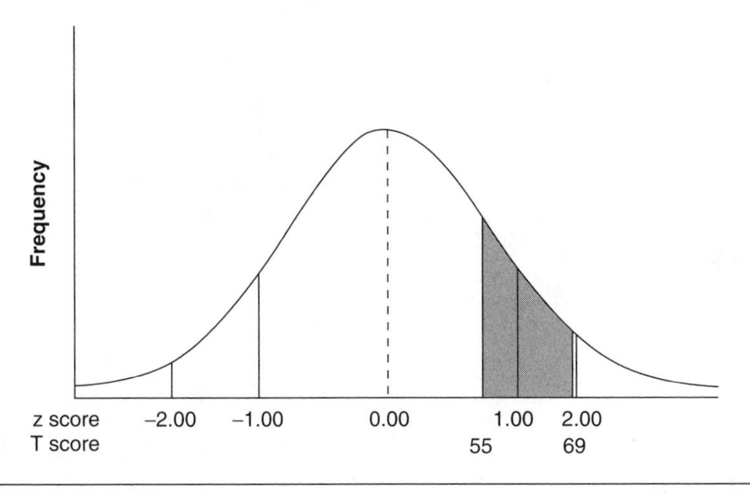

Figure 4.10

Our battle plan will be to convert both T scores to z scores; then we will look up the appropriate proportions and subtract the smaller proportion from the larger proportion.

So after we draw a picture, convert the two T scores into z scores:

$$z = \frac{x_i - \bar{x}}{S} = \frac{55 - 50}{10} = \frac{5}{10} = 0.50$$

and

$$z = \frac{x_i - \bar{x}}{S} = \frac{69 - 50}{10} = \frac{19}{10} = 1.90$$

What B areas are associated with these two values of z? The B area associated with the $z = 0.50$ is .1915. This value represents the proportion of scores from a z of 0.00 to a z of 0.50. The B area for $z = 1.90$ is .4713.

Note from the picture that the B area associated with $z = 1.90$ includes the B area of the $z = 0.50$. Therefore, if we subtract out the smaller B area from the larger, the result will be the proportion of scores between these two values. Thus, .4713 – .1915 = .2798, or 27.98% of the total number of T scores fall between 55 and 69.

Question: What percentage of IQ scores will be between an IQ score of 105 and an IQ score of 125?

Answer: First, draw a picture of the problem. See Figure 4.11.

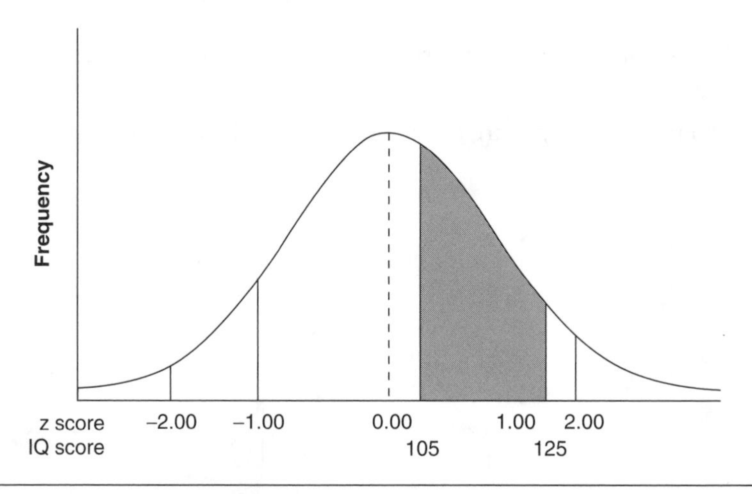

Figure 4.11

Our battle plan will be to convert both IQ scores to z scores; then we will look up the appropriate proportions and subtract the smaller proportion from the larger proportion.

So after we draw a picture, convert the two IQ scores into z scores:

$$z = \frac{x_i - \bar{x}}{S} = \frac{105 - 100}{15} = \frac{5}{15} = 0.33$$

and

$$z = \frac{x_i - \bar{x}}{S} = \frac{125 - 100}{15} = \frac{25}{15} = 1.67$$

The B area for the $z = 0.33$ is .1293. This value represents the proportion of scores from a z of 0.00 to a z of 0.33. The B area for $z = 1.67$ is .4525. Subtracting the smaller proportion from the larger proportion will yield the proportion of IQ scores between 105 and 125. Thus, .4525 – .1293 = .3232, or 32.32% of the total IQ scores will be between an IQ of 105 and 125.

Example 6: Finding the area in a z distribution that falls between two known scores where one known score is above the mean and one is below the mean

Question: What percentage of *T* scores will be between a *T* score of 43 and a *T* score of 72?

Answer: Again, draw a picture of the problem. See Figure 4.12.

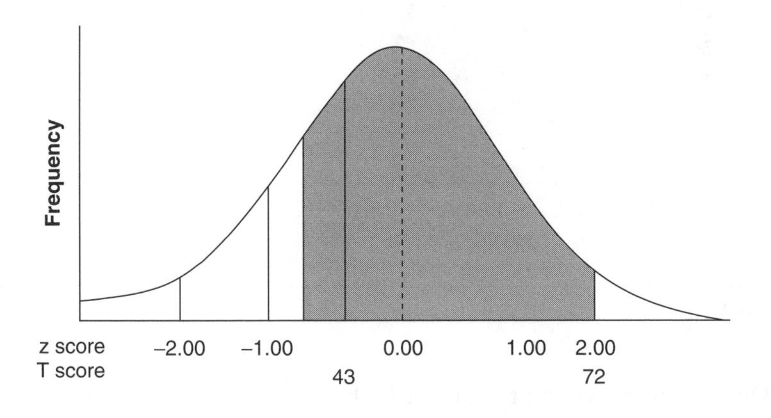

Figure 4.12

Our battle plan here will be to convert both *T* scores to z scores; then we will look up the appropriate proportions and add the two proportions together.

Convert the two *T* scores into z scores:

$$z = \frac{x_i - \bar{x}}{S} = \frac{43 - 50}{10} = \frac{-7}{10} = -0.70$$

and

$$z = \frac{x_i - \bar{x}}{S} = \frac{72 - 50}{10} = \frac{22}{10} = 2.20$$

Do we want the B areas or the C areas? For the z score of -0.70, the B area is needed because we want the proportion of scores between the $z = -0.70$ and the mean $z = 0.00$. This proportion is .2580. For the z score of 2.20, the B area is also needed because we want the proportion from the mean $z = 0.00$ to z score $= 2.20$. This proportion is .4861. Next, simply add the two proportions together to get the total proportion; thus, $.2580 + .4861 = .7441$, or 74.41% of all the T scores fall between the T scores of 43 and 72.

Example 7: Finding the area in a *z* distribution that falls between two known scores where both known scores are below the mean

Question: What percentage of *T* scores will be between a *T* score of 35 and 45?

Answer: Draw a picture of the problem! See Figure 4.13.

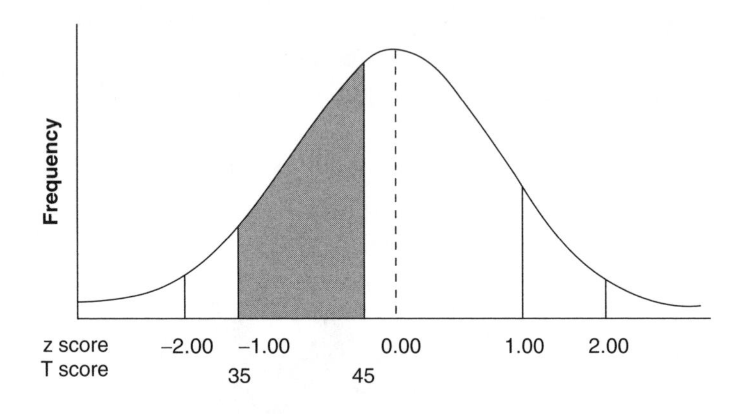

Figure 4.13

Our plan here will be to convert both T scores to z scores; then we will look up the appropriate proportions and subtract the smaller from the larger.

Convert the two T scores into z scores:

$$z = \frac{x_i - \bar{x}}{S} = \frac{35 - 50}{10} = \frac{-15}{10} = -1.50$$

and

$$z = \frac{x_i - \bar{x}}{S} = \frac{45 - 50}{10} = \frac{-5}{10} = 0.50$$

Do we want the B areas or the C areas? For the z score of -1.50, the B area is needed because we want the proportion of scores between the $z = -1.50$ and the mean $z = 0.00$. This proportion is .4332. For the z score of -0.50, the B area is also needed because we want the proportion from the mean $z = 0.00$ to z score $= 0.50$. This proportion is .1914. Next, subtract the smaller proportion from the larger to get the remaining proportion; thus,

.4332 – .1914 = .2418, or 24.18% of all the T scores fall between the T scores of 35 and 45.

Question: What percentage of IQ scores will be between an IQ score of 60 and an IQ score of 90?

Answer: First, draw a picture of the problem. See Figure 4.14.

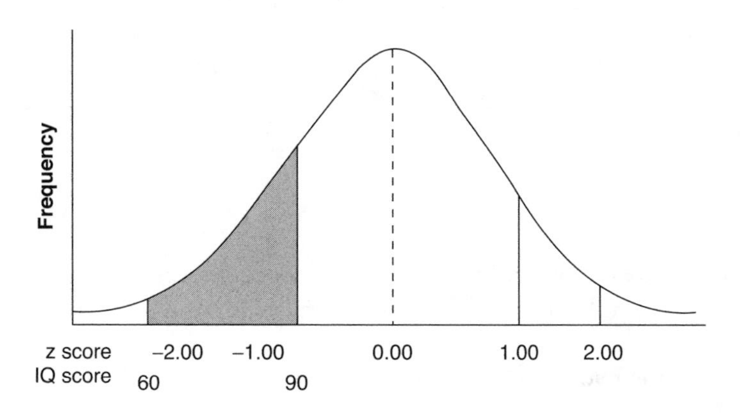

Figure 4.14

Our battle plan will be to convert both IQ scores to z scores; then we will look up the appropriate proportions and again subtract the smaller proportion from the larger proportion.

Convert the two IQ scores into z scores:

$$z = \frac{x_i - \bar{x}}{S} = \frac{60 - 100}{15} = \frac{-40}{15} = -2.67$$

and

$$z = \frac{x_i - \bar{x}}{S} = \frac{90 - 100}{15} = \frac{-10}{15} = -0.67$$

The B area for the $z = -2.67$ is .4962. This value represents the proportion of scores from a z of 0.00 to a z of -2.67. The B area for $z = -0.67$ is .2486. Subtracting the smaller proportion from the larger proportion will yield the proportion of IQ scores between 60 and 90. Thus, .4962 – .2486 = .2476, or 24.76% of the total IQ scores will be between an IQ of 60 and 90.

Question: What percentage of IQ scores will be between an IQ score of 85 and an IQ score of 115?

Answer: Draw a picture. See Figure 4.15.

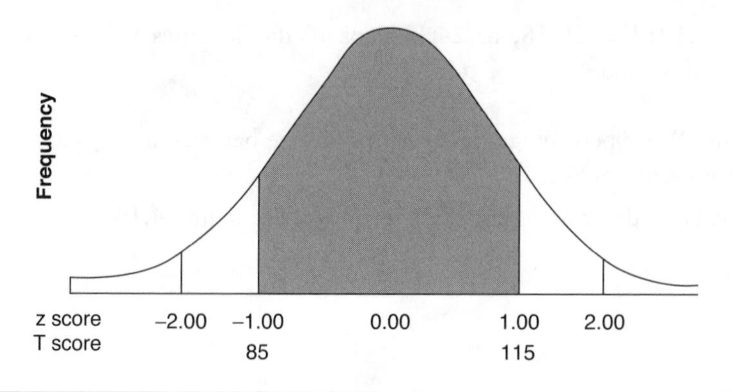

Figure 4.15

Here, our battle plan will be to convert both IQ scores to z scores; then we will look up the appropriate proportions and add the two proportions together.

Convert the two IQ scores into z scores:

$$z = \frac{x_i - \bar{x}}{S} = \frac{85 - 100}{15} = \frac{-15}{15} = -1.00$$

and

$$z = \frac{x_i - \bar{x}}{S} = \frac{115 - 100}{15} = \frac{15}{15} = 1.00$$

The B area for the $z = -1.00$ is .3413. This value represents the proportion of scores from a z of 0.00 to a z of -1.00. The B area for $z = 1.00$ is also .3413. Adding these two proportions together will yield the proportion of IQ scores between 85 and 115. Thus, $.3413 + .3413 = .6826$, or 68.26% of the total IQ scores will be between an IQ of 85 and 115.

Summarizing Scores Through Percentiles

People are fond of summarizing data with statistical parameters that are easily understood, such as **percentiles.** Percentiles provide a ranking in terms of a percentage compared to a larger group (presumably the population). One of the earliest uses of percentile ranking came from developmental psychology. In order to tell whether children were growing up normally, they were measured on height and weight charts. Parents were then told where their children stood compared to children the same age in terms of their height and weight. Let us practice determining percentiles using the previous examples in this chapter.

Question 1: A boy has a height of 4 feet, 4 inches at age 8. According to current height charts, the mean height at this age is 4 feet, 2 inches, with a standard deviation of 3 inches. What is this boy's height in percentile ranking?

Answer 1: First, convert the boy's height into a z score by the following formula:

$$z = \frac{x_i - \bar{x}}{S}$$
$$z = \frac{52 - 50}{3}$$
$$z = \frac{2}{3}$$
$$z = .67$$

The B area of the z distribution for this z score is .2486. Because this child's z score exceeds not only this area from the mean ($z = 0.00$) but also all scores below the mean (an area of .5000), the total proportion of the scores exceeded by this child's height is .2486 + .5000 = .7486. To convert this proportion to a percentage, move the decimal place over two places and round; thus, the child's height is at the 75th percentile. In other words, this child's height exceeds 75% of all children his age, and the child's height is exceeded by 25% of all children the same age.

Question 2: A girl weighs 37 lbs. at age 4. According to current weight charts, the mean weight at this age is 40 lbs., with a standard deviation of 4 lbs. What is this girl's weight as a percentile ranking?

Answer 2: First, convert the child's weight into a z score by the following formula:

$$z = \frac{x_i - \bar{x}}{S}$$
$$z = \frac{37 - 40}{4}$$
$$z = -\frac{3}{4} = -.75$$

The B area of the z distribution for this z score is .2734. In this case, we need the C area since the child's weight is less than the normative mean. The C area for this z score is .2266. Thus, the child's weight exceeds .2266, 22.66%, or places the child in the 23rd percentile of children the same age and gender. This child's weight exceeds 23% of children her age, and the child's weight is exceeded by 77% of children the same age.

History Trivia

Karl Pearson to Egon Pearson

Karl Pearson (1857–1936) has been called the founder of modern statistics. Although *z* scores and the *z* distribution were the creation of another English statistician, William S. Gosset (1876–1937), Pearson is credited (along with Francis Galton) with the inauguration of the first great wave of modern statistical thought.

Pearson received his college degree in mathematics and subsequently studied law and was admitted to the bar. He then traveled to Germany to study physics, metaphysics, religion, and socialism. In 1884, at the age of 27, he returned to London to take a university teaching position in applied mathematics and mechanics. In 1889, Francis Galton published his famous and influential book, *Natural Inheritance.* Pearson was "immensely excited" (Pearson's own words) by Galton's book, particularly by Galton's theoretical work in determining laws of inheritance and the measurement of variability.

In 1892, Pearson published his most provocative book, *The Grammar of Science,* in which he promoted goals and values of science, including its scope (every possible branch of knowledge) and the scientific method, including careful and accurate classification of facts and observations and appropriate statistical analysis. Pearson's subsequent contributions to the modern science of statistics are monumental. He developed the correlation coefficient, which measures the strength of association between two variables (and which bears his name, the Pearson product-moment correlation coefficient), and he developed a goodness-of-fit test, known as the chi-square, which determines how well a mathematical prediction fits some observed data. Even modern statistical language contains many words that were originally coined by Pearson, including the following: array, biserial, *r,* chi-square, coefficient of variation, kurtosis, leptokurtic, platykurtic, multiple correlation, partial correlation, normal curve, standard deviation, and the symbol *sigma* for the standard deviation.

Pearson also enjoyed the process of intellectual argument. He took on the medical establishment when he published a paper that said that contemporary medical treatment of tuberculosis did not reduce death rates. Of course, he made this argument only after examining real data. His recommendations may now be considered bizarre, in light of modern treatment methods. He thought people with a "pedigree" toward tuberculosis should be forbidden to marry or have children. His thoughts on racial intelligence are also now deemed racist and pernicious. However, the value he placed on the scientific method and statistical analysis are the very tools with which modern theorists might fight racism or determine whether a drug treatment for AIDS might be effective.

Alfred Adler (1870–1937) (who is not a famous statistician) was a neo-Freudian who was interested in how our earliest memories affected our later psychological development. Helen Walker (1891–1989), a statistician and historian, relates the following anecdotal story of Karl Pearson's earliest memory: "Well," he said, "I do not know how old I was, but I was sitting in a high chair and I was sucking my thumb. Someone told me to stop sucking it and said that unless I did so the thumb would wither away. I put my two thumbs together and looked at them a long time. 'They look alike to me' I said to myself, 'I can't see that the thumb I suck is any smaller than the other. I wonder if she could be lying to me'" (Walker, 1929/1975).

Helen Walker notes that in the anecdotal memory, Pearson rejected constituted authority, appealed to empirical evidence, had faith in his own interpretation of the observed data, and, finally, there was "imputation of moral obliquity to a person whose judgment differed from his own." Walker claims that these characteristics were prominent throughout Pearson's career.

Karl Pearson should also be touted for an additional contribution to statistics—his son, Egon S. Pearson (1895–1980), who was to make his own unique and important contributions to statistics. Finally, there is the anecdotal story of an American, with a recent Ph.D. degree, who went to study statistics in England and met and studied with Pearson. He asked Pearson how he had time to write, study, and compute so much. Pearson's reply was, "You Americans would not understand," he replied. "But I never answer a telephone or attend a committee meeting."

Key Terms and Definitions

Percentiles—A distribution that is divided into hundredths.

Standard score—A type of score whose mean and standard deviation are known or given.

***T* score**—A type of standard score with a mean of 50 and a standard deviation of 10.

***z* distribution**—A distribution of scores according to frequency that is perfectly normally distributed based on *z* scores.

***z* score**—The first standard score whose mean is zero and whose standard deviation is 1.0.

Chapter 4 Practice Problems

1. What is the definition of a standard score?
2. What are the advantages of using standard scores?
3. What is the mean and standard deviation of a *z* distribution?
4. What are the disadvantages of using *z* distributions?
5. Using the *z* distribution table, find the proportion of the total area for the following, draw a picture of each, and shade in the target area.

   a. between a *z* score of 0.00 and a *z* score of 1.64
   b. above a *z* score of 1.64
   c. between a *z* score of 0.00 and a *z* score of –1.00
   d. below a *z* score of –1.64

e. below a z score of 0.00
f. between a z score of −1.00 and a z score of 1.00
g. between a z score of −2.00 and a z score of −1.00

6. Using the z distribution table, find the z score(s) that most closely corresponds to the area described in each of the following, draw a picture of each, and shade in the target area.
   a. upper .2500
   b. upper .5000
   c. lower .2500
   d. lower .0500
   e. lower .5000
   f. lower .7500
   g. lower .9500
   h. lower .0100
   i. middle .5000
7. Convert the following compliance test scores into z scores and then into T scores: 37, 42, 51, 38, 45, 50, 31, 44, 34, 39, 52, 40, 41, 36, 32.
8. In the previous example, what proportion of the scores exceeds each of the following T scores?
   a. 70
   b. 60
   c. 50
9. In the previous example, determine what proportion of the scores is below each of the following T scores:
   a. 65
   b. 60
   c. 40
10. In the previous example, determine what proportion of the scores is between each of the following pairs of T scores:
   a. 40 and 60
   b. 30 and 70
   c. 45 and 55

Chapter 4 Test Questions

1. On the Horney-Coolidge Tridimensional Inventory, a student receives a raw score of 9 on the Compliance Dimension and a raw score of 12 on the Malevolence facet of Aggression. The student
   a. was more malevolent than compliant
   b. scored higher on the Malevolence facet than on the Compliance dimension
   c. scored nearly equally on the two measures given the possibility of chance
   d. it cannot be determined
2. Which of the following is not a standard score?
   a. SAT
   b. GRE
   c. IQ
   d. all of the above are standard scores
3. What essential properties are necessary to make a score a standard score?
   a. mean and standard deviation
   b. median and standard deviation

c. standard deviation and variance
c. a normative sample of at least 30 participants

4. A *T* score has a mean and standard deviation of
   a. 100 and 15
   b. 50 and 10
   c. 25 and 5
   d. 500 and 100

5. To create a single standard score, one must take any number in a set of numbers and __________ the mean of the set of numbers and __________ by the standard deviation of those numbers.
   a. add; divide
   b. subtract; add
   c. subtract; divide
   d. add; multiply

6. Which of the following scores has a *z* score of 0.00?
   a. 10
   b. 20
   c. 30
   d. 40
   e. 50

7. For the previous set of scores, which score is at the 50th percentile?
   a. 10
   b. 20
   c. 30
   d. 40
   e. 50

8. For a person with an IQ of 145, his or her score would be ________ standard deviations above the normative mean.
   a. 1
   b. 2
   c. 3
   d. 4

9. For a person with an IQ of 145, according to the *z* distribution, his or her score would be in the ________ percentile.
   a. 99th (or 100th)
   b. 95th
   c. 90th
   d. 145th

10. For a person with an SAT verbal score of 500, his or her score would be in the ________ percentile.
    a. 25th
    b. 50th
    c. 75th
    d. 100th

11. For a person with an SAT quantitative score of 600, his or her z score would be
    a. –1.00
    b. 0.00
    c. 1.00
    d. 2.00
12. What proportion of scores on a z distribution falls between a z score of 0.00 and a z score of 1.00?
    a. .2500
    b. .3000
    c. .3413
    d. .5000
13. What proportion of scores on a z distribution falls between a z score of –1.00 and a z score of 0.00?
    a. –.2500
    b. .2500
    c. –.3413
    d. .3413
14. What proportion of scores on a z distribution falls between a z score of –1.00 and a z score of 1.00?
    a. .3413
    b. .6826
    c. .7500
    d. .9500
15. What proportion of scores on a z distribution falls between a z score of –1.96 and 1.96?
    a. .3413
    b. .6826
    c. .7500
    d. .9500
16. According to the z distribution, what proportion of SAT verbal scores falls between 500 and 600?
    a. .3413
    b. .2500
    c. .5000
    d. .7500
17. According to the z distribution, what proportion of SAT verbal scores falls between 400 and 500?
    a. .3413
    b. .2500
    c. .6826
    d. –.3413
18. According to the z distribution, what proportion of SAT verbal scores falls between 400 and 600?
    a. .3413
    b. .2500

c. .6826
d. –.3413

19. What z score cuts off the upper 50% of the z distribution?
a. 0.00
b. 1.96
c. –1.00
d. –1.96

20. What z score cuts off the upper .0250 of the z distribution?
a. 0.00
b. 1.96
c. –1.00
d. –1.96

21. From what negative z score to what positive z score would cut off the middle .9500 of the z distribution?
a. ±1.64
b. ±1.96
c. ±2.00
d. ±1.00

22. From what negative z score to what positive z score would cut off the middle .6826 of the z distribution?
a. ±1.64
b. ±1.96
c. ±2.00
d. ±1.00

23. What proportion of scores on the z distribution would fall between a z of –1.00 and a z of 1.96?
a. .3413
b. .4750
c. .8163
d. .1683

24. According to the z distribution, what proportion of SAT verbal scores would fall between 400 and 696?
a. .3413
b. .4750
c. .8163
d. .1683

25. What SAT verbal score would convert to a z score of 2.00?
a. 400
b. 500
c. 600
d. 700

5 Inferential Statistics

The Controlled Experiment, Hypothesis Testing, and the z Distribution

Chapter 5 Goals

- Understand hypothesis testing in controlled experiments
- Understand why the null hypothesis is usually a conservative beginning
- Understand nondirectional and directional alternative hypotheses and their advantages and disadvantages
- Learn the four possible outcomes in hypothesis testing
- Learn the difference between significant and nonsignificant statistical findings
- Learn the fine art of baloney detection
- Learn again why experimental designs are more important than the statistical analyses
- Learn the basics of probability theory, some theorems, and probability distributions

Recently, when I was shopping at the grocery store, I became aware that music was softly playing throughout the store (in this case, the ancient rock group Strawberry Alarm Clock's "Incense and Peppermint"). In a curious mood, I asked the store manager, "Why music?" and "why this type of music?" In a very serious manner, he told me that "studies" had shown people buy more groceries listening to this type of music. Perhaps more businesses would stay in business if they were more skeptical and fell less for scams that promise "a buying atmosphere." In this chapter on inferential statistics, you will learn how to test hypotheses such as "music makes people buy more," or "HIV does not cause AIDS," or "moving one's eyes back and forth helps to forget traumatic events."

Inferential statistics is concerned with making conclusions about populations from smaller samples drawn from the population. In descriptive statistics, we were primarily concerned with simple descriptions of numbers by graphs, tables, and parameters that summarized sets of numbers such as the mean and standard deviation. In inferential statistics, our primary concern will be testing hypotheses on samples and hoping that these hypotheses, if true of the sample, will be true and generalize to the population. Remember that a population is defined as the mostly hypothetical group to whom we wish to generalize. The population is hypothetical for two reasons: First, we will rarely, if ever, have the time or money, or it will not be feasible to test everyone in the population. Second, we will attempt to generalize from a current sample to future members of the population. For example, if we were able to determine a complete cure for AIDS, we would hope that the cure would not only work for the current population of AIDS patients in the world but also any future AIDS patients.

The most common research designs in inferential statistics are actually very simple: We will test whether two different variables are related to each other (through correlation and the chi-square test) or whether two or more groups treated differently will have different means on a response (or outcome) variable (through *t* tests and analyses of variance). Examples of whether two different variables are related to each other are plentiful throughout science and its many disciplines. We may wish to know whether cigarettes are related to cancer, whether violent crime rates are related to crack cocaine use, whether breast implants are related to immunodeficiency disease, whether twins' IQs are more highly related than siblings' IQs, and so on. Note that finding a relationship between two variables does not mean that the two variables are causally related. However, sometimes determining whether relationships exist between two variables, such as smoking and rates of lung cancer, may give up clues that allow us to set up controlled experiments where causality may be determined. Controlled experiments, typically with two or more groups treated differently, are the most powerful experimental designs in all of statistics. Whereas correlational designs, which determine whether two variables are related, are very common and useful, they pale in comparison to the power of a well-designed experiment with two or more groups.

It is perhaps unfortunate (maybe statisticians should hire a public relations firm) that the most powerful experimental design is simply called a controlled experiment. There are other theories in science that have much better names, such as the big bang theory, which attempts to explain the origins of the universe. Nonetheless, for the present, we are stuck with the name *controlled experiment.* While some statisticians might argue which statistical tests best evaluate the outcome of some types of controlled experiments, few statisticians would argue about the powerful nature of the classic two-group controlled experiment.

The **controlled experiment** is a two-group experiment, consisting typically of an experimental group and a control group. In this experimental design, which allows the determination of causality, the independent variable is the factor that the experimenter is manipulating. For example, in a drug effectiveness experiment, the independent variable is the drug itself. One group receives the drug, and the other group receives a placebo. The experimenter then determines whether the drug has an effect on some outcome measure, response variable, or, as it is also known, the dependent variable. The dependent variable is the behavior, which is measured or observed to change as a function of the two levels of the independent variable (drug group and the placebo group). The experimenter wants to see if the independent variable changes the dependent variable. Some statisticians have compared this experimental process to signal detection theory. If a treatment really works, then the two groups treated differently should score differently on the dependent variable or response variable, and this difference is the *signal.* If the treatment does not work at all, then the two groups should score similarly on the response variable. However, due to random errors or pure chance, it is highly unlikely that the two groups (even if the drug does not work any differently than a placebo) will have exactly the same means on the response variable. If the independent variable or treatment does not work, the two groups' means should be close but not exactly equal. This difference between the two means is attributed to chance or random error, and it is called *noise.* Thus, inferential statistics can be likened to the **signal-to-noise ratio.** If the independent variable really works, then the signal should be much greater than the noise. If the independent variable does not work, then the signal will not exceed the background noise.

Hypothesis Testing in the Controlled Experiment

A hypothesis is an educated guess about some state of affairs. In the scientific world, researchers are usually conservative about their results, and they assume that nothing has been demonstrated unless the results (signal) can be clearly distinguished from chance or random error (noise). Usually, experiments are conducted with a research idea or hunch, which is typically called the **research hypothesis.** The research hypothesis is usually what the experimenter believes to be true. Despite this belief, however, in theory, all experiments are begun with a statement called the **null hypothesis** (abbreviated H_0), which states that there is no relationship between the independent variable and the dependent or response variable in the population.

In the drug effectiveness experiment, the null hypothesis would be that the drug has no effect on the dependent variable. Thus, frequently, the null hypothesis will be the opposite of what the scientist believes or hopes to be true. The prior research hunch or belief about what is true is called the **alternative hypothesis** (abbreviated H_a).

As noted earlier in the book, science must work slowly and conservatively. The repercussions of poorly performed science are deadly or even worse. Thus, the null hypothesis is usually a safe, conservative position, which says that there is no relationship between the variables or, in the case of the drug experiment, that the drug does not affect the experimental group differently on the dependent variable compared to the control group.

Hypothesis Testing: The Big Decision

All experiments begin with the statement of the null and alternative hypotheses (at least in the experimenter's mind, but not usually in the published article). However, the null hypothesis is like a default position: We will retain the null hypothesis (or we will fail to reject the null hypothesis) unless our statistical test tells us to do otherwise. If there is no statistical difference between the two means, then the null hypothesis is retained. If the statistical test determines that there is a difference between the means (beyond just chance differences), then the null hypothesis will be rejected.

In summary, when a statistical test is employed, one of two possible decisions must be made: (a) retain the null hypothesis, which means that there are no differences between the two means other than chance differences, or (b) reject the null hypothesis, which means that the means are different from each other well beyond what would be expected by chance.

How the Big Decision Is Made: Back to the *z* Distribution

A statistical test of the classic two-group experiment will analyze the difference between the two means to determine whether the observed difference could have occurred by chance alone. The z distribution, or a similar distribution, will be used to make the decision to retain or reject the null hypothesis.

To appreciate how this occurs, imagine a large vat of 10,000 ping-pong balls (see Figure 5.1).

Let us suppose that each ping-pong ball has a z score written on it. Each z score on a ball occurs with the same frequency as in the z distribution. Remember that the z distribution reveals that exactly 68.26% of the 10,000 balls will fall within ±1 standard deviation of the mean z score of 0.00. This means that 6,826 of the 10,000 ping-pong balls will have numbers ranging from −1.00 to +1.00. Also, 95.44% of all the balls will fall within ±2 standard deviations of the mean. Therefore, 9,544 of the ping-pong balls will range between −2.00 and +2.00. Finally, we know that 9,974 ping-pong balls will be numbered from −3.00 to +3.00.

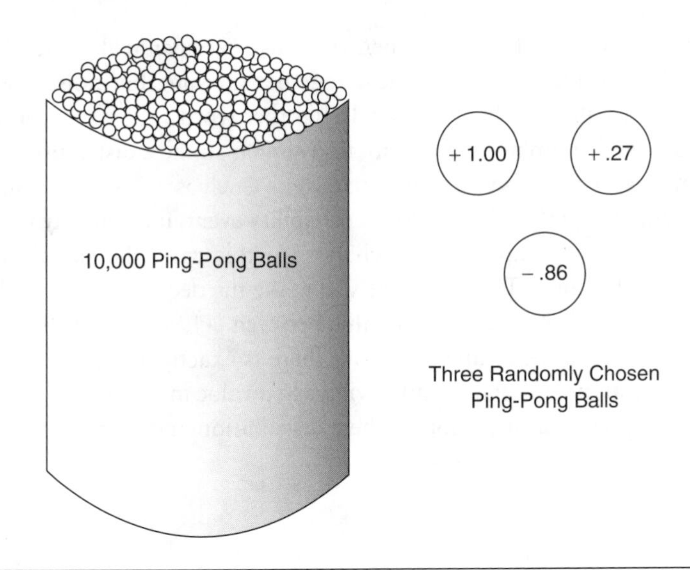

Figure 5.1 A Vat of 10,000 Ping-Pong Balls, Each With a Single Value of *z*, Occurring With the Same Frequency as in the *z* Distribution

Now, let us play a game of chance. If blindfolded and I dig into the vat of balls and pull out one ball in random fashion, what is the probability that it will be a number between –1.00 and +1.00? If I bet you $20 that the number would be greater than +1.00 or less than –1.00, would you take my bet? You should take my bet because the probability that the ball has a number between –1.00 and +1.00 is 68.26%. Therefore, you would roughly have a 68% chance of winning, and I would only have a 32% chance of winning.

How about if we up the stakes? I will bet you $100 that a *z* score on a randomly chosen ball is greater than +2.00 or less than –2.00. Would you take this bet? You should (and quickly) because now there is a 95.44% you would win and less than a 5% chance that I would win.

What would happen if we finally decided to play the game officially, and I bet a randomly chosen ball is greater than +3.00 or less than –3.00? You put your money next to my money. A fair and neutral party is chosen to select a ball and is blindfolded. What would be your conclusion if the resulting ball had a +3.80 on it?

There are two possibilities: Either we both have witnessed an extremely unlikely event (only 1 ball out of 10,000 has a +3.80 on it), or something is happening beyond what would be expected by chance alone (namely, that the game is rigged and I am cheating in some unseen way).

Now, let us use this knowledge to understand the big decision (retain or reject the null hypothesis). The decision to retain or reject the null hypothesis will be tied to the *z* distribution. Each of the individual subject's scores in the two-group experiment will be cast into a large and complicated formula, and a single *z*-like number will result. In part, the size of this single *z*-like number will be based on the difference between the two groups' means.

If the two means are far apart, then the *z*-like number will most likely be large. If the two means are very close together (nothing but noise), then the *z*-like number will more likely be small. In other words, the data will be converted to a single number in a distribution similar to the *z* distribution. If this *z*-like value is a large positive or negative value (such as +3.80 or −3.80), then it will be concluded that this is a low-probability event. It is unlikely that what has happened was simply due to chance. In this case, the signal is much greater than the noise. Therefore, we will make the decision to reject the null hypothesis. If the formula yields a value between +1.96 and −1.96, then the null hypothesis will be retained because there is exactly a 95.00% probability by chance alone that the formula will yield a value in that range. See Figure 5.2 for a graphic representation of the *z* distribution and these decisions.

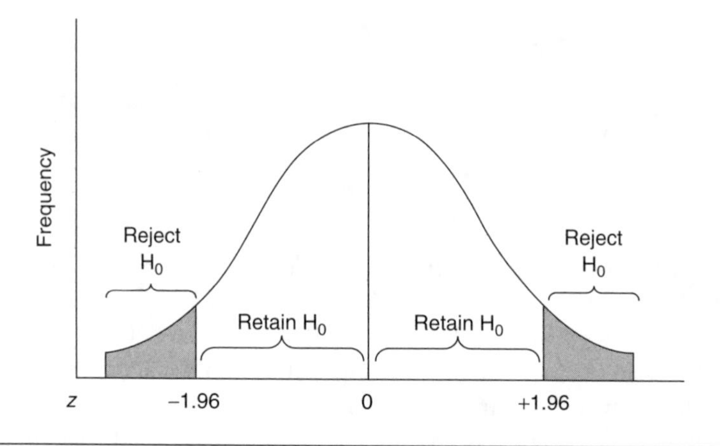

Figure 5.2 *z* Distribution With Retention and Rejection Regions for H_0

The Parameter of Major Interest in Hypothesis Testing: The Mean

In the classic two-group experiment, the means of the two groups are compared on the dependent variable. The null hypothesis states that there is no difference between the two populations' means:

$$H_0: \mu_1 = \mu_2$$

where μ_1 represents the mean for the first population, and μ_2 represents the mean for the second population. Because we will not be using the actual population means, we are going to be making inferences from our sample means, $\bar{x}_1$ and $\bar{x}_2$, to their respective population means. We hope that what we have concluded about the sample is true of the populations.

Another way of thinking about the two means is whether they were both drawn from the same population distribution (in other words, the treatment did not work to make one sample different from another) or whether the two means came from different populations (because the treatment did work on one group and made its mean much larger or much smaller than the other group's mean).

The alternative hypothesis is often what we hope is true in our experiment. The alternative hypothesis is most often stated as

$$H_a: \mu_1 \neq \mu_2$$

Note that the alternative hypothesis is stated as "Mean 1 does not equal Mean 2." This is its most common form, and it is called a **nondirectional alternative hypothesis.** Logically, the "does not equal" sign allows for two possibilities. One possibility is that Mean 1 is greater than Mean 2, and the other is Mean 1 can be less than Mean 2.

Because the controlled experiment involves making inferences about populations, the analysis of the experiment involves inferential statistics. Thus, the mean is an essential parameter in both descriptive and inferential statistics.

Nondirectional and Directional Alternative Hypotheses

An experimenter has a choice between two types of alternative hypotheses when hypothesis testing, a nondirectional or a directional alternative hypothesis. A **directional alternative hypothesis,** in the two-group experiment, states the explicit results of the difference between the two means. For example, one alternative hypothesis could be

$$H_a: \mu_1 > \mu_2$$

Here, the experimenter predicts that the mean for Group 1 will be higher than the mean for Group 2. Another possibility is that the experimenter predicts

$$H_a: \mu_1 < \mu_2$$

Here, the experimenter predicts that Mean 1 will be less than Mean 2. In practice, however, most statisticians choose a nondirectional alternative hypothesis. One of the reasons for this is that the nondirectional alternative hypothesis is less influenced by chance. Directional alternative hypotheses, however, are not all bad. They are more sensitive to small but real differences between the two groups' means. Most statisticians agree that the directional alternative hypothesis should be reserved for situations where the experimenter is relatively certain of the outcome. It is legitimate to wonder, however, why the experimenter was conducting the experiment in the first place, if he or she was so certain of the outcome.

A Debate: Retain the Null Hypothesis or Fail to Reject the Null Hypothesis

Remember that the classic two-group experiment begins with the statement of the null and the alternative hypotheses. Some statisticians are concerned about the wording of the decision that is to be made. Some say, "The null hypothesis was retained." Others insist that it should be worded, "The null hypothesis was not rejected." Although it may seem to be a trivial point, it has important implications for the entire meaning of the experiment.

After an experiment has been performed and statistically analyzed, and the null hypothesis was retained (or we failed to reject it), what is the overall conclusion? Does it really mean that your chosen independent variable has no effect whatsoever on your chosen dependent variable? Under any circumstances? With any kind of subjects? No! The conclusion is really limited to this particular sample of subjects. Perhaps the null hypothesis was retained because your sample of subjects (although it was randomly chosen) acted differently from another or larger sample of subjects.

There are other possibilities for why the null hypothesis might have been retained besides the sample of subjects. Suppose that your chosen independent variable does affect your subjects but you chose the wrong dependent variable. One famous example of this type of error was in studies of the effectiveness of Vitamin C against the common cold. Initial studies chose the dependent variable to be the number of new colds per time period (e.g., per year). In this case, the null hypothesis was retained. Does this mean that Vitamin C has no effect on the common cold? No! When the dependent variable was the number of days sick within a given time period, the null hypothesis was rejected, and it was preliminarily concluded that Vitamin C appears to reduce the number of days that people are sick with a cold.

It is important to remember that just because you do not find an effect does not mean it does not exist. You might be looking in the wrong place (using the wrong subjects, using the wrong experimental design) and/or you might be using the wrong dependent variable to measure the effect.

Thus, some statisticians recommend that it be stated, "The null hypothesis was not rejected." This variation of the statement has the connotation that there still may be a significant effect somewhere, but it just was not found this time. More important, it has the connotation that, although the null hypothesis was retained, it is not necessarily being endorsed as true. Again, this reflects the conservative nature of most statisticians.

The Null Hypothesis as a Nonconservative Beginning

The null hypothesis, however, is not always a conservative position. As in the case of side effects of many prescription drugs, the null hypothesis is that there are no side effects of the drugs! To correct for this unusual and nonconservative position, the experimenter might increase the regular dosage to exceptionally high levels. If no harmful side effects were observed at high levels, then it might be preliminarily concluded that the drug is safe. Perhaps you have heard that grilled steak fat contains known carcinogenic agents. The beef industry is quick to point out that the experiments to test this hypothesis use animals and levels of grilled fat that would be the equivalent of more than 100 steaks per day for a human being. However, in defense of statisticians and because of their conservative nature, they would be willing to conclude that steak fat does not cause cancer if the equivalent of 100 steaks per day did not cause cancer in experimental animals. Thus, if 100 grilled steaks per day did not seem to cause cancer, then it would be a relatively safe assumption that grilled steak fat is not carcinogenic.

Even in cases where high levels are shown to be safe, scientists are still conservative. Scientists will typically call for a **replication** of the experiment, which means that the experiment will be performed again by another experimenter in a different setting. As noted previously, in science, it is said that one cannot prove a hypothesis to be true or false. Even after high dosages are shown to have no side effects and after repeated experiments with other dosage levels, the drug's safety still has not been proven. Successful replication simply lends additional weight to the hypothesis that the drug is safe. It may still be found that the drug is not safe under other conditions or for other types of people. Thus, replication is important because it generally involves manipulations of other independent variables such as the types of subjects, their ages, and so on.

The Four Possible Outcomes in Hypothesis Testing

There are four possible outcomes in hypothesis testing: two correct decisions and two types of error.

1. Correct Decision: Retain H_0, When H_0 Is Actually True

In this case, we have made a correct decision. In an example involving the relationship between two variables, the H_0 would be that there is no relationship between the two variables. A statistical test (such as correlation) is performed on the data from a sample, and it is concluded that any relationship that is observed is due to chance. In this case, we retain H_0 and infer that there is no relationship between these two variables in the population from which the sample was drawn. In reality, we do not know whether H_0 is true. However, if it is true for the population and we retain H_0 for the sample, then we have made a correct decision.

2. Type I Error: Reject H_0, When H_0 Is Actually True

The **Type I error** is considered to be the more dangerous of the two types of errors in hypothesis testing. When researchers commit a Type I error, they are claiming that their research hypothesis is true when it really is not true. This is considered to be a serious error because it misleads people. Imagine, for example, a new drug for the cure of AIDS. A researcher who commits a Type I error is claiming that the new drug works when it really does not work. People with AIDS are being given false hopes, and resources that should be spent on a drug that really works will be spent on this bogus drug. The probability of committing a Type I error should be less than 5 chances out of 100 or $p < .05$. The probability of committing a Type I error is also called **alpha (α)**.

3. Correct Decision: Reject H_0, When H_0 Is Actually False

In this case, we have concluded that there is a real relationship between the two variables, and it is probably not due to chance (or that there is a very small probability that our results may be attributed to chance). Therefore, we reject H_0 and assume that there is a relationship between these two variables in the population. If in the population there is a real relationship between the two variables, then by rejecting H_0, we have made the correct decision.

4. Type II Error: Retain H_0, When H_0 Is Actually False

A **Type II error** occurs when a researcher claims that a drug does not work when, in reality, it does work. This is not considered to be as serious an error as the Type I error. Researchers may not ever discover anything new or become famous if they frequently commit Type II errors, but at least they have not misled the public and other researchers. The probability of a Type II error is also called **beta (β)**.

A summary of these decisions appears in Table 5.1.

Table 5.1

Our Decision	*In Reality*	*The Result*
Retain H_0	H_0 is true	Correct decision
Reject H_0	H_0 is true	Type I error (alpha = α)
Reject H_0	H_0 is false	Correct decision
Retain H_0	H_0 is false	Type II error (beta = β)

Significance Levels

A test of **significance** is used to determine whether we retain or reject H_0. The significance test will result in a final test statistic or some single number. If this number is small, then it is more likely that our results are due to chance, and we will retain H_0. If this number is large, then we will reject H_0 and conclude that there is a very small probability that our results are due to chance. The minimum conventional level of significance is p or $\alpha = .05$. This final test statistic is compared to a distribution of numbers, which are called critical values. The test statistic must exceed the critical value in order to reject H_0.

Significant and Nonsignificant Findings

When significant findings have been reported in an experiment, it means that the null hypothesis has been rejected. The word **nonsignificant** is the opposite of significant. When the word *nonsignificant* appears, it means that the null hypothesis has been retained. Do not use the word **insignificant** to report nonsignificant statistical findings. Insignificant is a value judgment, and it has no place in the statistical analysis section of a paper.

In the results section of a research paper, significant findings are reported if the data meet an alpha level of .05 or less. If the findings are significant, it is a statistical convention to report them significant at the lowest alpha level possible. Thus, although H_0 is rejected at the .05 level (or less), researchers will check to see if their results are significant at the .01 or .001 alpha levels. It appears more impressive if a researcher can conclude that the probability that his or her findings are due to chance is $p < .01$ or $p < .001$. It is important to note that this does not mean that results with alphas at .01 or .001 are any more important or meaningful than results reported at the .05 level.

Some statisticians also object to reporting results that are "highly significant." By this, they mean that their findings were significant not only at $p < .05$ but also at $p < .001$. These statisticians would argue that the null hypothesis is rejected at .05, and thus one's job is simply to report the lowest significance possible (e.g., $p < .01$ or $p < .001$). They find it inappropriate, therefore, to use the word *highly* before the word *significant*.

Trends, and Does God Really Love the .05 Level of Significance More Than the .06 Level?

Sometimes, researchers will report "trends" in their data. This usually means that they did not reject H_0 but that they came close to doing so. For example, computers do many of the popular statistics, and they commonly print out the exact alpha levels associated with the test statistic. A **trend** in the data may mean that the test statistic did not exceed the critical value at the .05 level, but the findings may be associated with an alpha of .06 or .10. In these cases, a researcher might say, "The findings approached significance." However, the American Psychological Association publication manual officially discourages reports of trends (American Psychological Association, 2001). The manual claims that if results do not meet the .05 level of significance, then they are to be interpreted as chance findings.

The decision to reject or retain the null hypothesis has been called dichotomous significance testing. Apparently, the need for dichotomous significance testing grew out of the early history of statistics, which developed in agriculture. Many of the statistical questions that an agriculturist might ask would be dichotomous in nature, for example, "Is the manure effective?" It is easy to see in this example how a yes or no answer is appropriate and practical. However, it has been noted, particularly in psychology, that dichotomous significance testing has no clear or early theoretical basis. Thus, two contemporary theoreticians have said "that surely, God loves the .06 nearly as much as the .05. Can there be any doubt that God views the strength of evidence for or against the null as a fairly continuous function of the magnitude of *p*?" (Rosnow & Rosenthal, 1989).

Directional or Nondirectional Alternative Hypotheses: Advantages and Disadvantages

Most statisticians use nondirectional alternative hypotheses. The advantage of the nondirectional alternative hypothesis is that it is less sensitive to chance differences in the data. Thus, the null hypothesis is less likely to be rejected, and a Type I error is less likely to be committed if a nondirectional alternative hypothesis is used. Because the Type I error is considered more serious than the Type II error, the nondirectional alternative hypothesis is more attractive to statisticians.

However, the nondirectional alternative hypothesis has one major disadvantage: It is less sensitive to real differences in the data compared to the directional alternative hypothesis. For example, in testing to see whether two groups' means are significantly different from each other, if there is a real difference but not a great difference, the nondirectional alternative hypothesis is less sensitive to this small but real difference between means.

Thus, it also follows that the directional alternative hypothesis has the advantage that it is more sensitive to real differences in the data. In other words, if there is a real difference between two groups' means, it is more likely to be detected with a directional alternative hypothesis. However, its major disadvantage is that it is also more sensitive to just chance differences between two groups' means.

Did Nuclear Fusion Occur?

In 1989, two chemists claimed that they produced nuclear fusion in a laboratory under "cold" conditions; that is, they claimed to have produced a vast amount of energy by fusing atoms and without having to provide large amounts of energy to do so. Their claims can still be analyzed in the hypothesis testing situation, although it is not absolutely known whether they did or did not produce fusion. However, most subsequent replications of their work were unsuccessful (see Park, 2000, for a fascinating discussion of the controversy).

The null and alternative hypotheses in this situation are as follows:

H_0: Fusion has not been produced.

H_a: Fusion has been produced.

Situation 1. If subsequent research supports their claims, then the two chemists made the correct decision to reject H_0. Thus, they will probably receive the Nobel Prize, and their names will be immortalized.

Situation 2. If subsequent research shows that they did not really produce fusion, then they rejected H_0 when H_0 was true, and thus they committed the grievous Type I error. Why is this a serious error? They may have misled thousands of researchers, and millions of dollars may have been wasted. The money and resources might have been better spent pursuing other lines of research to demonstrate cold fusion (because physicists claim cold fusion is theoretically possible) rather than these chemists' mistake.

What about the quiet researcher who actually did demonstrate a small but real amount of fusion in the laboratory but used a nondirectional alternative hypothesis? The researcher failed to reject H_0 when H_a was true, and thus the researcher committed a Type II error. What was the researcher's name? We do not know. Fame will elude a researcher if there is a continual commission of Type II errors because of an inordinate fear of a Type I error! Remember, sometimes scientists must dare to be wrong.

Baloney Detection

The late astronomer Carl Sagan, in his 1996 book *The Demon-Haunted World: Science as a Candle in the Dark,* proposed a baloney detection kit. The purpose of the kit was to evaluate new ideas. The primary tool in the kit was simply skeptical thinking, that is, to understand an argument and to recognize when it may be fallacious or fraudulent. The baloney detection kit would be exceptionally useful in all aspects of our lives, especially in regards to our health, where sometimes the quest for profit may outweigh the dangers of a product or when the product is an outright fraud. In the traditional natural sciences, the baloney detection kit can help draw boundaries between real science and pseudoscience. Michael Shermer, publisher of *Skeptic* magazine (www.skeptic.com), has modified Sagan's baloney detection kit. Let's use some of Sagan's and Shermer's suggestions to investigate three claims: (a) magician David Copperfield's recent announcement that he predicted Germany's national lottery numbers 7 months before the drawing; (b) mangosteen, a South Asian fruit, cures cancer, diabetes, and a plethora of other diseases and illnesses, and it works as well or better than more than 50 prescription drugs; and (c) therapeutic touch (TT), a therapy in which a medical patient is not actually touched but the patient's negative energy aura is manipulated by a trained TT therapist in order to relieve pain.

How Reliable Is the Source of the Claim?

A corollary of this criterion would be, Does the claimant have a financial (or fame) interest in the outcome? Pseudoscientists may, on the surface, appear to be reliable, but when we examine their facts and figures, they are often distorted, taken out of context, or even fabricated. Often, the claims are merely based on a desire for money and/or fame. Copperfield is a professional magician. He specializes in illusions such as making large jet planes disappear. How reliable is his claim to have predicted lottery numbers in advance? Not very. Would his claim advance his fame (and fortune)? Of course!

The chief promoter of mangosteen is identified as a prominent medical doctor and medical researcher. In reality, the doctor is a Georgia family physician who has not published even a single clinical study in any medical journal. He has written a self-published book on mangosteen, touting near-miraculous cures for a variety of diseases with his patients. We noted earlier in Chapter 1 that books, particularly self-published and those published by commercial presses, have no scientific standards to meet; therefore, they often fail to supply us with any acceptable scientific evidence *whatsoever!* Claiming something is true or saying something is true does not make it so. Mangosteen is being marketed for $37 a bottle. Distributorships are being sold. Mangosteen's proponents are clearly interested in financial gain. The latter is not a heinous crime, but it becomes one if its proponents know there are no clinical studies with humans that support their outlandish claims.

TT was developed by a nursing professor and a "fifth-generation sensitive" in the 1970s. While the nursing professor's academic credentials are credible, a self-proclaimed "sensitive" is someone who can perceive "energies" not normally perceived by other people. Thus, TT claims are initially tainted by their association with a less than scientific credible source. This, of course, does not automatically make all TT claims false, but the claims would have been more scientifically credible had they been made by another medical practitioner, scientist, or professor. In addition, those who now control the training of TT therapists do so for profit. They have a strong financial interest in maintaining their claims about TT.

Does This Source Often Make Similar Claims?

Although this may be the first time Copperfield claims to have predicted the winning numbers in a lottery, he has made many similar claims in the past. It becomes very difficult to believe in his present boast when his entire life consists of making magic and creating illusions. One of the current leading sellers of mangosteen was previously involved in the selling (multilevel marketing) and excessive hype of another "miracle" juice, which was supposedly from Tahiti. With regard to TT therapy, Dora Kunz had already claimed to be a "sensitive." Thus, she had already claimed to have extrasensory powers. Such previous claims seriously diminish her stature as a reliable expert with regard to TT.

Have the Claims Been Verified by Another Source?

Pseudoscientists often do not have their claims verified by anyone other than themselves or close followers. With regard to Copperfield, he said he "predicted the numbers 7 months ago." It would have been a much more impressive demonstration if he had produced the winning ticket 7 months earlier. In this case, Copperfield simply announced *after the lottery* that he had predicted these numbers 7 months earlier. He offered no tangible proof. His only "proof" was his claim! In addition, it is probably dangerous for any scientist or statistician without training in illusions or magic to investigate Copperfield's claim. Magicians and illusionists are typically very good at what they do, and a scientist or statistician, no matter how sophisticated their scientific knowledge, could be as easily fooled as anyone.

The literature regarding TT follows a similar course. Successful demonstrations of TT (despite severe methodological design flaws) cite other successful demonstrations of TT. Meta-analyses (a summary study of other studies) of TT cannot take into account design flaws. Interestingly, there are meta-analytic studies that have given some credence to TT claims. However, all of the positive TT studies have failed to control for the placebo effect, and thus meta-analytic studies will include these flawed studies in their overall analysis.

If TT therapists were serious about the scientific establishment of TT, they would employ acceptable scientific standards in their research. They would show that the results of TT are not due to the placebo effect. They would demonstrate scientifically that trained TT therapists can detect energy fields. To date, only one published TT study has attempted to determine whether TT therapists can detect energy auras better than chance. That study was published by a 9-year-old girl as a fourth-grade science fair project, and she found that experienced TT therapists could do no better than chance in detecting which hand she held over one of the therapist's hands (when the TT therapists could not see their hands). TT proponents tend to seek out other proponents. They cite research with positive outcomes. They ignore or deny claims to the contrary.

How Does the Claim Fit With Known Natural Scientific Laws?

A corollary of this criterion would be, Does the finding seem too good to be true? Copperfield claims his lottery prediction was not a trick. He said it was more like an experiment or a mental exercise. If it was, how does it fit into any known or replicated scientific principle? It simply does not. We would have to create a new principle to explain his mind/matter experiment or use an old one that is without any scientific merit (such as clairvoyance). There is no accepted scientific principle that explains how one would predict lottery numbers in advance. That is simply too good to be true.

The health claims for mangosteen actually pass this criterion but not its corollary. The fruit does appear to contain known antioxidants called xanthones. Xanthones from mangosteen do appear to have some antibacterial and antiviral properties *in test tubes only!* Where mangosteen fails to live up to its excessive hype is that there has not been one human clinical study to date that has demonstrated that the xanthones in mangosteen have helped or cured a disease.

TT proponents propose that humans have an energy field, which can be detected by other "trained" humans. They propose that imbalances in the patient's energy field cause disease and pain. TT therapists claim they can restore these imbalances by sweeping their hands about 3 inches over the patients' bodies and in order to get rid of their excess negative energy. Does this fit with any scientifically supported natural laws? No. Does it seem too good to be true? Yes.

This is a common ploy in pseudoscience: Concoct exaggerated claims around a kernel of scientific truth. Some fruits (those containing Vitamin C) do appear to aid physical health. Some cancer drugs have been created from plants. But it is not scientifically ethical to claim that mangosteen prevents and cures cancer, as well as lowers cholesterol and prevents heart disease, without acceptable scientific proof, and theoretical proof (i.e., mangosteen has xanthones, xanthones have antioxidant properties, and antioxidants are thought to aid physical health) is not sufficient. Its power to prevent and cure disease must be demonstrated in empirical studies with humans.

The same is true of TT. For example, there is some evidence that humans can interact with energy fields. For example, have you ever noticed that when straightening an antenna, you can sometimes get better reception when you are holding the antenna? However, it is a severe stretch (and pseudoscientific) to claim humans generate energy fields, that imbalances in these fields cause pain, and that restoring balance by eliminating negative energy is a skill that can be learned.

Sagan noted that we tell children about Santa Claus, the Easter Bunny, and the Tooth Fairy, but we retract these myths before they become adults. However, the desire to believe in something wonderful and magical remains in many adults. Wouldn't it be wonderful if there were super-intelligent, super-nice beings in spaceships visiting the Earth who might give us the secrets to curing cancer and Alzheimer's disease? Wouldn't it be great if we only had to drink 3 ounces of mangosteen twice a day to ward off nearly all diseases and illnesses? Wouldn't it be great if playing a classical CD to a baby boosted his or her IQ? Wouldn't it be amazing if a person could really relieve pain without touching someone else? But let us return to the essential tool in the baloney detection kit—skeptical thinking. If something seems too good to be true, we should probably be even more skeptical than usual. Perhaps we should demand even higher scientific standards of evidence than usual, especially if the claims appear to fall outside known natural laws. It has been said that extraordinary claims should require extraordinary evidence. An extraordinary claim, however, might not always have to provide extraordinary evidence if the evidence for the claim was *ordinary but plentiful.* A preponderance of ordinary evidence will suffice to support the scientific credibility of a theory. Thus, the theory of evolution has no single extraordinary piece of evidence. However, a plethora of studies and observations help to support it overall. I tell my students not to be disappointed when wonderful and magical claims are debunked. There are plenty of real wonders and magic in science yet to be discovered. We do not have to make them up. Francis Crick, Nobel Prize winner for unraveling DNA, reportedly told his mother when he was young that by the time he was older, everything will have been discovered. She is said to have replied, "There'll be plenty left, Ducky."

Can the Claim Be Disproven or Has Only Supportive Evidence Been Sought?

Remember, good scientists are highly skeptical. They would always fear a Type I error, that is, telling people something is true when it is not. Pseudoscientists are not typically skeptical. They believe in what they propose without any doubts that they are wrong. Pseudoscientists typically seek only confirmatory evidence, and they ignore or even reject vehemently any contradictory evidence. They often seek out only people who support their theories and often castigate and demean those who do not. Good scientists check their claims, check their data, recheck their claims and data, verify their findings, seek replication, and encourage others to replicate their findings.

In the case of mangosteen, test tube studies are certainly the first step in establishing scientific credibility. However, before mangosteen can be touted as a cancer preventative or curative agent, human trials must be conducted.

In a recent study of TT's "qualitative" effectiveness, the results were reported of treating 605 patients who were experiencing discomfort. There was no control condition. Only 1 patient out of the 605 rated the TT treatment as poor. All of the others rated it from excellent (32%), very good (28%), good (28%), or fair (12%). However, the design of this study does not allow for the evaluation of the placebo effect. Are the patients simply responding to some special, caring time that a nurse spends with his or her patient regardless of the treatment? Studies such as these do not advance the scientific credibility of TT. These studies are simply seeking supportive evidence for their claims. Interestingly, the study also reported a "small sampling" of 11 written patient responses, and all were very positive. If the authors were truly interested, as they initially stated, in improving the quality of TT, shouldn't they have interviewed the ones who said it was only fair or poor? How does interviewing only the patients who said positive things about TT improve the quality of TT? It does not. The crux of the problem in TT is the lack of studies that demonstrate that it is working for reasons other than the placebo effect. "Qualitative studies," such as the one previously cited, are not interested in disconfirming evidence. They are only interested in confirming what they already believe.

Do the Claimants' Personal Beliefs and Biases Drive Their Conclusions or Vice Versa?

As Shermer noted, all scientists have personal beliefs and biases, but their conclusions should be driven by the results of their studies. It is hoped that their personal beliefs and biases are formed by the results of their studies. The title of the qualitative TT study was "Large Clinical Study Shows Value of Therapeutic Touch Program." Does it sound like the authors had a preconceived bias as to the outcome of their study? Perhaps the authors' findings helped form their personal beliefs in TT's effectiveness, but I doubt it. Even their initial paragraph cites only studies of the positive effects of TT. It cites no negative outcome studies or any studies that are critical of TT.

Recently, a highly influential religious leader (with a Ph.D. in psychology) claimed that one's sexual orientation was completely one's choice. When asked whether his conclusion was based on his religious beliefs or based on scientific evidence, he firmly stated it was definitely not based on his religious beliefs but on the lack of even a "shred" of scientific evidence that sexual orientation is biologically determined. Because there is clear and increasing empirical evidence that sexual identity and sexual orientation are highly heritable and biologically based (e.g., Bailey, Pillard, Neale, & Agyei, 1993; Bailey et al., 1999; Bailey, Dunne, & Martin, 2000; Coolidge, Thede, & Young, 2002), it might be concluded that this religious leader is woefully ignorant of such studies, he is unconsciously unaware that his religious beliefs are driving his conclusions, or he is lying about his religious beliefs not biasing his conclusions.

Conclusions About Science and Pseudoscience

As noted earlier, skeptical thinking helps to clear a boundary between science and pseudoscience. As Shermer noted, it is the nature of science to be skeptical yet open-minded and flexible. Thus, sometimes science seems maddeningly slow and even contradictory. Good scientists may even offer potential flaws or findings that would disconfirm their own hypotheses! Good science involves guessing (hypothesizing), testing (experimentation), and retesting (replication). The latter may be the most critical element in the link. Can the results of a particular experiment be duplicated (replication) by other researchers in other locations? There may not always be a very clear boundary between science and pseudoscience, but the application of the skeptical thinking offered by the principles in the baloney detection kit may help to light the way.

The Most Critical Elements in the Detection of Baloney in Suspicious Studies and Fraudulent Claims

In my opinion, there are two most critical elements for detecting baloney in any experiment or claim. The first and most important in the social and medical sciences is, Has the placebo effect been adequately controlled for? For example, in a highly controversial psychotherapeutic technique, eye movement desensitization reprocessing (EMDR), intensively trained therapists teach their patients to move their eyes back and forth while discussing their traumatic experience (see Herbert et al., 2000, for a critical review). Despite calls for studies to control for the placebo effect—in this case, the therapist's very strong belief that the treatment works—there are few, if any, EMDR studies in which the placebo effect has been adequately controlled. In addition, there are obviously *demand characteristics* associated with the delivery of EMDR. **Demand characteristics** are the subtle hints and cues in human interactions (experiments, psychotherapy, etc.) that prompt participants to act in ways consistent with the beliefs of the experimenter or therapist. Demand characteristics usually operate below one's level of awareness. Psychologist Martin Orne has repeatedly demonstrated that demand characteristics can be very powerful. For example, if a devoted EMDR therapist worked for an hour on your traumatic experience and then asked you how much it helped (with a very kind and expectant facial expression), would you not be at least slightly inclined to say "yes" or "a little bit" even if in reality it did not help at all because you do not wish to disappoint the EMDR therapist? Controlling for the gleam, glow, and religiosity of some devotees of new techniques and the demand characteristics of their methods can be experimentally difficult. However, as Sagan and Shermer have noted, often these questionable studies do not seek evidence to disconfirm their claims, and only supporting evidence is sought. As I have already stated, this is particularly true where strong placebo effects are suspected.

The second most important element of baloney detection for your author is Sagan's and Shermer's fourth principle: How does the claim fit with known natural scientific laws? In the case of EMDR, its rationale relies on physiological and neurological processes such as information processing and eye movements somehow related to rapid eye movement (REM) sleep. Certainly, there is good support for cognitive behavioral models of therapy, and there is a wealth of evidence for REM sleep. However, the direct connection between information-processing models, cognitive behavior techniques, and eye movements in the relief of psychological distress arising from traumatic experiences has not been demonstrated. In each case where I hear of a new technique that seems too good to be true, I find that the scientific or natural explanation for why the technique works is unstated, vague, or questionable. In the case of new psychotherapeutic techniques, I always think that the absence of clear scientific explanations with specifically clear sequences for how the therapy works makes me wonder about how strong the placebo effect plays in the therapy's outcome. In their defense, I will state that it is sometimes difficult to explain how some traditionally accepted therapies, such as psychoanalysis, work. However, that defense is no excuse for not searching for reasonable and scientific explanations for how a new therapy works. *It is also absolutely imperative, in cases where scientific explanations for the therapeutic mechanism are somewhat difficult to demonstrate, that the placebo effects are completely and adequately controlled for and that disconfirming evidence has been sincerely and actively sought.*

Can Statistics Solve Every Problem?

Of course not! In fact, I must warn you that sometimes statistics may even muddle a problem. It has been pointed out that it is not statistics that lie; it is people who lie. Often, when I drive across the country, I listen to "talk radio." I've heard many extremely sophisticated statistical arguments from both sides of the gun control issue. I am always impressed at the way each side manages to have statistics that absolutely show that guns are safe or guns are dangerous. As a statistician myself, if I get muddled by these statistical arguments, I imagine it must be difficult for almost anyone to see his or her way through them. Thus, I firmly believe that statistics can help us solve most of our problems in most aspects of our lives. However, in some extremely contentious issues (e.g., religious beliefs, gun control, abortion rights, etc.), both proponents and opponents of these issues prepare themselves so well with statistical studies that it will become virtually impossible to make a decision based on statistical evidence. In these situations, I recommend that you vote with your heart, beliefs, or intuition. If you believe that you and your children are safer in your house with a gun, then by all means, keep a gun in your house. If you believe your children are safer without a loaded gun in your house, then do not have one.

Probability

The Lady Tasting Tea

An all-time classic statistics book, *The Design of Experiments,* was written by Sir Ronald A. Fisher (1890–1962) and first published in 1935. It became a standard textbook and reference book for statisticians and their students, and it was continuously published until 1970. Curiously, Chapter 2 begins with the following:

> A Lady declares that by tasting a cup of tea made with milk she can discriminate whether the milk or the tea was first added to the cup. (p. 11)

In his book, Fisher proceeded to design an experimental test of the null hypothesis that the "Lady" cannot discriminate between the two types of tea/milk infusion, and he also demonstrated how to apply a test of significance to the **probability** associated with the null hypothesis. Although Fisher's example appears to be hypothetical, a statistician, Professor H. Fairfield Smith of the University of Connecticut, revealed that Fisher's example was not hypothetical but actually occurred in the late 1920s at an afternoon tea party in Cambridge, England. Apparently, after the Lady voiced her opinion, there was a consensus from the professors in attendance that it was impossible to distinguish whether tea had been added to the milk or the milk was added to the tea. Professor Smith said that he had attended the tea party, and at the moment the Lady made her declaration and the professors voiced their august and oppositional opinions, Ronald Fisher (in his 30s at the time) came forward to suggest there was an applied mathematical analysis (called statistics) and an associated experimental method to discern whether she could or could not taste the difference in the two tea infusions. Fisher proposed that she be told she would taste a series of cups of tea and voice her opinion after each. In his 1935 book, he gave the example of a series of eight cups of tea, four with milk added to tea and four with tea added to milk. He proposed that the order of the presentation of the infusions be randomized and that the Lady be told that half of the eight cups would contain tea added to milk and half would be milk added to tea.

Probability is the science of predicting future events. The probability of a specified event is defined as the ratio of the number of ways the event can occur to the total number of equally likely events that can occur. Probabilities are usually stated numerically and in decimal fashion, where 1 means that the event will certainly happen, 0 means the event will not happen, and a probability of .5 means that the event is likely to happen once in every two outcomes.

As your intuition might already tell you, if only one cup of tea was presented to the Lady for testing, then the probability of her being right would be ½ or $p = .50$. The probability of her being wrong is also $p = .50$, and the total probability sums to 1.00; that is, after she tastes the tea, the probability of her being right or wrong is 1.00. If we label the tea tasting as an *event* and the Lady's opinion as an *outcome,* then we have established that an event with two equally likely outcomes will have each of those outcomes equal to $p = .50$.

The Definition of the Probability of an Event

If an event has n equally likely outcomes, then each individual outcome will have a probability of $1/n$ or (p_i), where n is the number of equally occurring outcomes. Thus, if a rat has a choice of entering three equally appearing doors in search of food, the probability of one door being chosen is $p_i = .33$ because $p_i = 1/n$ or $p_i = 1/3 = .33$. Also, the overall probability is $p_1 + p_2 + p_3 = 1.00$. For a larger number of events (also called mutually exclusive events), it holds that

$$p_1 + p_2 + \ldots + p_n = 1.00$$

where p_n is the final event.

The Multiplication Theorem of Probability

If there are two independent events (i.e., the outcome of one event has no effect on the next event) and the first event has m possible outcomes, and the second event has n possible outcomes, then there are $m \times n$ total possible outcomes. Now, in the case of the Lady tasting tea, if there are eight events (eight cups of tea in succession), and each event has two outcomes, then her probability of choosing correctly for all eight cups of tea is $1/2^8$ or $\frac{1}{2} \times \frac{1}{2} \times \frac{1}{2} \times \frac{1}{2} \times \frac{1}{2} \times \frac{1}{2} \times \frac{1}{2} \times \frac{1}{2} = (\frac{1}{2})^8 = p = 1/256$ or $p = .0039$. The latter holds because there are two outcomes (right or wrong) for each event, and there are eight events (the eight cups of tea); thus, there are 256 total outcomes (2^8), but only one can be correct (identifying all eight cups of tea correctly).

However, remember Fisher said that the Lady should be informed that half of the eight cups would have tea added to milk and half would have milk added to the tea. Thus, the Lady knows that if she is forced to guess, her best strategy would be to make half her guesses one way (e.g., tea added to milk) and half her guesses the other way (e.g., milk added to tea). This reduces her overall probability of being wrong on at least one guess and increases her overall probability of being right on all 8 guesses to 1/70 or $p = .014$ because of the combinations theorem of probability.

Combinations Theorem of Probability

This theorem states that if a selection of objects is to be selected from a group of n objects and the order of the arrangement selected is not important, then the equation for obtaining the number of ways of selecting r objects from n objects is

$$C = \frac{n!}{r!(n-r)!}$$

where

C = the total number of combinations of objects,

$n!$ = (is read n factorial or $n(n-1)(n-2)(n-3)$ etc.) so $3! = 3 \times 2 \times 1 = 6$.

Example 1. Thus, for the Lady tasting tea, there were 70 possible outcomes because

$$C = \frac{8!}{4!(8-4)!}$$

$$C = \frac{40320}{24(24)}$$

$$C = 70$$

Fisher had already established the concept of significance and p values in his 1925 book. He began the practice of accepting an experiment's outcome as significant if it could not be produced by chance more frequently than once in 20 trials or $p = .05$. Importantly, he noted in his 1935 book that it was the experimenter's choice to be more or less demanding of the **p level** acceptable for demonstrating significance. However, he cautioned that even for the Lady tasting tea, who might correctly identify every tea infusion at a probability of 1 in 70 or .014 (well below the conventional level of significance of $p = .05$),

> no isolated experiment, however significant in itself, can suffice for the experimental demonstration of any natural phenomenon; for the "one chance in a million" will undoubtedly occur, with no less and no more than its appropriate frequency, however surprised we may be that it should occur to *us*. (Fisher, 1935, pp. 13–14)

Example 2. Suppose it is suspected that a combination of two drugs might interact to improve attention and reduce hyperactivity in children with attention-deficit hyperactivity disorder (ADHD). Let us propose that there are four available drugs, and label them A, B, C, and D. How many combinations are there?

$$C = \frac{n!}{r!(n-r)!}$$

where

$r = 2$ drug objects from $n = 4$ drugs,

$C =$ the total number of combinations of 2 drugs.

$$C = \frac{4!}{2!(4-2)!}$$
$$C = \frac{4!}{2!(2)!}$$
$$C = \frac{24}{4}$$
$$C = 6$$

Intuitively, you can see they would be AB, AC, AD, BC, BD, and CD.

Permutations Theorem of Probability

This theorem states that if the order of selection of r objects is important from a group of n objects, then

$$P = \frac{n!}{(n-r)!}$$

where

$P =$ the total number of permutations of objects,

$n! =$ is the factorial number of objects,

$r =$ the number of objects selected.

Note: It is not possible to give a relevant example for the Lady tasting tea because the order in which she gets the teacups correctly identified is not relevant here.

Example 1. Suppose it is suspected that a combination of two drugs might interact to help children with ADHD, but the order of presentation of the drugs might be important. Thus, taking Drug A in the morning and taking Drug B in the afternoon might produce a different response than taking Drug B in the morning and Drug A in the afternoon. If there are four available drugs (A, B, C, and D), how many different drug permutations are there?

$$P = \frac{n!}{(n-r)!}$$

where

P = the total number of permutations of objects,

$n!$ = is the factorial number of objects,

r = the number of objects.

$$P = \frac{4!}{(4-2)!}$$
$$P = \frac{24}{2}$$
$$P = 12$$

Because there are a limited number of possibilities, we can actually list them all: AB, BA, AC, CA, AD, DA, BC, CB, BD, DB, CD, and DC.

Let us revisit these rules briefly with another intuitive example. A couple wishes to have three children. Assuming that the probability for this couple is the same for having a girl (G) as having a boy (B) (*Note:* This is probably not true for all couples), then there are k^n different outcomes or $2 \times 2 \times 2 = 8$ permutations (GGG, GGB, GBB, GBG, BBB, BBG, BGG, BGB) and four combinations corresponding to the total number of girls, irrespective of order (three girls, two girls, one girl, no girls). Now, let us summarize the probability of these outcomes in a table and a graph (Table 5.2 and Figure 5.3), where order is not important (a summary of combinations and not permutations). The results will be known as a probability distribution.

Table 5.2 Probability Distribution Table

Event	*Fractional Probability*	*Probability*
Three girls	1/8	.125
Two girls, one boy	3/8	.375
One girl, two boys	3/8	.375
Three boys	1/8	.125
Totals	8/8	1.000

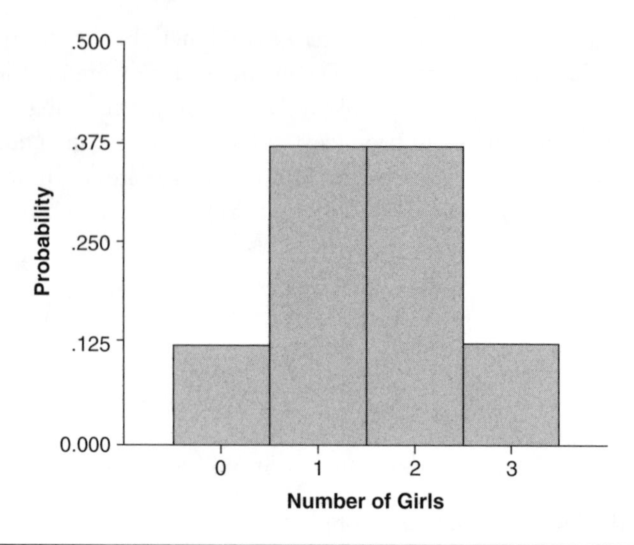

Figure 5.3 Probability Distribution Graph

Probability distributions are the heart of statistical tests of significance. One of the more popular and useful probability distributions is the *binomial distribution.* In this type of distribution, the outcome has only two choices (it is said to be dichotomous), such as boy or girl, heads or tails, or at risk or not at risk. Probability distributions may appear as tables or graphs, as previously noted.

As a final example of the application of probability theory to statistics, who among us has not dreamed of winning a lottery? Certainly, enough states and countries have discovered that lotteries are a consistently powerful means of generating income. Many state lotteries quote that the odds of winning the big grand prize is said to be about 1 in 5 million. How are these probabilities generated?

In my state lottery (Colorado), there are six different balls drawn from numbers 1 to 42. If there was only 1 number drawn (from 1 to 42), then it is easy to see that my odds of winning would be 1 in 42 or about $p = .0238$. If only two balls were drawn, my odds of picking one of the two winning balls with my first number would be 2 in 42 or about $p = .0476$, and now there are only 41 balls left to be drawn, so my odds of picking each of the two balls is the product of the two individual probabilities, that is, $2/42 \times 1/41$, which is $p = .0012$. Now, we can see that my odds of winning have dropped dramatically from about 2 chances in 100 to a little more than 1 chance in 1000. In order to pick six correct numbers (where order does not matter), we will use the following formula for combinations:

$$C = \frac{n!}{r!(n-r)!}$$

$$C = \frac{42!}{6!(42-6)!}$$

$$C = \frac{42!}{6!(36!)}$$

$$C = \frac{42 \cdot 41 \cdot 40 \cdot 39 \cdot 38 \cdot 37 \cdot 36 \cdot 35 \cdot \ldots 2 \cdot 1}{6 \cdot 5 \cdot 4 \cdot 3 \cdot 2 \cdot 1 \cdot 36 \cdot 35 \cdot 34 \cdot 33 \cdot \ldots 2 \cdot 1}$$

$$C = \frac{42 \cdot 41 \cdot 40 \cdot 39 \cdot 38 \cdot 37}{6 \cdot 5 \cdot 4 \cdot 3 \cdot 2 \cdot 1}$$

$$C = \frac{3776965920}{720}$$

$$C = 5245786$$

Thus, my chances of winning are about 1 in 5 million! Of course, I also once got a fortune from a fortune cookie that read, "Without a dream, no dream comes true," so despite these odds, I continue to support Colorado parks (which receive half the losing lottery money each week).

Gambler's Fallacy

A word of caution must be issued. Humans have an innate ability to search for meaning and understanding when an unusual or improbable event takes place. People who win lotteries often attribute their win to special numbers, a higher authority, or being born lucky. They are making a causal connection where none exists. If 10 million or 20 million lottery tickets are sold in Colorado every few days and the odds of a winning ticket are about 1 in 5 million, then eventually we expect that at least one ticket will win. Although the winning ticket is a low-probability event, it will eventually occur due to the law of very large numbers (which, simply stated, means that with a very large sample, any outrageous thing is likely to occur). Gambler's fallacy occurs when people attribute a causal relationship to truly independent events. For example, in tossing a fair coin, if I obtain two heads in a row, some might believe that tails are slightly more likely on the next toss because they are "due." However, the odds of tossing heads or tails on the next toss are equally likely events. The coin has no memory for the previous two tosses. Also, remember obtaining three heads in a row, although it has a probability of only 1/8, is still likely to occur sometime, if I engage in long strings of coin tossing. Probability is about predicting outcomes in a long series of future events.

Coda

The Lady Tasting Tea is also the name of a wonderful book about how statistics revolutionized science in the 20th century (Salsburg, 2001). It is in this book that the story of the Lady tasting tea is recounted by Professor Smith. Fisher did not reveal the outcome of his example in his 1935 book because he presented it as hypothetical, but Professor Smith reported that the Lady correctly identified every single cup of tea!

History Trivia

Egon Pearson to Karl Pearson

Egon S. Pearson (1895–1980) was the son of the famous statistician Karl Pearson. Egon was born and reared in England, and he studied under his father at University College in London in the Department of Applied Statistics. In 1922, Egon began to publish articles that would establish him as an important theoretical figure in modern statistics. In the 1920s, Egon Pearson also began an important collaboration with Jerzy Neyman (1894–1981), who would help shape one of Egon's most important contributions, statistical hypothesis testing.

Neyman grew up in the Ukraine. He studied mathematics early in his college training, and he moved to Poland in 1921 and lectured in mathematics and statistics. He received his doctorate in 1924 at the University of Warsaw. In 1925, he received a fellowship to study statistics in England at the University College with Karl Pearson. Here, Neyman would meet William S. Gosset, who would ultimately make his own unique contributions to statistics, the development of statistical analysis with small samples. Gosset introduced Neyman to Ronald Fisher (who, among his other contributions, developed the analysis of variance). Neyman was also to meet Egon Pearson, who worked as an assistant in his father's statistics laboratory.

Egon Pearson, with the mathematical help of Neyman, further developed the notion of hypothesis testing, including the testing of a simple hypothesis against an alternative; developed the idea of two kinds of errors in hypothesis testing; and proposed the likelihood ratio criterion, which can be used to choose between two alternative hypotheses by comparing probabilities. With respect to the interactions between Gosset, Egon Pearson, and Neyman, it has been said that Gosset asked the question, Egon Pearson put the question into statistical language, and Neyman solved it mathematically.

In 1933, when Karl Pearson retired, the Department of Applied Statistics was split in two. Ronald Fisher became the Galton Professor, succeeding Karl Pearson. Egon Pearson was appointed reader and, later, professor of statistics. In 1936, Egon took over the editorship of the famous journal *Biometrika* which was founded by his father to study mathematical and statistical contributions to life sciences. He remained editor for 30 years until his own retirement in 1966.

According to Helen Walker (1929/1975), a statistician and statistics historian, the modern history of statistics could be summarized thusly. The first great wave of theoretical contributions came from Francis Galton and Karl Pearson. They promoted the idea that statistical analysis would provide important information, heretofore unknown, about people, plants, and animals. With their far-reaching direction and influence, even medicine and society would be positively changed by the science of statistics. Their contributions also included the invention of measures of association such as the correlation coefficient and chi-square analysis, as well as the construction and publication of tables of statistics that were needed by statisticians and biometricians.

The second wave was begun by Gosset and completed by Ronald Fisher. According to Walker, this period was characterized by the development of statistical methods with small samples, initial development of hypothesis testing, design of experiments, and the development of criteria to choose among statistical tests.

The third wave was led by Egon Pearson and Jerzy Neyman. During their 10-year collaboration, the science of statistics enjoyed an ever-increasing popularity and appreciation. Hypothesis testing was refined, and the logic of statistical inference was developed. The notion of confidence intervals was created, and ideas for dealing with small samples were further advanced and refined.

Although Karl Pearson held some controversial views, his contributions to the philosophy of science and statistics are tremendous. His idea that the scientific method and statistical analysis should be considered part of the "grammar of science" is a watershed in the history of statistics.

Key Terms, Symbols, and Definitions

Alpha (α)—The probability of committing the Type I error. In order to consider findings significant, the probability of alpha must be less than .05.

Alternative hypothesis (H_a)—Most frequently, what the experimenter thinks may be true or wishes to be true before he or she begins an experiment; also called the research hypothesis. It can also be considered the experimenter's hunch.

Beta (β)—The probability of committing the Type II error. A Type II error can occur only when the null hypothesis is false and the experimenter fails to reject the null hypothesis.

Controlled experiment—A two-group experiment, with one group designated as the experimental group and one as the control group. The parameter of statistical interest is the difference between the two groups' means.

Demand characteristics—Subtle hints and cues that guide participants to act in accordance with the experimenter's wishes or expectations.

Directional alternative hypothesis—Also called a one-tailed test of significance where the alternative hypothesis is specifically stated beforehand; for example, Group 1's mean is greater than Group 2's mean.

Insignificant—A value judgment, such as deciding between good and evil, worthless and valuable. It typically has no place in statistics.

Nondirectional alternative hypothesis—Also called a two-tailed test of significance where the null hypothesis will be rejected if either Group 1's mean exceeds Group 2's mean, or vice versa, or where the null hypothesis will be rejected if a relationship exists, regardless of its nature.

Nonsignificant—Findings are considered statistically nonsignificant if the probability that we are wrong is greater than .05. Nonsignificant findings indicate that the null hypothesis has been retained, and the results of the experiment are attributed to chance.

Null hypothesis (H_0)—The starting point in scientific research where the experimenter assumes there is no effect of the treatment or no relationship between two variables.

***p* level**—The probability of committing the Type I error; that is, rejecting H_0 when H_0 is true.

Probability—The science of predicting future events or the likelihood of any given event occurring.

Replication—A series of experiments after an initial study where the series of experiments varies from the initial study in types of subjects, experimental conditions, and so on. Replication should be conducted not only by the initial study's author but also by other scientists who do not have a conflict of interest with the eventual outcome.

Research hypothesis—Also called the alternative hypothesis. It is most frequently what the experimenter thinks may be true or wishes to be true before he or she begins an experiment. It can also be considered the experimenter's hunch.

Signal-to-noise ratio—Borrowed from signal detection theory, in which the effect of a treatment is considered the signal, and random variation in the numbers is considered the noise.

Significance—Findings are considered statistically significant if the probability that we are wrong (where we reject H_0 and H_0 is true) is less than .05. Significant findings indicate that the results of the experiment are substantial and not due to chance.

Trend—Frequently reported when the data do not reach the conventional level of statistical significance (.05) but come close (e.g., .06 or .07). The American Psychological Association's (2001) publication manual officially discourages reports of trends.

Type I error—When an experimenter incorrectly rejects the null hypothesis when it is true.

Type II error—When an experimenter incorrectly retains the null hypothesis when it is false.

Chapter 5 Practice Problems

1. State H_0 and H_a for the following problems:
   a. a new drug cures the AIDS virus
   b. the relationship between drinking milk and longevity

2. Be able to recognize and state the four possible outcomes in hypothesis testing.

3. Be able to state why statisticians prefer NOT to say "insignificant findings," "highly significant," and "the null was retained."

Problems 4–10. True/False

4. A Type II error is considered less serious than a Type I error.
5. The probability of making a Type I error is referred to as alpha.
6. In comparison to a directional alternative hypothesis, a nondirectional alternative hypothesis is more sensitive to real differences in data.
7. The null hypothesis always represents the conservative position.
8. The conventional level of significance is $p < .01$.
9. An experimenter should report the lowest alpha level possible.
10. The probability of a Type II error is called beta.

Chapter 5 Test Questions

1. Which of the following is true about correlational designs?
   a. Finding a relationship between two variables does not mean that one causes the other but could give clues to set up experiments where causality may be determined.
   b. Finding a relationship between two variables NEVER means that one causes the other.
   c. Correlational designs are not very common.
   d. Correlational designs are more powerful than a controlled experiment.
2. An independent variable in an experiment can be likened to the __________ in the signal-to-noise ratio.
   a. signal
   b. to
   c. noise
   d. ratio
3. The difference between two means in a controlled experiment that is just attributed to chance or random error is called __________ in the signal-to-noise ratio.
   a. signal
   b. to
   c. noise
   d. ratio
4. Which of the following is a Type I error?
   a. rejecting H_0 when H_0 is false
   b. rejecting H_0 when H_0 is true
   c. not rejecting H_0 when H_0 is true
   d. not rejecting H_0 when H_0 is false
5. Because it is said in science "one experiment does not prove anything," statisticians rely on __________ to test the usefulness of theories.
   a. a controlled experiment
   b. at least two experiments

c. parametric and nonparametric tests
d. replication

6. The probability of committing the Type I error is also called __________.
   a. alpha
   b. beta
   c. delta
   d. omega
7. The probability of committing the Type II error is also called __________.
   a. alpha
   b. beta
   c. delta
   d. omega
8. The minimum conventional level of statistical significance is __________.
   a. .01
   b. .05
   c. .10
   d. .50
9. When findings in an experiment do not reach the conventional level of statistical significance, they are reported to be __________.
   a. nonsignificant
   b. insignificant
   c. significant
   d. unworthy
10. A trend in the data means that the experimenter
    a. rejected the null hypothesis at $p < .05$
    b. rejected the null hypothesis at $p < .01$
    c. did not reject the null hypothesis but came close to doing so
    d. none of the above
11. In the section "Trends, and Does God Really Love the .05 Level of Significance More Than the .06 Level?" Rosnow and Rosenthal have argued that the strength of evidence for or against a null hypothesis should be a fairly continuous function of the size of
    a. p (the significance level)
    b. beta
    c. the sample size
    d. the size of the standard deviations for each group
12. If a magician claims that he can bend spoons with his mind, what did Carl Sagan propose to test the claim?
    a. a lie detector test
    b. the baloney detection kit
    c. the salami detection kit
    d. the Atkins detection kit

13. In the section "Can Statistics Solve Every Problem?" your course author argues
    a. yes
    b. no

14. In the classic two-group controlled experiment, what parameter is of central interest between the two groups?
    a. the means
    b. the standard deviations
    c. the sample size
    d. the median

15. Which of the following is more likely to end up being too sensitive to chance differences?
    a. a nondirectional alternative hypothesis
    b. a directional alternative hypothesis
    c. a nondirectional research hypothesis
    d. all of the above

6 An Introduction to Correlation and Regression

Chapter 6 Goals

- Learn about the Pearson product-moment correlation coefficient (r)
- Learn about the uses and abuses of correlational designs
- Learn the essential elements of simple regression analysis
- Learn how to interpret the results of multiple regression
- Learn how to calculate and interpret Spearman's r, point-biserial r, and the phi correlation

Does listening to classical music increase intelligence? Does drinking milk have a relationship to getting cancer? If a woman smokes and drinks, does it increase her probability of being abused? Correlation can help answer these questions, but there is no statistical technique more used or more abused than correlation. **Correlation** is a statistical method that determines the degree of relationship between two different variables. It is also known as a "bivariate" statistic, with *bi-* meaning two and *variate* indicating variable. The two variables are usually a pair of scores for a single participant. The relationship between any two variables can vary from strong to weak or none. When a relationship is strong, this means that knowing a participant's score on one variable helps to predict his or her score on the second variable. In other words, if a participant has a high score on variable A (compared to all the other participants' scores on A), then he or she is likely to have a high score on variable B (compared to the other participants' scores on B). This example would be considered a strong positive correlation. If the correlation or relationship between variables A and B is a weak one, then knowing a participant's score on variable A does not help to predict his or her score on variable B.

One very nice feature of the correlation coefficient is that it can only range from -1.00 to $+1.00$. Any values outside this range are invalid. The following is a graphic representation of correlation's range. Note that the correlation coefficient is represented in a sample by the value r.

Strong Negative Relationship	Weak Relationship	Strong Positive Relationship
−1.00 to −.50	−.30 to +.30	+.50 to +1.00

When the correlation coefficient approaches $r = +1.00$ (or greater than $r = +.50$), it means there is a **strong positive relationship** or high degree of relationship between the two variables. This also means that the higher the score of a participant on one variable, the higher the score will be on the other variable. Also, if a participant scores very low on one variable, then his or her score will also be low on the other variable. For example, there is a positive correlation between years of education and wealth. Overall, the greater the number of years of education a person has, the greater his or her wealth. A strong correlation between these two variables also means the lower the number of years of education, the lower the wealth of that person. If the correlation was a perfect one ($r = +1.00$), then there would not be a single exception in the entire sample to increasing years of education and increasing wealth. It would mean that there would be a perfect *linear* relationship between the two variables. We will revisit the term *linear,* but preliminarily, it means that a straight line would best fit a graphic plot of variable A on the x-axis and variable B on the y-axis. However, perfect relationships do not exist between two variables in the real world. Thus, a strong but not perfect relationship between education and wealth in the real world would mean that the relationship holds for most people in the sample, but there are some exceptions. In other words, some highly educated people are not wealthy, and some uneducated people are wealthy.

When the correlation coefficient approaches $r = -1.00$ (or less than $r = -.50$), it means that there is a **strong negative relationship**. This means that the higher the score of a participant on one variable, the lower the score will be on the other variable. For example, there might be a strong negative relationship between cigarette smoking and longevity. In other words, the more a person smokes, the shorter his or her longevity, and the less one smokes, the longer one might live.

A correlation coefficient that is close to $r = 0.00$ (note that the typical correlation coefficient is reported to two decimal places) means knowing a participant's score on one variable tells you nothing about his or her score on the other variable. For example, there might be a zero (or close to zero) correlation between the number of letters in a person's last name and the number of miles he or she drives per day. If you know the number of letters in a last name, it tells you nothing about how many miles that person drives per day. There is no relationship between the two variables; therefore, there will be approximately zero correlation.

It is also important to note that there are no hard rules about labeling the size of a correlation coefficient. Statisticians generally do not get excited about a correlation until it is greater than $r = .30$ or less than $r = -.30$. They most often call a correlation of $r \geq .50$ a strong positive relationship and an $r \leq -.50$ a strong negative relationship. A moderate positive relationship might fall between $r = .30$ and $r = .49$, and a moderate negative relationship might fall between $r = -.30$ and $r = -.49$, although again there are no firm rules for labeling the strength of a correlation coefficient.

A correlational statistical analysis usually accompanies correlational designs. In a correlational design, the experimenter typically has little or no control over the variables to be studied. The variables may be statistically analyzed long after they were initially produced or measured. Such data are often called **archival data.** The experimenter no longer has any experimental power to control the gathering of the data. The data have already been gathered, and the experimenter now has only statistical power in his or her control. Lee Cronbach (1916–2001), a Stanford Professor Emeritus, stated well the difference between the experimental and correlational techniques: "The experimentalist [is] an expert puppeteer, able to keep untangled the strands to half-a-dozen independent variables. The correlational psychologist is a mere observer of a play where Nature pulls a thousand strings" (Cronbach, 1957, p. 679).

One of the potential benefits of a correlational analysis is that *sometimes* a strong (either + or −) correlation between two variables may provide *clues* about possible cause-effect relationships. However, some statisticians claim that a strong correlation *never* implies a cause-effect relationship. As much as correlational designs and statistical techniques are abused in this regard, I can understand the conservative statisticians' concerns. I think that correlational designs and techniques may allow a researcher to develop ideas about potential cause-effect relationships between variables. At that point, the researcher may conduct a controlled experiment and determine whether his or her cause-effect hunch between two variables has some support. Indeed, after a controlled experiment, a researcher may claim a cause-effect relationship between two variables. Because correlational designs and techniques *may* yield clues for future controlled experimental investigations of cause-effect relationships, correlational designs and correlational statistical analyses are probably the most ubiquitous in all of statistics. Their mere frequency, therefore, may help to contribute to their continued abuse, yet it is also something about their very nature that contributes to their misinterpretation.

Correlation: Use and Abuse

The crux of the nature and the problem with correlation is that, just because two variables are correlated, it does not mean that one variable *caused* the other. We mentioned earlier about a governor who wanted to supply every parent of a newborn child in his state with a classical CD or tape to boost the child's IQ. The governor supported his decision by citing studies that have shown a positive relationship between listening to classical music and intelligence. In fact, the controversy has grown to the point where it is referred to as the Mozart effect. The governor is making at least two false assumptions. First, he is assuming a **causal relationship** between classical music and intelligence; that is, classical music causes intelligence to rise. Correlational studies have long noted that when college students were asked what music they preferred, the ones with higher IQs favored classical music. However, the technique of correlation does not allow the implication of causation. It cannot be assumed that IQ can increase from listening to classical music. Yes, there may be a correlation between the two variables, but one cannot assume causation.

For example, it may be that intelligent people are inherently attracted to classical music and their minds are genetically predispositioned against rap or rock. Thus, playing classical music for someone without the genetic predisposition to be intelligent will do nothing for that person's IQ. In this example, intelligence may be a consequence of a predisposition to have brains prewired for an affinity for classical music. Classical music will not increase intelligence; it is just a sign or a symptom of an intelligent person.

Another scenario, however, is more likely. Classical music and intelligence may have been correlated in these early studies because of a third factor: socioeconomic status. Thus, intelligence might increase when children are given wonderful (and expensive) resources for their education (e.g., finest of private schools, access to books and computers, etc.). Thus, children growing up in families with higher socioeconomic status may have higher intelligence and may have been exposed to classical music because their families could afford to take them to the symphony. Of course, a host of additional factors may also have an effect on the relationship, and this is why it is dangerous to assume a cause-effect relationship in correlation.

The second reason the governor should have been leery of the relationship between classical music and intelligence is that it was also based on an initial experiment where college students were played classical music, and the raised intelligence lasted *only for 15 minutes!* Besides the facts of this ephemeral intelligence increase, including the fact that it produced only one kind of increase in intelligence, the study was not universally replicated by others. As noted earlier in the book, I believe *the* most important principle of experimentation is replication. Before we begin to believe in the usefulness of any statistical finding, we must be extremely cautious until the finding has been replicated a number of times and by scientists who do not have a relationship to the original author and who are financially independent from the outcome of the study. I recently saw a study where participants lost 70% more weight than any other group when they drank at least 12 ounces of milk a day. The author's work, not surprisingly, was sponsored by a dairy council. It is possible that increased milk in diets *may* cause dieting people to lose more weight, but before we "bet the farm," we should wait for replication by other scientists who are not sponsored by dairy associations.

If x and y are correlated, then x is related to y, and y is related to x. Therefore, as noted earlier, it may not be that classical music increases intelligence (x is related to y), but maybe more intelligent people listen to classical music (y is related to x). Correlation does not distinguish or give us any guidance *whatsoever* about when x is correlated with y, whether it is x that is related to y or whether y is related to x.

Let's apply some of the principles of Sagan's baloney detection kit. Have the claims been verified by another source? At this point, the effect has received little support by researchers other than those who first claimed it. How does this claim fit in the world as we know it? It does not fit very well. It would require new and undiscovered brain mechanisms. Does it seem too good to be true? Yes! Wouldn't it be wonderful if just playing a Mozart CD boosted every baby's IQ? Of course it would, but with such a wonderful claim as this, we must be very skeptical. We must ask for high standards of research excellence. Scientists must always be cautious. Findings must be replicated through experiments in a variety of settings with a variety of people by a variety of different researchers. Interestingly, the Mozart effect may be another uncommon example of the benign result of rejecting a true null hypothesis. What are the consequences of a Type I error in this case? Parents are exposing their children to classical music when it really does not boost their children's IQs. I also know some highly educated and highly skeptical parents and grandparents who buy their children and grandchildren classical music toys that are marketed directly because of the probably unreal Mozart effect. These parents and grandparents are fully aware there is little probability that the Mozart effect is real, but it is a high-risk but low-cost and benign consequence situation. If there is even a 1 in 10,000 chance that the Mozart effect is real, the musical toys do not cost that much (because some kinds of toys will be purchased anyway), and exposure to classical music is, at the very, very worst, harmless. It has been suggested that one way to counter believing in things we like to believe is to ask ourselves what the consequences would be if it were really true. For example, if intelligence is boosted by listening to classical music, shouldn't everyone be listening to classical music nearly all the time? Wouldn't there be laws against playing *any* other kind of music in nurseries, kindergartens, and schools? Wouldn't we make our own children constantly listen to classical music? Or would we want our children to end up dumber than the kids next door?

A Warning: Correlation Does Not Imply Causation

A major caution must be reiterated. Correlation does not imply causation. Because there is a strong positive or strong negative correlation between two variables, this *does not* mean that one variable is caused by the other variable. As noted previously, many statisticians claim that a strong correlation *never* implies a cause-effect relationship between two variables. Yet there are daily published abuses of the correlational design and statistical technique, not only in newspapers but major scientific journals! A sampling of these misinterpretations follows.

1. Marijuana Use and Heroin Use Are Positively Correlated

Some drug opponents note that heroin use is frequently correlated with marijuana use. Therefore, they reason that stopping marijuana use will stop the use of heroin. Clear-thinking statisticians note that even a higher correlation is obtained between the drinking of milk in childhood and later adult heroin use. Thus, it is just as absurd to think that if early milk use is banned, subsequent heroin use will be curbed, as it is to suppose that banning marijuana will stop heroin abuse. It is strongly suspected that if marijuana laws become absolutely draconian, heroin abuse would increase, not decrease.

2. Milk Use Is Positively Correlated to Cancer Rates

Although this is not a popular finding within the milk industry, there is a moderately positive correlation with drinking milk and getting cancer (Paulos, 1995). Could drinking milk cause cancer? Probably not. However, milk consumption is greater in wealthier countries. In wealthier countries, people live longer. Greater longevity means people live long enough to eventually get some types of cancer. Thus, milk and cancer are correlated, but drinking milk does not cause cancer (nor does getting cancer cause one to drink more milk).

3. Weekly Church Attendance Is Negatively Correlated With Drug Abuse

A recent study demonstrated that adolescents who attended church weekly were much less likely to abuse illegal drugs or alcohol. Does weekly church attendance cause a decrease in drug abuse? If the federal government passed a law for mandatory church attendance, would there be a decrease in drug abuse? Probably not, and there might even be an increase in drug abuse. This study is another example of the abuse of the interpretation of the correlational design and statistical analysis.

4. Lead Levels Are Positively Correlated With Antisocial Behavior

A 1996 correlational study (Needleman, Reiss, Tobin, Biesecker, & Greenhouse, 1996) examined the relationship of lead levels in the bones of

301 children and found that higher levels of lead were associated with higher rates of antisocial behavior. "This is the first rigorous study to demonstrate a significant association between lead and antisocial behavior," said one environmental health professor about the study. Although the study's authors may have been very excited, the study is still correlational in design and analysis; thus, implications of causation should have been avoided. Perhaps antisocial children have a unique metabolic chemistry such that their bodies do not metabolize lead like normal children. Perhaps lead is not a *cause* of antisocial behavior but the *result* of being antisocial. Therefore, the reduction of lead exposure in early childhood may not reduce antisocial behavior at all. Also, note that there was a statistically "significant" relationship. As you will learn later in this chapter, with large samples (such as 301 children in this study), even very weak relationships can be statistically significant with correlational techniques. With 301 children, significance could be obtained with a correlation as weak as $r = .12$!

5. The Risk of Getting Alzheimer's Dementia Is Negatively Correlated With Smoking Cigarettes

In studies funded by the tobacco industry, it was found that the risk of getting Alzheimer's dementia was negatively correlated with smoking cigarettes (yes, the risk went *down* with an increased use of cigarettes!). The implication of these findings for the tobacco industry was that increases in smoking (probably from increases in nicotine) would prevent the onset of Alzheimer's dementia. This serves as another good example of the error in assuming a causal relationship because a correlation exists between two variables, as well as a demonstration of how a third variable may be controlling the other two variables. The risk of getting Alzheimer's dementia increases with longevity. While about 10% of people older than age 65 are diagnosed with Alzheimer's, nearly 50% of those older than age 90 are diagnosed with Alzheimer's. Heavy smokers die at a rate of about 500,000 a year. Smokers do not get the chance to get Alzheimer's dementia because they do not live long enough.

6. Sexual Activity Is Negatively Correlated With Increases in Education

A 1997 report based on 10,000 interviews found that those with less education reported more sexual contacts per year than those who had been to postgraduate schools (Ph.D. programs, law school, medical school, etc.). Again, this is another example of a correlational design with a correlational analysis; therefore, causation cannot be inferred in these findings. And once again, age may be the mitigating factor in this study. People with less than a baccalaureate degree tend to be younger than those with postgraduate degrees. Younger people tend to be more sexually active than older people. The study did not completely control for the age of the participants.

7. An Active Sex Life Is Positively Correlated With Longevity

The 1997 newspaper headline for this study, published in a British medical journal, read, "Study suggests frequent sex equals long life." The study was conducted on a sample of 918 Welsh men divided into three groups: sex twice or more a week, an intermediate group, and those who had sex less than once a month. In a 10-year follow-up, the sexually inactive group had the highest death rate. The authors said the results could not be attributed to age or health. The authors, however, did not control for psychological variables such as depression, which could have been the precursor of physical disease. Thus, depression may have lowered sexual interest, and subsequent physical disease may have accounted for the increase in death rates. Imagine doctors prescribing or ordering their patients to have frequent sex in order to increase longevity. It is quite possible we might witness a sudden increase in death rates. To change this correlational design to an experiment, the authors should have randomly assigned the men to one of the three sexual frequency groups, ordered the men to have sex according to the group they were assigned, and then assessed their longevity 10 years later. Of course, this experimental design is not feasible, but any causal inferences from the original correlational design are equally unfeasible.

8. Coffee Drinking Is Negatively Correlated With Suicidal Risk

In this 1996 study, published in an American medical journal, 86,626 registered nurses were evaluated over a 10-year period. Increased coffee drinking was associated with lower rates of suicide. The authors concluded that the caffeine in the coffee might enhance mood and well-being, resulting in lowered suicide risk. While the number of nurses studied is impressive, once again this study is a correlational design, and causation cannot be inferred. A stronger argument might have been made for the positive mood hypothesis had the authors randomly assigned depressed patients to groups, then prescribed various levels of caffeine. If the groups receiving the highest levels of caffeine had a subsequently higher sense of well-being and lowered suicide rates, then the hypothesis might be more plausible.

9. Excessive Drinking and Smoking Causes Women to Be Abused

In a 2005 study, a group of authors concluded that, in women, excessive use of alcohol and addiction to smoking were associated with a much greater risk of being involved in an abusive relationship. One touted implication of this study was that if the women stopped drinking and smoking, their likelihood of abuse would go down. This is a nearly absurd conclusion. There is also a tendency for women who live in trailers to be abused more than women who own houses. If we banned women from trailers, would spousal abuse diminish? The latter conclusion is as absurd as the former *just* based on a correlational relationship.

Another Warning: Chance Is Lumpy

The contemporary Yale statistician Robert Abelson (1995) postulated Abelson's first law of statistics: **Chance is lumpy.** By this, he means that if 86,626 nurses were measured for hundreds of variables over a 10-year period, it would be highly surprising if dozens of significant relationships were *not* found! What Abelson is concerned with is that people fail to appreciate that long runs of occurrences can often be attributed to pure chance or random processes. Often, people will attribute some unusual finding or run of luck to a mysterious or a systematic process when, in fact, only chance is operating. As Abelson notes, "Attributing a data set to mere chance is deflating" (p. 21).

Abelson (1995) also notes that there is the psychological tendency to minimize the great variability that exists in small samples. Thus, we tend to overestimate the generalizability of small samples when, in reality, our conclusions may have varied widely across other samples.

One solution to the chance-is-lumpy problem is to try to replicate our findings. If this study is our first, then we should keep in mind that we should be somewhat conservative in our initial conclusions and be mindful of the repercussions of our findings and equally mindful of the tricks that can occur when we measure hundreds of variables in huge samples or a few variables in small samples.

Correlation and Prediction

The correlation coefficient may also be used as an indicator of prediction. If a strong positive or negative correlation is obtained, then the relationship between the two variables may be likened to a predictive relationship. For example, in the previous strong negative relationship between smoking and longevity, it would be suspected that longer lives would be associated with less smoking. It could be said that smoking predicts longevity and vice versa. If there was no relationship between the variables, or the correlation coefficient is close to or equal to zero, then no predictions can be made with any reliability.

The Four Common Types of Correlation

1. *Pearson's r.* A measure of the strength of a relationship between two continuous variables.

2. *Spearman's r.* A measure of the similarity between two ordinal rankings of a single set of data.

3. *Point-biserial* r. A measure of the strength of a relationship between one continuous variable and one dichotomous variable (a two-level-only variable such as gender).

4. *Phi (ϕ) correlation.* A measure of the strength of a relationship between two dichotomous variables.

The Pearson Product-Moment Correlation Coefficient

The single most common type of correlation is the **Pearson product-moment correlation coefficient,** which measures the degree of relationship between two continuous variables. A continuous variable is a variable that can be measured along a line scale. For example, smoking can be measured continuously because the number of cigarettes smoked per week, month, or year(s) can be measured along a line scale from zero to a large number. (*Note:* A clothing designer recently said he smoked 140 cigarettes a day for years.) Gender (male or female) is not considered a continuous variable because if numbers (e.g., 1 or 2) were assigned to the two categories, a person could not be a 1.3 or a 1.7. The story of the naming of the correlation coefficient appears at the end of this chapter in the History Trivia section.

Pearson's product-moment correlation (coefficient r) is obtained for a sample drawn from a population. The population value of Pearson's coefficient is called rho (ρ), and thus r is an estimate of ρ.

The formula for r is as follows:

$$r = \frac{N\sum xy - (\sum x)(\sum y)}{\sqrt{[N\sum x^2 - (\sum x)^2][N\sum y^2 - (\sum y)^2]}}$$

Note: In this formula, N is equal to the number of pairs of scores, and $\sum xy$ is called the *sum of the cross-products.* Let us see how the formula works in the following example.

A tobacco company statistician wishes to know whether heavy smoking is related to longevity. From a sample of recently deceased smokers, the number of cigarettes (estimated per day for their last 5 years after visits with their surviving relatives) is paired with the number of years that they lived.

We will arbitrarily name one variable x and the other variable y. The results of Pearson's r will be exactly the same no matter which variable is labeled x or y.

Subject	*Cigarettes*	*Years Lived*
1	25	63
2	35	68
3	10	72
4	40	62
5	85	65
6	75	46
7	60	51
8	45	60
9	50	55

Step 1. First, obtain $\sum x$, $\sum y$, $\sum x^2$, $\sum y^2$, $\left(\sum x\right)^2$, and $\left(\sum y\right)^2$.

$$\sum x = 25 + 35 + 10 + 40 + 85 + 75 + 60 + 45 + 50 = 425$$

$$\sum x^2 = 25^2 + 35^2 + 10^2 + 40^2 + 85^2 + 75^2 + 60^2 + 45^2 + 50^2 = 24{,}525$$

$$\sum y = 63 + 68 + 72 + 62 + 65 + 46 + 51 + 60 + 55 = 542$$

$$\sum y^2 = 63^2 + 68^2 + 72^2 + 62^2 + 65^2 + 46^2 + 51^2 + 60^2 + 55^2 = 33{,}188$$

$$\left(\sum x\right)^2 = (25 + 35 + 10 + 40 + 85 + 75 + 60 + 45 + 50)^2 \leftarrow \text{HINT : square } \Sigma x$$

$$\left(\sum x\right)^2 = 425^2 = 180{,}625$$

$$\left(\sum y\right)^2 = (63 + 68 + 72 + 62 + 46 + 51 + 60 + 55)^2 \leftarrow \text{HINT : square } \Sigma y$$

$$\left(\sum y\right)^2 = 542^2 = 293{,}764$$

Step 2. Obtain the sum of the cross-products ($\Sigma\ xy$) by multiplying each x score by its paired y score.

$$\sum xy = (25 \times 63) + (35 \times 68) + (10 \times 72) + (40 \times 62) + (85 \times 65) + (75 \times 46) + (60 \times 51) + (45 \times 60) + (50 \times 55)$$

$$\sum xy = 1575 + 2380 + \cdots + 2750$$

$$\sum xy = 24{,}640$$

Step 3. Substitute the values derived above into the formula for r. Remember, N is equal to the number of pairs of scores. In this example, there are nine pairs of scores.

$$r = \frac{N\sum xy - (\sum x)(\sum y)}{\sqrt{[N\sum x^2 - (\sum x)^2][N\sum y^2 - (\sum y)^2]}}$$

$$r = \frac{(9)(24,640) - (425)(542)}{\sqrt{[(9)(24,525) - (425)^2][(9)(33,188) - (542)^2]}}$$

$$r = \frac{221,760 - 230,350}{\sqrt{[220,725 - 180,625][298,692 - 293,764]}}$$

$$r = \frac{-8590}{\sqrt{[40,100][4928]}}$$

$$r = \frac{-8590}{14,057.482}$$

$$r = -.6111$$

$$r = -.61$$

Note that the Pearson r is usually rounded off to two decimal places. Thus, $r = -.61$ means that there is a strong negative correlation between smoking and longevity. This indicates that the higher the number of cigarettes smoked in the past 5 years, the lower the number of years lived. And the lower the number of cigarettes, the higher the number of years lived. Remember, this relationship between these two variables *does not* mean that heavy smoking causes one to live a shorter life. It may, however, give clues as to further research ideas for experiments. In this case, an experiment might be set up (perhaps with animals) with an experimental group and a control group to determine whether cigarette smoking actually has a causal relationship with early morbidity.

Testing for the Significance of a Correlation Coefficient

A correlation coefficient may be tested to determine whether the coefficient significantly differs from zero. The value r is obtained on a sample. The value rho (ρ) is the population's correlation coefficient. It is hoped that r closely approximates rho. The null and alternative hypotheses are as follows:

$$H_0: \rho = 0$$
$$H_a: \rho \neq 0$$

The value of r and the number of pairs of scores are converted through a formula into a distribution (similar to the z distribution) called the ***t* distribution** (in Appendix B). The t formula can only be used to test whether r is equal to zero. It cannot be used to test to see whether r might be equal to some number other than zero. It is also important to note that the t distribution may be used to test other types of inferential statistics. Therefore, if someone says that a t test is being used, it would be a legitimate question to ask "why?" The t distribution is most commonly used to test whether two means are significantly different, but it may also be used to test the significance of the correlation coefficient. The t distribution also has other uses. Interestingly, the t distribution becomes the z distribution when the data are infinite, but they are also strikingly visually similar when there are only several hundred numbers in the set of data.

The t test formula, in order to test the null hypothesis for a correlation coefficient, is the following:

$$t = \frac{r}{\sqrt{\frac{1 - r^2}{N - 2}}}$$

In the previous example on smoking, the research question was whether heavy smoking was related to longevity. An r of $-.61$ was obtained, which meant that there was a strong negative relationship between smoking and longevity. To test whether this obtained r significantly differs from zero, the t formula is used.

$$t = \frac{r}{\sqrt{\frac{1 - r^2}{N - 2}}}$$

where N = the number of pairs of scores.

$$t = \frac{-.6111}{\sqrt{\frac{1 - (-.6111)^2}{9 - 2}}}$$

$$t = \frac{-.6111}{\sqrt{\frac{1 - .3734}{7}}}$$

$$t = \frac{-.6111}{\sqrt{\frac{.6266}{7}}}$$

$$t = \frac{-.6111}{\sqrt{.0895}}$$

$$t = \frac{-.6111}{.2992}$$

$$t = -2.042$$

Obtaining the Critical Values of the *t* Distribution

We will now obtain the critical values of t, which are obtained from the t distribution in Appendix B.

Step 1: Choose a One-Tailed or Two-Tailed Test of Significance

The alternative hypothesis establishes whether we will use a one-tailed or two-tailed significance test. If the alternative hypothesis is nondirectional, as is the case in most studies, then a two-tailed test of significance is required.

Step 2: Choose the Level of Significance

The conventional level of significance is $p = .05$. Only in rare circumstances would one ever depart from $p = .05$ as a starting point.

Step 3: Determine the Degrees of Freedom (df)

The df is an advanced statistical concept related to sampling. We will keep things simple: The formula for this t test statistic is $df = N - 2$, where N is the number of pairs of scores. In this example, there were nine pairs of scores, so $df = 9 - 2$ or $df = 7$.

Step 4: Determine Whether the t From the Formula (Called the Derived t) Exceeds the Tabled Critical Values From the t Distribution

For a two-tailed test of significance at $p = .05$ with $df = 7$, the critical values of t are $t = +2.365$ and $t = -2.365$. If the derived t is greater than $t = +2.365$ or less than $t = -2.365$, then the null hypothesis will be rejected. In this example, the derived $t = -2.042$ is not less than $t = -2.365$; therefore, the null hypothesis is not rejected, and it will be concluded that $r = -.61$ indicates a nonsignificant relationship. Curiously, although the strength of the relationship was strong ($r = -.61$), the test of significance indicated that the obtained relationship was likely due to chance or there was greater than 5 chances out of 100 that the relationship was due to chance.

In a research paper, the results might be reported as follows:

> There was a strong negative correlation found between smoking and longevity, although the correlation was not statistically significant, $r(7) = -.61$, $p > .05$. Note that the degrees of freedom appear in the parentheses to the right of r.
>
> To demonstrate the effect of the number of pairs of scores upon the significance of the correlation coefficient, had the number of pairs been 12, the correlation $r = -.61$ would have been statistically significant. Thus, it is ironic that we found a strong negative relationship between smoking and longevity, although the relationship is not statistically significant (i.e., there is a greater than 5% probability that the finding could

have been due to chance). In the world of conservative statisticians, this finding is not good enough to be called statistically significant.

If the Null Hypothesis Is Rejected

Remember also that if the null hypothesis is rejected, the experimenter would report the lowest p level possible. In that case, the derived t would be compared to the critical values of t at .01 and .001. If the derived t exceeded the critical values at both .01 and .001, then the experimenter would report the r as significant at $p < .001$. It is more impressive (whether valid or not) for statisticians to conclude a finding significant with $p < .01$ or $p < .001$. In general, statisticians report the lowest p level possible (which only occurs when H_0 has been rejected).

Representing the Pearson Correlation Graphically: The Scatterplot

A **scatterplot** is a graphic presentation of the pairs of scores involved in a correlation coefficient. It is also sometimes called a bivariate distribution. Let us produce a scatterplot of the example about the number of cigarettes smoked and longevity. To construct a scatterplot, prepare to graph each of the variables along one of the graph's axes. Ultimately, it does not matter which variable is chosen for which axis, just as long as you prepare each axis for one of the variables (see Figure 6.1).

The order in which the pairs are graphed is not important. Thus, it does not matter if you begin with the last pair of scores or the first pair of scores. However, each *pair* of scores is very important, so be sure to plot each participant's score on one variable with his or her *corresponding score* on the other variable. For example, if we began with the participant who smoked an average of 85 cigarettes per day and lived 65 years, we would first locate the 85 cigarettes value on the cigarette axis and draw a horizontal line across the graph along this value. Next, locate this participant's longevity score on the longevity axis and draw a line vertically up the graph. The intersection of the two lines is the graphic representation of the pair of values in mathematical two-dimensional space. Figure 6.2 shows a completed scatterplot.

Fitting the Points With a Straight Line: The Assumption of a Linear Relationship

When using the Pearson correlation coefficient, it is assumed that the cluster of points is best fit by a straight line. Look at the cluster in Figure 6.2, and imagine how a straight line would pass through the points with as little

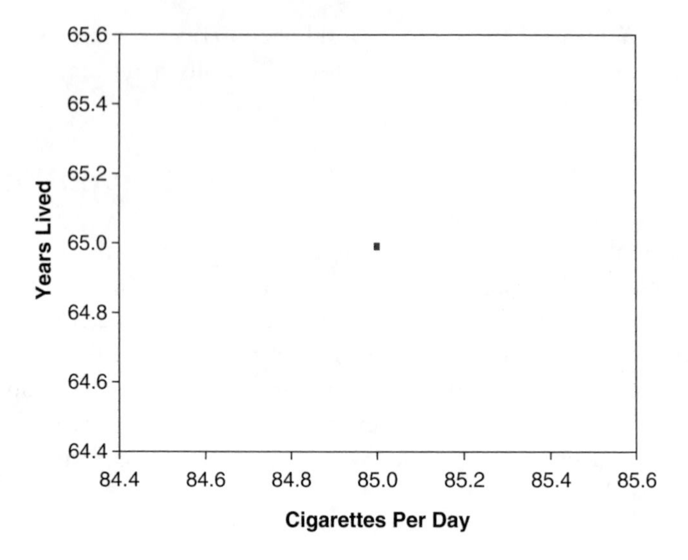

Figure 6.1

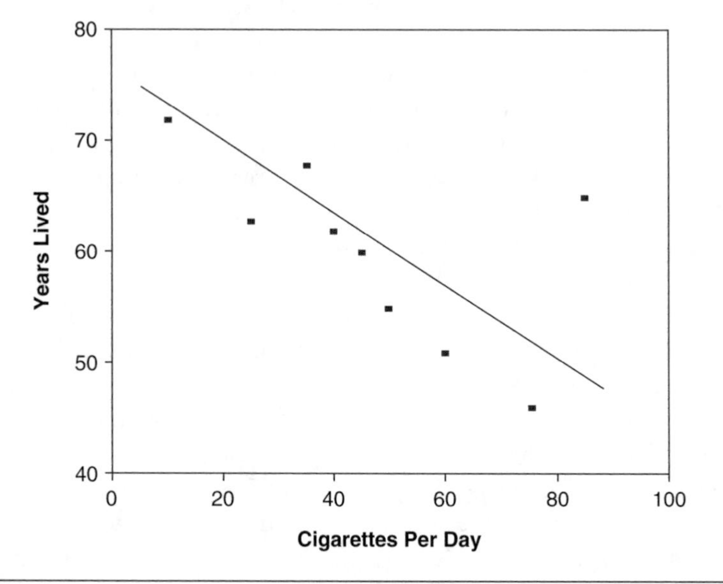

Figure 6.2

overall distance between all the points and the straight line as possible. Obviously, there is no single straight line that would pass through all the points. There is only a best-fitting line that minimizes all of the distances between the points and the line. If a straight line, as opposed to a curved line, best fits the points, then the relationship between the two variables is said to be a **linear relationship.** If a curved line best fits the points, then the relationship between the two variables is said to be a **curvilinear relationship.**

Remember, it is an assumption of the correlation coefficient that the best-fitting line is linear, or in other words, it is assumed that the relationship between the two variables is linear. The violation of this assumption is typically not harmful, at least in terms of committing a Type I error. If we assume a relationship is linear when it is really curvilinear, it will result in a lower r value, and statistical significance is less likely to be attained. Of course, it could conceivably be harmful in subtle ways to the experimenter if no relationship was found where one actually exists (Type II error). Figure 6.3 presents a curvilinear relationship. Curvilinear relationships are not uncommon. For example, there is a curvilinear relationship between strength and age: When we are younger, we are weaker; when we are older, we are stronger; but when we become very old, we are weaker again.

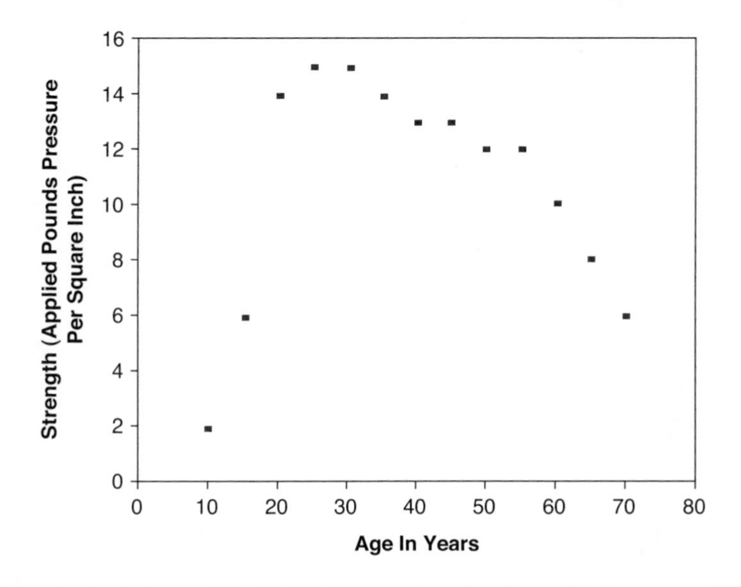

Figure 6.3

Interpretation of the Slope of the Best-Fitting Line

The best-fitting line may also be used to interpret the nature and strength of the relationship between two variables. In the previous example, the best-fitting line slopes from the upper left of the graph to the lower right. This indicates that there is a negative correlation between the two variables. The relatively small amount of overall distance between the points and the line indicates that this negative relationship is also strong; that is, it will be a large negative number.

A positive relationship will produce a best-fitting line that slopes from the lower left of the graph to the upper right. A weak or no relationship will produce a seemingly random cluster of points. There will be no best-fitting line, or it could be said that a straight horizontal line through the cluster is as good as any.

A perfect correlation ($r = 1.00$) would produce a scatterplot where the best-fitting straight line passes through all of the points, and thus it is an extremely unlikely event in the real world of data. See Figure 6.4 for a graphic representation of positive, negative, weak, and perfect correlations.

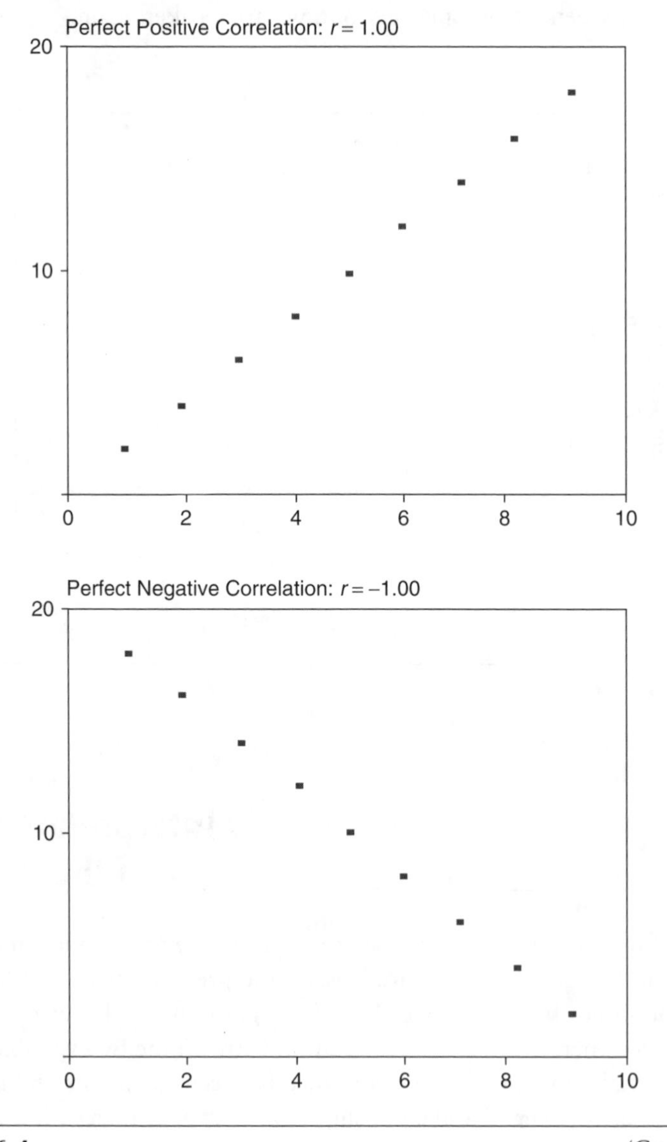

Figure 6.4 (Continued)

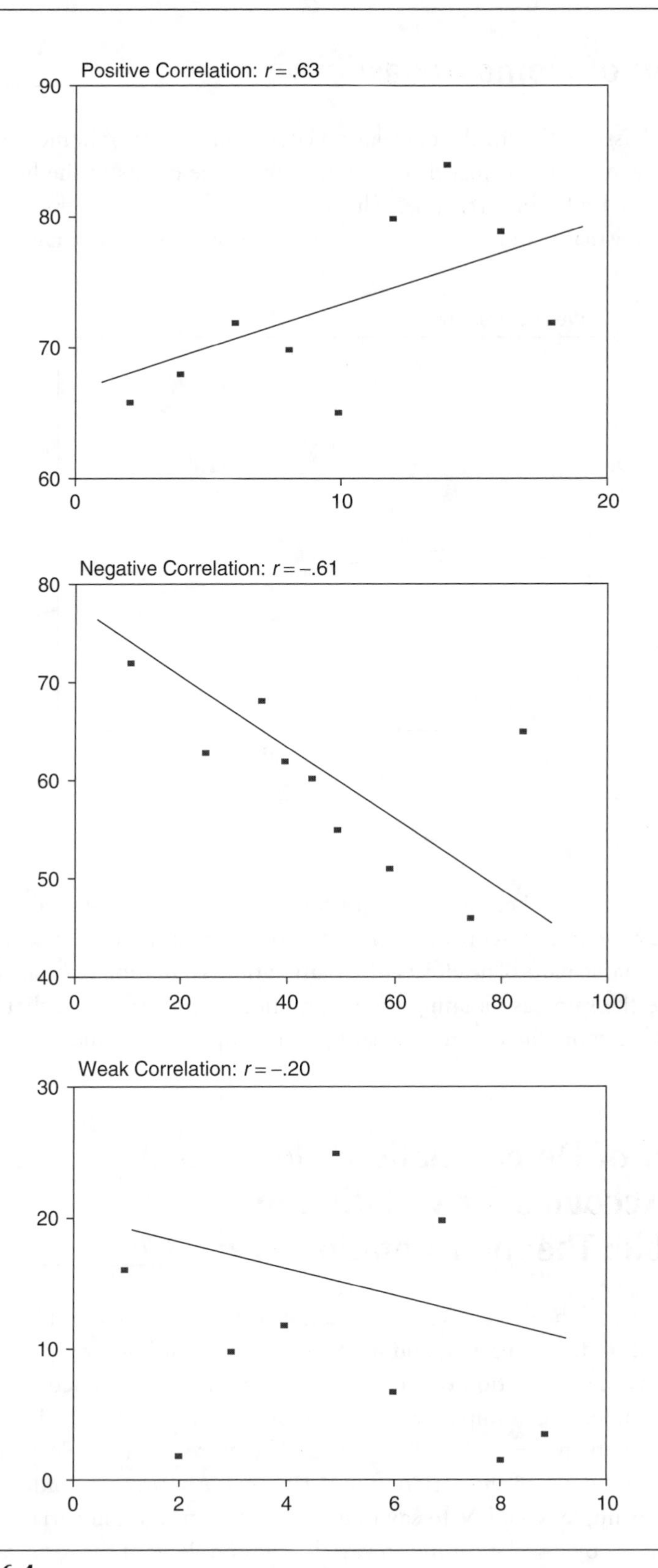
Positive Correlation: r = .63
90
80
70
60
0
10
20
Negative Correlation: r = –.61
80
70
60
50
40
0
20
40
60
80
100
Weak Correlation: r = –.20
30
20
10
0
0
2
4
6
8
10

Figure 6.4

The Assumption of Homoscedasticity

A second assumption of the correlation coefficient is that of **homoscedasticity.** This assumption is met if the distance from the points to the line is relatively equal all along the line. The violation of the assumption is called **heteroscedasticity,** and a graphic representation is presented in Figure 6.5.

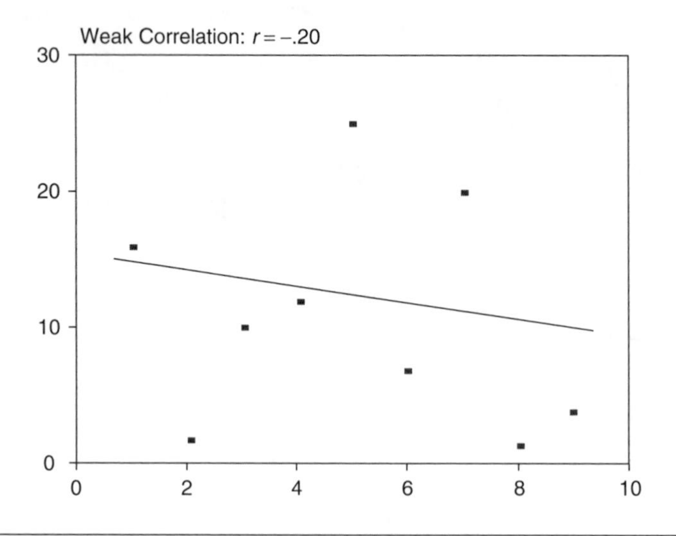

Figure 6.5

There is greater heteroscedasticity for the four largest x-variable scores with their y-variable scores than for the four lowest x-variable scores and their y-variable pairs. The effects of violating the assumption of homoscedasticity are the same as violating the assumption of linearity, and that is, the value of r is more likely to underestimate the population value of r.

The Coefficient of Determination: How Much One Variable Accounts for Variation in Another Variable: The Interpretation of r^2

Another use of the correlation coefficient is its squared value. r^2 is called the **coefficient of determination,** and it has two important interpretations. First, it explains the proportion of variance in one variable accounted for by the other variable. For example, if the correlation between two variables A and B is $r = .25$, then $r^2 = (.25)(.25) = .0625$. Therefore, the variable A explains approximately 6% of the variation in variable B. Another way of looking at r in this example would be to say that 6% of the variation in variable B can be explained by its relationship to variable A, and 94% of the total variance between variables *A* and *B* remains unexplained. The former variance is called *shared variance,* and the latter variance is called *uncommon, unshared,* or *unexplained variance.*

A second interpretation of r^2 is its use as a measure of strength between two or more r values. For example, if given $r = .25$ and $r = .50$, it is clear that the second r value is twice as great as the first r value. However, the r^2 values are 6.25% and 25%, respectively. Thus, $r = .50$ actually explains four times as much variance as does $r = .25$. The moral here is to be careful when interpreting and comparing r values, particularly smaller values of r where any $r < .31$ will explain less than 10% of the variance in two variables.

The same causality caution that was applied to the interpretation of the simple correlation coefficient r is applied to the interpretation of r^2. The coefficient of determination does not provide an *explanation* for the observed relationship, nor does it *account* for the *reason* for the relationship. The coefficient of determination simply provides another way of viewing the relationship between two variables.

It is important to note, however, that the coefficient of determination has its critics. Rosnow and Rosenthal (1996) warn against squaring the value of r since r^2 tends to underestimate the practical importance of the observed outcome. For example, a study of a drug, AZT, used in the treatment of AIDS, found the value of $r = .23$ (Barnes, 1986). The coefficient of determination yields $r^2 = .05$, which would be about half the value required for statisticians to get excited about the relationship ($r^2 \geq .10$). However, in the AZT study, the difference in survival rates between the AZT group and the placebo group was about 23%, which is a substantial practical difference. Also in the medical and health worlds, outright cures are relatively rare. Modern medical researchers are often quite happy with the management of diseases such as AIDS and cancer. Thus, improvements of more than 20% are judged to be of considerable practical importance. In summary, there is no accepted standard for the use or nonuse of the coefficient of determination. It is commonly used and often criticized.

Quirks in the Interpretation of Significant and Nonsignificant Correlation Coefficients

There are some serious quirks in the interpretations of the correlation coefficients. We have witnessed one already in our example on smoking and longevity. The interpretation of the *strength* of the relationship between these two variables was actually *independent* of the significance testing. The derived $r = -.61$ indicated that there was a moderately strong negative correlation between smoking and longevity. However, we found that this relationship was not statistically significant at $p < .05$. The statistical quirk is that significance tests are artifactually dependent on sample size: Larger sample sizes will more likely produce significance than smaller samples. In our example, $N = 9$ was an exceptionally small sample; thus, despite a moderately strong r value, we were not able to reject the null hypothesis. If we had had 30 pairs of scores, we would have obtained significance.

At this point, the coefficient of determination helps us to interpret this sample size quirk. When we square $r = -.61$, we obtain .37, or 37% of the variance is shared by the two variables. This is not a small amount of variance, all things being equal. In a 1988 prejudice study, it was found that $r = .06$ was significant for the relationship between anti-Semitism and prejudice against smoking. How could the latter r be statistically significant when our $r = -.61$ was not? In the prejudice study, there were 5977 pairs of scores. When we square $r = .06$, we obtain $r^2 = .0036$, or about one third of 1% of the total variance is explained variance! This also indicates that well over 99% of the variance remains unexplained between the two variables. In these two examples, the coefficient of determination has helped clear away some of the confusion in the interpretation of significance or a lack of significance.

Linear Regression

Now that you have learned many practical aspects of the Pearson product-moment correlation coefficient, let us delve into some other theoretical aspects. It can be said that in correlation to this point, we have used the x and y variables symmetrically; that is, the correlation between x and y was the same as the correlation between y and x. Now, let us consider them asymmetrically, where the Y variable will be called the dependent variable, and X will be labeled the independent variable. Our interest will be in seeing how various values of the independent variable predict corresponding values in the dependent variable. The dependent variable could also be called the response variable, and the independent variable could also be called the explanatory variable. This statistical technique is called **regression analysis,** and it is probably the most common statistical technique in business and economics, but it is quite popular in the social sciences such as psychology, medical sciences such as nursing, and natural sciences such as anthropology. Because virtually no one calculates regression analyses by hand any longer, we will focus on its uses, meaning, and interpretation.

Regression analysis deals with the way one variable (Y) changes based on how one or more other variables (X_1 and X_2, etc.) change. If we are interested in how schizoid behavior (Y) might vary as a function of gender (X_1), then this kind of relationship is called **simple regression** or simple linear regression. If we are interested in how schizoid behavior (Y) varies as a function of gender (X_1), age (X_2), and, perhaps, additional variables, then the investigation of these relationships is called **multiple regression analysis.** Notice that regression analyses can involve either dichotomous or continuous variables, or both.

In simple linear regression, a regression equation is used to plot a straight line through the middle of the scatterplot. The formula is the following:

$$Y = a + bX$$

where Y is the dependent variable or the value we are trying to predict:

- a is the Y intercept or the point at which the straight line crosses the ordinate or y-axis when X is 0,
- b defines the slope or the angle of the straight line.

The regression equation attempts to choose among an infinite number of straight lines to produce the single best-fitting line that predicts a Y score given an X score. The **least squares method** is used to produce the best-fitting line. This method involves measuring the square of the distance from each point to a potential best-fitting line and then choosing the line that produces the smallest value or "least squares." The values are squared to control for the positive and negative distances that result from points above and below the best-fitting line. If these values were added without squaring them, then the positive and negative distances would cancel each other out. By squaring the distances, negative distances are turned into positive values. The statistical interest is not in the real distance of points to the best-fitting line but in a measure that estimates the best-fitting line.

Reading the Regression Line

In regression, we use X scores to predict Y scores. If there is a correlation between X and Y (if there was no correlation between X and Y, then X scores could not predict Y scores), then each value of X predicts a different value of Y. To find a specific value of Y given a specific value of X, find the X value on the horizontal x-axis (abscissa) and draw a line parallel to the vertical y-axis (ordinate). When that straight line meets up with the regression line, another line is drawn at a right angle (parallel to the x-axis) until it meets the vertical or y-axis. The value at which this line intersects the y-axis is the predicted value of Y given that value of X (see Figure 6.6).

The slope of the regression line indicates how many units the line rises (positive slope) or falls (negative slope) on the y-axis for every unit moved to the right on the x-axis. For example, if $b = .75$, then the regression line would rise .75 units of Y with every successive unit of X. Also, a positive slope means that there is a positive correlation, and a negative slope means there is a negative correlation. If there is no correlation, then the regression line runs parallel to the x-axis (a flat line), and the slope is 0. Interestingly, when $r = 0$, the slope of the line $b = 0$, and the regression line intersects the y-axis at the mean value of Y for all values of X.

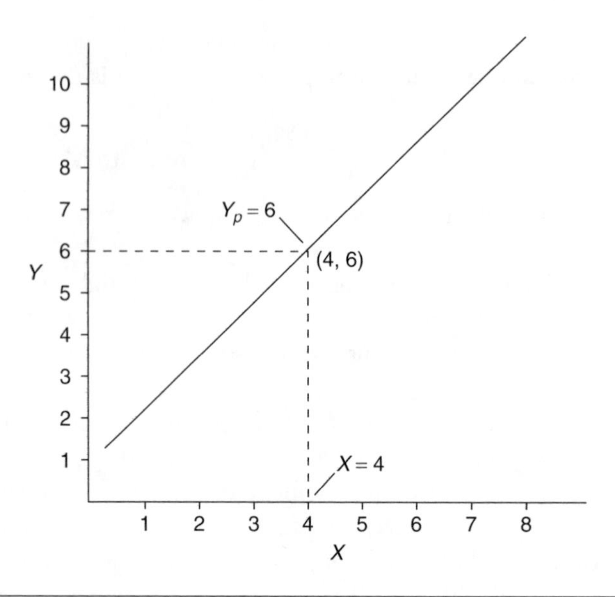

Figure 6.6

The world, however, is complex. It is much more common that the single variable or score is dependent on multiple independent variables; thus, multiple regression analysis will often be more useful. Let us see how multiple regression analysis can be applied to understanding the schizoid personality disorder (people with this disorder are extreme loners). A psychologist wishes to know what demographic and personality variables affect schizoid behavior. The psychologist suspects that the diagnosis is affected by the age of the patient, evidence of delayed maturation, and obsessive-compulsive traits. I am going to use the statistical software Statistical Packages for the Social Sciences (SPSS). I have entered the data for 183 children brought to an outpatient clinic for psychological problems. In the *Analyze* menu, I chose *Regression.* In the submenu for *Regression,* I chose *Linear.* An options menu appeared, and I entered the *T* score for the Schizoid Personality Disorder scale as my dependent variable, and then I entered the variables age, delayed maturation score, and the *T* score for the Obsessive-Compulsive Personality Disorder scale as the independent variables. I accepted the default method of regression, which is *enter.* As a point of information, there are many types of linear regression, including *enter, backward, forward, hierarchical,* and so on, and interestingly, most types yield similar results. Consult a multivariate statistics book or a book specifically on regression analyses (e.g., Tabachnick & Fidell, 2001) to learn further about multiple regression and the differences between its various procedures.

The following is the initial part of the SPSS output.

Variables Entered/Removed(b)

Model	Variables Entered	Variables Removed(a)	Method
1	Obsessive-Compulsive Personality Disorder scale T score, Age, Delayed Maturation scale T score		Enter

a. All requested variables entered.

b. Dependent Variable: T-score for Schizoid Personality Disorder scale

This table demonstrates three things. First, under variables entered, it tells us that the three independent variables are Obsessive-Compulsive Personality Disorder scale *T* score, age, and Delayed Maturation scale *T* score. Second, it tells us that the dependent variable is the *T* score for the Schizoid Personality Disorder scale. Third, it shows which method of multiple regression was used, and it was the default in SPSS, the *enter* method. The next table in the printout presents three of the most important statistical parameters in multiple regression analyses: *R*, *R*-square, and adjusted *R*-square.

Model Summary

Model	R	R-Square	Adjusted R-Square	Std. Error of the Estimate
1	.405(a)	.164	.150	12.78

a. Predictors: (Constant), Obsessive-Compulsive Personality Disorder T score, age, Delayed Maturation scale T score

R

The coefficient *R* in this model summary can be interpreted *similarly* to the interpretation of a Pearson product-moment correlation coefficient. In this case, $R = .41$ indicates that there is a very strong positive *relationship* between the dependent variable (schizoid personality disorder) and the set of three independent variables (Obsessive-Compulsive Personality Disorder scale *T* score, age, and Delayed Maturation scale *T* score). I emphasize the word *relationship* because the coefficient *R* is not theoretically a measure of correlation but a measure of relationship. Although the distinction appears minor, it is not. In general, statisticians view any values of $R \geq .30$ as reflecting the beginnings of a meaningful relationship between the dependent and independent variables.

R-Square

The *R*-square value represents the percentage of variance accounted for in the dependent variable (schizoid personality disorder) by the set of three independent variables (Obsessive-Compulsive Personality Disorder scale *T* score, age, and Delayed Maturation scale *T* score). In other words, approximately 16% (.164 in the table) of the variance or changes in Schizoid Personality Disorder scale scores can be accounted for by the confluence of the three independent variables. In general, statisticians view any values of *R*-square $\geq .10$ as reflecting the beginnings of an important amount of the variance explained between the dependent and independent variables.

Adjusted *R*-Square

The adjusted *R*-square value also represents the percentage of variance accounted for in the dependent variable by the set of three independent variables. However, the adjusted *R*-square value is always slightly lower than the *R*-square value. This lowering of the estimate is similar to a penalty paid for having more than one independent variable. Typically, the larger the number of independent variables, the lower the adjusted *R*-square value will be than the *R*-square value. As in any distribution of samples, chance fluctuations may be larger with smaller sample sizes. There is a tendency, then, for *R*-square to be overestimated, so the adjusted *R*-square is the adjustment made for this expected inflation in a sample. Thus, 15% of the variance has been explained in the dependent variable by the set of three independent variables, and again, adjusted *R*-square values $\geq .10$ are interpreted as the beginnings of an important proportion of the variance explained between the dependent and independent variables.

The next table in the output is labeled ANOVA, which stands for analysis of variance. This ANOVA tells us whether the three main regression coefficients were significantly different from zero. Of primary interest is the *F* value and its significance level. If the three main regression coefficients were close to zero, then the *F* value would be approximately 1.00. To be statistically significant, the significance level must be less than the conventional level of statistical significance (i.e., .05). In the present case, the obtained $F = 11.68$ is statistically significant. Statistical convention dictates that one reports the lowest significance level possible. In a write-up of the present results so far, it would appear like this: The regression equation indicated that Obsessive-Compulsive Personality Disorder *T* score, age, and Delayed Maturation scale *T* score were significantly related to the Schizoid Personality Disorder scale *T* score, $R = .41$, *R*-square = .16, and adjusted *R*-square = .15, $F(3, 179) = 11.68$, $p < .001$ (or $p < .0005$).

ANOVA(b)

Model		Sum of Squares	df	Mean Square	F	Sig.
1	Regression	5719.777	3	1906.592	11.682	.000(a)
	Residual	29212.989	179	163.201		
	Total	34932.766	182			

a. Predictors: (Constant), Obsessive-Compulsive Personality Disorder scale T score, age, Delayed Maturation scale T score

b. Dependent Variable: T-score for Schizoid Personality Disorder scale

Note that the significance level in the SPSS printout was .000 (or $p = .000$). It is standard convention not to report $p = .000$ but less than some whole number. Because SPSS calculations round from the fourth decimal place, we therefore know that $p = .000$ actually means that we could report $p < .0005$, although $p < .001$ is more common.

Now that we have established the significance of the regression equation, we need to examine the strength of the individual independent variables in the prediction of the dependent variable. The following is the final section of the multiple regression output.

Coefficients(a)

Model		Unstandardized Coefficients		Standardized Coefficients		
		B	Std. Error	Beta	t	Sig.
1	(Constant)	17.459	6.840		2.552	.012
	age	1.134	.293	.270	3.870	.000
	Delayed Maturation scale T score	.126	.064	.139	1.973	.050
	Obsessive-Compulsive Personality Disorder scale T score	.359	.084	.295	4.252	.000

a. Dependent Variable: TSZ

The *B* column under Unstandardized Coefficients gives the actual value that would be multiplied against the specific values of the independent variables to make a prediction of the dependent variable. However, the *B* weights are not very useful in understanding the relative importance of the independent variables. These weights will be more interpretable if the dependent and independent variables have been standardized into *z* scores with a mean of 0.00 and a standard deviation of 1.00. These standardized values appear in the beta column under Standardized Coefficients. The beta weights can be interpreted as correlation coefficients. It can be seen that Obsessive-Compulsive Personality Disorder scale *T* score's beta weight of .295 appears to be the strongest of the three in the prediction of schizoid behavior. Next, examine the significance of the *t* values to see whether the Obsessive-Compulsive Personality Disorder scale *T* score's beta weight is significantly different from zero. The $t = 4.252$ and the significance level is $p < .001$ (or $p < .0005$), which is less than the conventional level of significance (.05). Thus, it can be concluded that the Obsessive-Compulsive Personality Disorder scale *T* score is a significant predictor of the Schizoid Personality Disorder scale score. Also, because the beta weight is positive, it means that as the score on the Schizoid Personality Disorder scale increases, so does the score on the Obsessive-Compulsive Personality Disorder scale. The second highest beta weight occurs for age, and it is .270 ($t = 3.870$, $p < .001$ or $p < .0005$). Because it is also positive, this indicates that as age increases, so does the Schizoid Personality Disorder scale score. Finally, the Delayed Maturation scale score's beta = .139 is also a significant contributor ($t = 1.973$, $p = .05$). Although a significant contributor to the prediction of the Schizoid Personality Disorder scale scores, the beta for Delayed Maturation scale was about half as strong as the other two predictors. Also of interest was the finding of a significance level exactly equal to the conventional level of significance, $p = .05$. Although not less than the conventional level of significance, at a significance level of $p = .05$, most statisticians would still reject the null hypothesis and would declare the results statistically significant.

Final Thoughts About Regression Analyses

As noted earlier, regression analyses are common throughout the sciences. They can also become highly complex, and their interpretation can sometimes be controversial. If you wish to learn more about these procedures, consult an advanced statistical text. I would recommend *Using Multivariate Statistics* by Tabachnick and Fidell (2001).

Spearman's Correlation

Spearman's correlation coefficient (sometimes referred to as Spearman's rho [ρ] or r_s, where the sub *s* is in honor of Spearman) determines the degree of relationship for two sets of ranked data. **Spearman's correlation** is also called the rank-order correlation coefficient. Although Spearman's correlation is far less common than Pearson's *r*, occasionally variables are ordered according to rank (e.g., 1st through 10th), or variables may be subsequently ranked on the basis of a continuous variable. The formula for Spearman's *r* is the following:

$$r_s = 1 - \frac{6\sum D^2}{N(N^2 - 1)}$$

where N = the number of pairs of scores.

Spearman's might be most typically used in situations where there are a number of variables and they are all ranked by two independent judges. For example, a psychologist wishes to determine how alike husbands and wives are in their vegetable preferences. The members of the couple were independently asked to rate their preference for seven vegetables from most preferred (#1 rank) to least preferred (#7). Their data are as follows:

	Husband	*Wife*	D *Score*	D^2
Broccoli	4	3	1	1
Cauliflower	3	1	2	4
Brussels sprouts	6	7	−1	1
Okra	1	2	−1	1
Cabbage	5	5	0	0
Spinach	2	4	−2	4
Turnips	7	6	1	1
				$\Sigma D^2 = 12$

Note that the D score is the difference between the pairs of ranks on the first variable ranked and so forth. The number "6" in the formula is a **constant** and remains "6" regardless of the number of ranked variables. N is the number of pairs of ranks (or the number of variables that is ranked).

In this example, $N = 7$.

$$r_s = 1 - \frac{6\sum D^2}{N(N^2 - 1)}$$

$$r_s = 1 - \frac{6(12)}{7(49 - 1)}$$

$$r_s = 1 - \frac{72}{7(48)}$$

$$r_s = 1 - \frac{72}{336}$$

$$r_s = 1 - .214$$

$$r_s = .786$$

$$r_s = .79$$

Note that Pearson's r and Spearman's r are most typically reported to two decimal places.

Spearman's r may be interpreted as a measure of the linear correlation between ranks. Pearson's r will produce the same value as Spearman's r on the same set of ranked data. In the case where the variables are expressed in their original form as continuous measures, Pearson's r will not equal Spearman's r after they have been converted to ranks, but they will have similar values.

Significance Test for Spearman's *r*

Spearman's r cannot be tested for significance in the same manner as Pearson's r. Refer to the significance table for Spearman's r in Appendix C. This table presents the minimum size of Spearman's r in order to reject the null hypothesis at $p < .05$ and $p < .01$. This table is unique because the **degrees of freedom** do not have to be calculated because N is used directly in the table. In the previous example, the null and alternative hypotheses are as follows:

$$\mathrm{H_0}: \rho = 0$$
$$\mathrm{H_a}: \rho \neq 0$$

Spearman's significance table reveals that an r value of at least .786 is necessary to reject H_0 at the $p < .05$ level. The obtained r_s value equals this level exactly; therefore, H_0 can be rejected at $p = .05$. According to American Psychological Association format, the derived r_s value may be reported as

$$r_s(7) = .79, p = .05$$

In conclusion, with respect to the previous example, the results indicate that there is a strong positive correlation between the husband's and the wife's preferences for the seven vegetables. The probability that r_s would equal .79 by chance alone is equal to five chances out of a hundred or .05. Thus, the $r_s = .79$ may be reported as statistically significant.

Ties in Ranks

The following example comes from a study (Coolidge, 1983) of the Wechsler Intelligence Scale for Children–Revised (WISC-R) profiles of emotionally disturbed children (EDC) and learning disabled children (LDC). The WISC-R contains 10 separate subtests. The focus of the study was whether the relative strengths and weaknesses within each group of children were similar between the two groups. The subtests were ranked from highest (#1) to lowest (#10) in terms of overall group performance. The data are tabled as follows:

Subtest	*EDC Rank*	*LDC Rank*	D	D^2
1	10	10	0	0
2	7	5.5*	1.5	2.25
3	9	9	0	0
4	8	8	0	0
5	3	1	2	4
6	6	7	−1	1
7	5	5.5*	−.5	.25
8	1	2	−1	1
9	2	4	−2	4
10	4	3	1	1
				$\Sigma D^2 = 13.5$

Note: Because there is a tie between two subtests at the 5th rank, positions 5 and 6 are added together and divided by 2 for a 5.5 average rank for both subtests. An asterisk was added to indicate the tie. Note that since the 5th and 6th ranked places are now taken, the next lowest subtest is given the 7th-place rank. Had a three-way tie occurred at the 5th-rank place, then places 5, 6, and 7 would have been added together and divided by 3. Thus, all three tied subtests would be given a 6. The next subtest would be 8th ranked.

$$r_s = 1 - \frac{6\sum D^2}{N(N^2 - 1)}$$
$$r_s = 1 - \frac{6(13.5)}{10(100 - 1)}$$
$$r_s = 1 - \frac{81}{990}$$
$$r_s = 1 - .0818$$
$$r_s = .918$$
$$r_s = .92$$

The null and alternative hypotheses are the following:

$$H_0: \rho = 0$$
$$H_a: \rho \neq 0$$

According to the Spearman significance table, H_0 is rejected at $p < .01$. It may be concluded that there is a significant, strong positive correlation between the two sets of ranks, and the relative strengths and weaknesses within the groups are similar between the two groups. This means that knowing the rank of a subtest in one group will predict the rank of the same subtest in the other group $r_s = .92$, $p < .01$.

Point-Biserial Correlation

The **point-biserial correlation** (r_{pb}) gives an estimate of the degree of relationship between a dichotomous variable and a continuous variable. Before the advent of modern calculators, students who had to use slow, noisy, mechanical ones would dichotomize one of two continuous variables (e.g., using the median score on the variable to be dichotomized) and then run the point-biserial correlation instead of Pearson's because the formula was simpler. Interestingly, Pearson's correlation coefficient yields the same value at the point-biserial correlation formula.

One typical use of r_{pb} is correlating a single test item, which is dichotomous (e.g., yes-no), with the overall test score, which is continuous. This might tell the researcher whether an individual item is a good predictor of the overall test score. The formula is as follows:

$$r_{pb} = \frac{\bar{x}_1 - \bar{x}_2}{S} \cdot \sqrt{pq}$$

where

$\bar{x}_1$ = the mean score on the continuous variable of just the participants in Level 1 of the dichotomous variable,

$\bar{x}_2$ = the mean score on the continuous variable of just the participants in Level 2 of the dichotomous variable,

S = the standard deviation of all the participants on the continuous variable,

p = the proportion of persons in Level 1 of the dichotomous variable,

$q = 1 - p$.

For example, a neuropsychologist wanted to determine whether fine motor performance, as measured by speed of finger tapping, was related to gender in an older sample of males and females. In this case, gender is inherently a dichotomous variable, while finger tapping (number of taps in 5 seconds) is measured as a continuous variable. The data are tabled as follows:

Participants	*Finger Taps*
Male	10
Female	12
Female	14
Male	9
Male	11
Female	13
Female	14
Female	10
Male	8
Female	11

Arbitrarily, consider the males as Level 1 and females as Level 2 of the dichotomous variable.

However, caution is in order! As previously noted, it is entirely arbitrary which dichotomous group ends up in Level 1 or 2. However, the interpretation of the final test statistic very much depends on those two levels. It is an artifactual quirk, but Level 1 is considered the "higher" level and Level 2 is considered as "lower" on the dichotomous scale. Thus, if there is a positive correlation, then higher scores on the dichotomous variable (meaning Level 1) are associated with higher scores on the continuous variable. If there is a negative correlation, then higher scores on the dichotomous variable (Level 1) are associated with lower scores on the continuous variable.

Step 1. Calculate the mean score on the continuous variable of the Level 1 group ($\bar{x}_1$):

$$\bar{x}_1 = \frac{10 + 9 + 11 + 8}{4} = 9.50$$

Step 2. Calculate the mean score on the continuous variable of the Level 2 group ($\bar{x}_2$):

$$\bar{x}_2 = \frac{12 + 14 + 13 + 14 + 10 + 11}{6} = 12.33$$

Step 3. Calculate the standard deviation of the continuous variable (S):

$$S = \sqrt{\frac{\sum x^2 - \frac{(\sum x)^2}{N}}{N - 1}}$$

$$S = \sqrt{\frac{1292 - \frac{(112)^2}{10}}{10 - 1}}$$

$$S = 2.044$$

Step 4. Calculate the proportion of the participants in Level 1 to the total number of participants (p):

$$p = \frac{4}{10} = 0.4$$

Step 5. Calculate the proportion of the participants in Level 2 to the total number of participants (q):

$$q = 1 - p = 1 - 0.4 = 0.6$$

Step 6. Finally, plug the values derived above into the formula and derive r_{pb} to two decimal places (the conventional standard):

$$r_{pb} = \frac{\bar{x}_1 - \bar{x}_2}{S} \cdot \sqrt{pq}$$
$$r_{pb} = \frac{9.50 - 12.33}{2.044} \cdot \sqrt{(0.4)(0.6)}$$
$$r_{pb} = -.679$$
$$r_{pb} = -.68$$

Testing for the Significance of the Point-Biserial Correlation Coefficient

The significance of the point-biserial correlation coefficient is tested the same way as Pearson's r. The null and alternative hypotheses are as follows:

$$\mathrm{H}_0: \rho = 0$$
$$\mathrm{H}_a: \rho \neq 0$$

The value of r and the number of pairs of scores are converted to a t distribution with $N - 2$ degrees of freedom, where N is the number of pairs of scores. The t statistic can only be used to test whether r is equal to zero. The formula is as follows:

$$t = \frac{r}{\sqrt{\dfrac{1 - r^2}{N - 2}}}$$

where N = the number of pairs of scores, and $df = N - 2$.

Thus,

$$t = \frac{-.6790}{\sqrt{\dfrac{1 - 0.4611}{8}}}$$
$$t = \frac{-.6790}{\sqrt{0.0674}}$$
$$t = \frac{-.6790}{0.2595}$$
$$t = -2.617$$

The critical value of t is obtained from the t distribution in Appendix B. In this case, the critical value of t with $df = 8$ and $p = .05$ is ± 2.306. Our formula-derived t of -2.617 exceeds this critical value; therefore, we reject H_0 and conclude that our $r_{pb} = -.68$ is statistically significant. This means that there is a significant relationship between gender and finger tapping. Since males were Level 1, the interpretation of the negative correlation would be that males are associated with lower levels of the continuous variable (finger tapping), and the lower level of the dichotomous variable (females) is associated with higher levels of the continuous variable (finger tapping).

In a research paper, the r_{pb} might be reported as follows:

> There was a strong and significant negative correlation found between gender and fine motor movement $r_{pb}(8) = -.68, p < .05$; in other words, older males are associated with slower fine motor movements, and older females appear to have faster fine motor movements.

Phi (ϕ) Correlation

The phi (pronounced like "fee") correlation gives an estimate of the degree of relationship between two dichotomous variables. The value of the phi (ϕ) correlation coefficient is interpreted just like the Pearson *r;* that is, it can vary from -1.00 to $+1.00$.

For example, attention-deficit/hyperactivity disorder (ADHD) in children has often been found to be comorbid with oppositional defiant disorder (ODD). A group of 183 children with behavioral problems were evaluated for a diagnosis of ADHD and/or ODD. After an array of psychological tests, clinical interviews, and behavioral observations, the children were then classified as having ADHD, ODD, both, or neither. The frequencies of these diagnostic combinations are tabled as follows:

		ODD	
		No	*Yes*
ADHD	*No*	110[a]	19[b]
	Yes	22[c]	32[d]

The individual cells in this matrix of numbers have been labeled *a* through *d* to identify the cells in the formula. The formula for ϕ is

$$\phi = \frac{ad - bc}{\sqrt{(a+b)(c+d)(a+c)(b+d)}}$$

$$\phi = \frac{(110)(32) - (22)(19)}{\sqrt{(110+19)(22+32)(110+22)(19+32)}}$$

$$\phi = \frac{3520 - 418}{\sqrt{(129)(54)(132)(51)}}$$

$$\phi = \frac{3102}{\sqrt{46,895,112}}$$

$$\phi = \frac{3102}{6848.000584}$$

$$\phi = .453$$

$$\phi = .45$$

Testing for the Significance of Phi

The **phi correlation** can be tested for significance by converting the value of phi into a chi-square statistic (χ^2) and comparing it to the **chi-square distribution.** The null and alternative hypotheses are as follows:

$$H_0: \phi = 0$$
$$H_a: \phi \neq 0$$

The formula for the conversion of phi to chi-square is

$$\chi^2 = N(\phi)^2$$

where N = the number of participants in the correlation, and the *df* is always equal to 1.

The critical values of chi-square are in Appendix D. The critical value of chi-square with $df = 1$ at $p = .05$ is 3.84. The obtained value of chi-square is $(183)(.453)^2 = 37.55$. The derived value of $\chi^2 = 37.55$ exceeds the tabled critical value of $\chi^2 = 3.84$ with $df = 1$ at $p = .05$. Therefore, H_0 is rejected, and it is concluded that there is a significant positive relationship between a diagnosis of ADHD and ODD, χ^2 (1, N = 183) = 37.55, $p < .001$. (Note that the lowest p level was reported in the last statement because the derived value of chi-square also exceeded the critical value at $p = .001$. In a research article, the ϕ correlation might be reported as follows:

> There was a strong and significant positive correlation found between diagnoses of ADHD and ODD in a sample of 183 children with behavioral problems, $\phi = .45$, $p < .001$. In other words, almost 60% of the children with a diagnosis of ADHD also had a diagnosis of ODD, and nearly 63% of the children with a diagnosis of ODD also had a diagnosis of ADHD.

History Trivia

Galton to Fisher

Francis Galton (1822–1911) is credited with the first formal presentation of a statistical relationship. He was profoundly influenced by Charles Darwin, and Galton's work was devoted to prediction as a tool for the study of inheritance. For part of his research, Galton used sweet pea seeds, and he ranked the size of the parent and offspring seeds. He also studied height in fathers and sons. From this research, he noted that the offspring at their mature height showed less variability and fewer extremes than the parents. He called this phenomenon "reversion," and thus was derived the symbol r that appears now as the correlation coefficient.

Karl Pearson (1857–1936) was a friend of Galton, and he is likewise famous in the early history of statistics. In 1893, Pearson presented the term *standard deviation*. In 1895, he published an article deriving the current correlation coefficient formula and its test statistic. In the 1920s, Pearson's son, Egon, developed the idea of hypothesis testing, and his work led to the present definitions of the null and alternative hypotheses.

In 1904, Charles Spearman (1863–1945) published an article in which he stated a formula for rank-order correlation. However, the present formula was actually derived by Karl Pearson in 1907. In addition, Galton was probably the first to develop the concept of correlation with ranks when he rank-ordered his sweet pea seeds. Furthermore, it was Pearson who proposed to use rho (ρ) as the symbol for the rank-order correlation coefficient. Most statisticians use rho (ρ) as the symbol for the population correlation coefficient for the Pearson product-moment correlation coefficient and for the rank-order correlation coefficient. The symbol r_s (where the s occurs in honor of Spearman) is currently used as the sample rank-order correlation coefficient.

In 1958, Ronald Fisher, an avid pipe smoker, came to the defense of the tobacco industry. He was concerned that there was only correlational evidence that smoking caused lung cancer. In part, his objections were fueled by the use of government money to promote its agenda, whatever that agenda might be (even the health of its citizens). Also, in part, he objected to the lack of a clearly demonstrated mechanism for exactly how smoking caused lung cancer. In this regard, Fisher even presented evidence that the tendency to smoke was genetically inherited, and he challenged other researchers to show whether lung cancer might not be caused by smoking but by an underlying genetic heritability. In 1960, an influential meta-analysis (which is a statistical overview of a large number of studies) was published, clearly establishing an undeniable link between smoking and lung cancer. Despite a lack of controlled experimental evidence, there were too many lines of other types of evidence to ignore some causal relationship between smoking and lung cancer, even if the cause-effect relationship was unspecified. The medical community, based on this meta-analysis, adopted the position from that time forward that smoking was harmful.

Thus, it was not a single, definitive, controlled experiment that established the medical community's position against smoking, but it was the nearly unanimous consistency across many different kinds of studies that finally established the usefulness of the hypothesis that smoking was causally related to lung cancer. Fisher, a very strong proponent of the controlled experiment, failed to appreciate that one cannot ethically randomly assign groups of young adults to start smoking or not and then follow their health for 30 years. There are some situations where the controlled experiment cannot be employed. Fisher, probably because he was a very happy pipe smoker, was overly influenced by his addiction, his wish that there were no consequences of his addiction, and his adamant refusal to accept nothing less than controlled experiments as a proof of causation. In fact, Fisher said nothing could be proved without a randomized experimental design or a controlled experiment. Ironically, this interesting debate continues today as the effects of second-hand smoke on health are debated. Again, however, it must be recognized that there will never be any controlled experiments, for example, of second-hand smoke's effects on children's health. Thus, we must be willing to accept other lines of evidence, even if they are correlational in nature.

Key Terms, Symbols, and Definitions

Archival data—A type of retrospective study on data that have already been collected. The experimenter only has statistical power and no longer has any means of changing the original experimental design.

Causal relationship—A relationship in which one factor can be said to be the cause of another. Causal relationships should never be inferred from a correlational design or a correlational test statistic, although correlational relationships may suggest causal relationships, which then might be tested with controlled experiments.

Chance is lumpy—A rule of thumb in statistics where a long string or an unusual occurrence in data should first be attributed to randomness or chance.

Chi-square (χ^2) distribution—A theoretical sampling distribution that can be used to test the significance of a phi correlation and also to test nonparametric frequency data.

Coefficient of determination (r^2)—A statistic that explains the proportion of variance in one variable accounted for by the other variable and is obtained by squaring the Pearson's coefficient r.

Constant—A specific number that always stays the same in a statistical formula.

Correlation—The degree of relationship between two variables.

Curvilinear relationship—An assumption that a curved line best fits the graphic representation between two variables.

Degrees of freedom—A parameter that is equal to the number of observations or groups in a study minus some value(s) that limit the observations' or groups' freedom to vary.

Heteroscedasticity—The condition where the bivariate distribution has greater variance for some values of one variable compared to others. It is the opposite of homoscedasticity.

Homoscedasticity—An assumption of the Pearson product-moment correlation coefficient and linear regression that the distributions have similar variances for all values of the y scores along the corresponding x scores.

Least squares method—A method in regression that produces the single best-fitting line that predicts a Y score given an X score.

Linear relationship—A relationship in which a straight line best fits the bivariate distribution of two continuous variables.

Multiple regression analysis—A statistical procedure that measures the strength of a relationship between multiple independent variables and a single dependent variable.

Pearson's product-moment correlation (coefficient *r*)—A measure of the strength of a relationship between two continuous variables. It is represented by the coefficient r.

Phi correlation—A measure of the strength of a relationship between two dichotomous variables. It is represented by the coefficient ϕ.

Point-biserial correlation—A measure of the strength of a relationship between one continuous variable and one dichotomous variable. It is represented by the coefficient r_{pb}.

Regression analysis—A statistical procedure that measures the strength of a relationship between an independent variable and a dependent variable.

Scatterplot—A graphic representation of the relationship of two continuous variables in correlational designs. It is the graphic representation of a bivariate distribution.

Simple regression—A statistical procedure where the focus is on the way one variable (Y) changes based on how one other variable (X) changes.

Spearman correlation—A measure of the relationship between two ordinal rankings of the same set of data. It is represented by the coefficient r_s.

Strong negative relationship—A relationship in which a high score on variable x will be associated with a low score on variable y, and a low score on variable x will be associated with a high score on variable y.

Strong positive relationship—A relationship in which a high score on variable x will be associated with a high score on variable y, and a low score on variable x will be associated with a low score on variable y.

***t* distribution**—A table of critical values used to test correlation hypotheses.

Chapter 6 Practice Problems

1. A consumer psychologist wants to know if shoe size and weight are correlated in female adults. She measures the shoe size and asks the weight of 12 consecutive customers. They are as follows: size 6, 116 lbs.; size 5.5, 121 lbs.; size 10, 165 lbs.; size 3, 101 lbs.; size 9.5, 148 lbs.; size 6, 150 lbs.; size 10, 167 lbs.; size 5.5, 134 lbs.; size 7.7, 145 lbs.; size 8, 201 lbs.; size 8.5, 138 lbs.; and size 5.5, 123 lbs.

   a. Conduct the Pearson correlation on these data.
   b. Test for statistical significance.
   c. Graph a scatterplot of the data.

2. Two sportswriters ranked 10 pro football teams for best team in the 1990s. The team and rankings by the two writers respectively are as follows:

49ers	#1	#2
Broncos	#2	#3
Cowboys	#3	#5
Packers	#4	#1
Raiders	#5	#4
Colts	#6	#8
Dolphins	#7	#6
Titans	#8	#10
Jaguars	#9	#9
Bills	#10	#7

Conduct a Spearman correlation on these data, and test for statistical significance.

Chapter 6 Test Questions

1. A Pearson correlation above $r = .50$ may be labeled
   a. weak
   b. significant
   c. strong
   d. nonsignificant

2. Given a strong negative correlation, a score that is relatively high on variable x will be relatively _____ on variable y.
   a. low
   b. medium
   c. high
   d. significant

3. Variables may be analyzed and studied long after the data have been gathered. Such data are known as _______ data.
   a. correlation
   b. experimental
   c. archival
   d. archetypal

4. According to the text, a correlational analysis _________
   a. sometimes provides clues about cause-effect relationships
   b. often provides clues about cause-effect relationships
   c. never provides clues about cause-effect relationships
   d. is relatively useless compared to controlled experiments

5. According to the text, what are the consequences of the Type I error in the Mozart effect?
   a. benign
   b. horrible

c. could lead to mental disturbances
d. more classical musicians

6. Milk has been found to be positively correlated to cancer because
   a. milk causes cancer
   b. drinking milk provides an environment for cancer
   c. milk drinkers clearly live longer and clearly get more cancers
   d. there are suspicions but a correlation design does not allow one to say definitely why two variables are correlated

7. Abelson's first law of statistics is
   a. correlation never implies causation
   b. correlation sometimes implies causation
   c. chance is lumpy
   d. if something is too good to be true, it probably is too good to be true

8. _______________ correlation measures the strength of a relationship between two continuous variables.
   a. Pearson
   b. Spearman
   c. Point-biserial
   d. Phi

9. The ___________ hypothesis establishes whether a one-tailed or two-tailed significance test will be used.
   a. null
   b. alternative
   c. general
   d. specific

10. In the study of 30 pairs of scores in a Pearson correlation, what are the degrees of freedom?
    a. 30
    b. 29
    c. 28
    d. 15

11. A _____________ is a graphic representation of the pairs of scores plotted along the *x*- and *y*-axes in a correlation coefficient.
    a. bivariate distribution
    b. scatterplot
    c. homoscedast
    d. heteroscedast

12. The coefficient of determination explains
    a. the relationship between two variables
    b. the proportion of variance in one variable accounted for by another variable and vice versa
    c. the cause-effect relationship between two variables
    d. all of the above

13. The __________ assumption is met if, in a scatterplot, the distance from the points to the line is relatively equal all along the line.
    a. homoscedasticity
    b. linearity
    c. curvilinearity
    d. heterolinearity

14. The strength of a relationship between two variables in correlation is ____________ of the significance testing.
    a. highly dependent
    b. actually independent (not completely, but mostly)
    c. highly independent
    d. the obverse

15. The interest in ____________ analysis is seeing how various values of the independent variable predict corresponding values in the dependent variable.
    a. correlational
    b. curvilinear
    c. topographical
    d. regression

16. In the formula $Y = a + bX$, the letter b defines
    a. the intercept of the x-axis
    b. the intercept of the y-axis
    c. the slope or angle of the straight line
    d. all of the above

17. In multiple regression, the coefficient R can be interpreted as
    a. the percentage of variance accounted for in the dependent variable by the set of independent variables
    b. the percentage of variance accounted for in the dependent variable by a single independent variable
    c. the strength of a relationship between the dependent variable and a set of independent variables
    d. all of the above

18. In multiple regression, the R-square can be interpreted as
    a. the percentage of variance accounted for in the dependent variable by the set of independent variables
    b. the percentage of variance accounted for in the dependent variable by a single independent variable
    c. the strength of a relationship between the dependent variable and a set of independent variables
    d. the percentage of variance accounted for in the dependent variable by the set of independent variables minus an estimate penalty

19. In multiple regression, the adjusted R-square can be interpreted as
    a. the percentage of variance accounted for in the dependent variable by the set of independent variables

b. the percentage of variance accounted for in the dependent variable by a single independent variable
c. the strength of the relationship between the dependent variable and the set of independent variables
d. the percentage of variance accounted for in the dependent variable by the set of independent variables minus an estimate penalty

20. In multiple regression, the ANOVA table tells us whether the three main regression coefficients (*R*, *R*-square, adjusted *R*-square) are significantly different from

a. zero
b. 1.00
c. .50
d. 1.50

21. In a multiple regression analysis, the final section of the output contains the coefficients. Which of these coefficients is of primary concern?

a. unstandardized *B*
b. standard error of *B*
c. standardized coefficient beta
d. standard error of beta

22. The __________ correlation gives an estimate of the degree of relationship between two dichotomous variables.

a. Pearson
b. Spearman
c. point-biserial
d. phi

23. To test for the significance of a Pearson correlation, one must use the ______________ distribution.

a. t
b. F
c. χ^2
d. z

24. To test for the significance of the phi correlation, one must use the ______________ distribution.

a. t
b. F
c. χ^2
d. z

25. The formula for Spearman's rank-order correlation was actually derived by

a. Francis Galton
b. Charles Darwin
c. Karl Pearson
d. Egon Pearson

Problems 26–30. A consumer psychologist wants to know if shoe size and weight are correlated in adult males. She measures the shoe size and asks the weight of 14 consecutive customers. They are as follows: size 9, 176 lbs.; size 7.5, 141 lbs.; size 10, 185 lbs.; size 12, 202 lbs.; size 9.5, 174 lbs.; size 10, 150 lbs.; size 10, 193 lbs.; size 10.5, 237 lbs.; size 13, 248 lbs.; size 8, 159 lbs.; size 8.5, 136 lbs.; size 9.5, 174 lbs.; size 9, 172 lbs.; and size 11, 183 lbs.

26. The Pearson correlation on these data is $r =$

    a. .61
    b. .71
    c. .81
    d. .91

27. The r value can be reported significant (lowest possible) at p

    a. $> .05$
    b. $< .05$
    c. $< .01$
    d. $< .001$

28. The *df* for this problem are

    a. 14
    b. 12
    c. 10
    d. 8

29. The coefficient of determination for this problem is

    a. .65
    b. .81
    c. .809
    d. .99

30. The percentage of variance in shoe size that can be accounted by weight is

    a. 65
    b. 81
    c. 90
    d. 99

7 The *t* Test for Independent Groups

Chapter 7 Goals

- Learn the statistical analysis of a controlled experiment
- Learn the assumptions of the *t* test for independent groups
- Learn how to analyze and interpret the *t* test for independent groups
- Learn about statistical power and the nature of a power analysis
- Learn how to test for effect size in the *t* test (correlation of effect size)
- Learn how to construct confidence intervals

The Statistical Analysis of the Controlled Experiment

We previously learned that the correlation coefficient is probably the most often used inferential statistic. It has the limitation, however, that we can never imply a causal relationship between two correlated variables. We also mentioned earlier the very powerful **controlled experiment,** where a large group ($N \geq 30$) of participants is randomly assigned to either an experimental group or a control group. The controlled experiment does allow us to assume a causative relationship between the independent variable and the dependent variable. This experimental design and procedure is known and statistically analyzed by the **independent groups *t* test.**

The focus of the *t* test is to determine whether there is a significant difference between the experimental and control groups' means on the dependent variable beyond mere chance differences. To understand the focus of the *t* test, let us return to the signal detection theory analogy. If the independent variable really works, it is similar to creating a large signal because there will be a large difference between the experimental group's mean and the control group's mean. The magnitude of the difference between the two groups' means is considered to be the signal. If the independent variable really works (the signal is large), then the two groups' means will be very different on the dependent variable; thus, the magnitude of the difference between the two groups' means will be large. This variation and difference in scores between the two groups because of the independent variable or treatment is called the between-groups variance.

Let us also imagine the situation where the independent variable does not work: In other words, nothing more than chance is having an influence on the participants' mean scores. In this case, there should not be much of a difference between the experimental group's mean and the control group's mean beyond chance differences. These chance differences are similar to noise in signal detection theory. Notice also that each participant in the experimental group will vary from the group's mean. This variance is considered to be the within-group noise or within-group variance. In any experiment, the participants in a group will always vary from each other, even though they were all in the same group and even though they all received the same treatment. Statisticians refer to within-group variance as error or within-group error. In reality, it is rather bizarre to label any variation from participant to participant as an "error," but it is considered to be an "error" in name only. It is a verbal anachronism from scientific times before about 1700, when it was thought that all of the sciences would be governed by invariant rules and laws. Any observations or measures that varied from these rules and laws were thought to be errors.

In summary, the signal (the difference between the two groups' means) in a controlled experiment must exceed the noise (within-group variation) for us to determine that the independent variable had a genuine effect on the dependent variable. In terms of hypothesis testing, the null and alternative hypotheses for the controlled experiment would be

$$H_0: \mu_1 = \mu_2$$
$$H_a: \mu_1 \neq \mu_2$$

These mean population values will actually be tested with sample means drawn from these population values.

One *t* Test But Two Designs

There are actually two popular experimental designs that can be analyzed by the *t* test for independent groups. The first design is the controlled experiment where the sample of participants is randomly selected from the population and then the participants are randomly assigned to one of the two experimental groups, an experimental group and a control group. The experimental group receives the treatment, and the control group is treated exactly like the experimental group, except that it receives a placebo instead of the treatment. For example, the effects of caffeine levels on mood states (good to bad mood) might be evaluated. Both groups' moods might be measured after giving one group caffeinated coffee and another group decaffeinated coffee. Some period of time (let us say 10 minutes) after drinking the caffeinated or decaffeinated coffee, each participant would be measured on his or her current mood state, perhaps using some continuous variable ranging from 1 (bad mood) to 10 (good mood). The independent *t* test would tell us whether the mean mood level of the caffeinated coffee group was significantly different from that of the decaffeinated coffee group. Next, we could preliminarily and cautiously conclude that the participants' moods were differentially affected by the caffeine or lack thereof.

The second experimental design is sometimes called the **in situ design.** *In situ* is Latin and means "in its original place." In this design, the participants come as they are, preassigned by God or nature, to one of the two groups. For example, we might measure the mood states of everyone in a coffee shop who is drinking caffeinated coffee and those who are drinking decaffeinated coffee. The *t* test, in this case, would measure the difference in the mean mood scores between the two groups. If we found a significant difference between the two means, we would not be able to imply that caffeine or its absence was causing the difference in moods because the participants were not randomly assigned to the two groups.

The in situ experimental design does result in two independent groups and can be statistically analyzed with a *t* test exactly the same as the controlled experiment. Again, however, the limitation of the in situ experimental design is that we are much less certain about causation since a plethora of other variables might have accounted for the differences between the two groups. Despite this limitation, in situ studies are frequently the only way some relationships between variables can be studied. For example, we cannot ethically randomly assign people to receive a head injury or not in order to determine whether a helmet prevents memory loss after a head injury from a motorcycle accident. However, we could measure the memory loss of two groups after a motorcycle accident, comparing those who wore a helmet and those who did not. In situ experimental designs are popular despite the problem of not being able to determine causation.

Assumptions of the Independent *t* Test

Specific assumptions must be met to use the *t* test appropriately. They are as follows:

Independent Groups

The participants must be different in each group; that is, no participant is allowed to serve in both groups. There is another form of the *t* test, the dependent groups *t* test, that does assume that the participants are the same in each group, and it will be discussed in Chapter 8.

Normality of the Dependent Variable

The *t* test and its critical values are based on the assumption that the sample dependent variable values come from a population of values that is normally distributed. However, the *t* test's value is enhanced because it is still a fairly reliable measure even for nonnormal but still mound-shaped distributions. An interesting characteristic of the *t* test is that it is robust. The **robustness** of a statistical test means that its assumptions may be violated to some extent, yet the correct statistical decision will still be made, which is to correctly reject or fail to reject the null hypothesis.

Homogeneity of Variance

While we may expect the means to be different between two groups in a *t* test, the assumption is made that the variances of the two groups about their respective means will be equal or approximately equal no matter whether the two groups' means are different or not. In reality, this assumption is often a safe one. It is actually rare that the variances of two groups are radically unequal. Also, violations of the assumption matter more when the samples are small than when they are large. Thus, using larger sample sizes (e.g., $N > 15$ or 20 in each group) helps to minimize unequal variances. Another way to reduce the effect of the **heterogeneity of variance** is to use an equal number of participants in each group (also called equal *N*). The use of an equal number of participants in each group has other beneficial statistical properties as well. Finally, the *t* test is robust against the violation of the assumption of homogeneity of variance and that means that, despite violations of the assumption of **homogeneity of variance,** we are still likely to make the correct statistical decision.

The Formula for the Independent *t* Test

The formula for a *t* test between two different groups of scores is as follows:

$$t = \frac{\bar{x}_1 - \bar{x}_2}{\sqrt{\left[\frac{\sum x_1^2 - \frac{(\sum x_1)^2}{N_2} + \sum x_2^2 - \frac{(\sum x_2)^2}{N_2}}{N_1 + N_2 - 2}\right]\left[\frac{1}{N_1} + \frac{1}{N_2}\right]}}$$

where

$\bar{x}_1$ = the mean of the scores of the first group,

$\bar{x}_2$ = the mean of the scores of the second group,

Σx_1^2 = the sum of the squares of the first group,

Σx_2^2 = the sum of the squares of the second group,

$(\Sigma x_1)^2$ = the square of the sum of the scores of the first group,

$(\Sigma x_2)^2$ = the square of the sum of the scores of the second group,

N_1 = the total number of scores in the first group,

N_2 = the total number of scores in the second group.

You Must Remember This! An Overview of Hypothesis Testing With the *t* Test

If the *t* value obtained by the formula exceeds the tabled critical value at the .05 significance level, then the null hypothesis will be rejected. It will then be concluded that one mean is significantly different from the other mean. If the *t* value obtained by the formula does not exceed the tabled critical value, then the null hypothesis will not be rejected, and it will be concluded that there is no significant difference between the two means (other than just chance differences).

What Does the *t* Test Do? Components of the *t* Test Formula

Let us examine the components of the *t* test formula. Mathematically, there are three major parts: the numerator, the left half of the denominator, and the right half of the denominator.

The numerator contains the difference between the two groups' means. What is the effect on the final *t* value of the size of the numerator? It is easily seen that as the magnitude of the difference between the two means gets larger, the *t* value will get larger. In general, if the difference between the means is small, the *t* value will be small.

Now let us examine the left half of the denominator. After some of you recover from the shock, you will realize that this part of the formula looks vaguely familiar. It is the computational formula for the standard deviation. However, in this formula, we have joined the variances for the two groups together. This procedure results in what is called **pooled variance.** This means that the size of the variances for the two groups is approximately the same (or at least not radically different from each other), and therefore the variances can be combined in a single formula. What is the effect of the size of the pooled variances on the final *t* value? Because the variances are in the denominator, it means that, as they get larger, the *t* value gets smaller. If this is not intuitively obvious, imagine a pizza being divided among a family. If the family is large, the resulting pieces of pie will be small. If the family is smaller, then the pieces of pie will be larger. Thus, a large variance or standard deviation in the denominator will reduce the size of the final t value. A small variance or standard deviation in the denominator will make the t value larger.

The final component of the t formula is the right half of the denominator. It contains the reciprocals of the sizes of the two groups. Why do we take the reciprocals? Taking the reciprocals reverses the effects of large and small values in the denominator. Normally, a large value in the denominator reduces the value of the final t. In this case, by taking the reciprocal, a large sample size increases the value of t, and a small sample size reduces the value of t. It is as if a larger sample increases our confidence that two means are reliably different.

In summary:

1. A large difference between the means creates a larger t value; therefore, the null hypothesis is more likely to be rejected.
2. Small variances or small standard deviations for the two groups create a larger t value; therefore, the null hypothesis is more likely to be rejected.
3. Large sample sizes create a larger t value; therefore, the null hypothesis is more likely to be rejected.

What If the Two Variances Are Radically Different From One Another?

One way to test statistically the assumption of homogeneity of variance is **Levene's test of homogeneity of variance.** In Levene's test, if the F value is large enough to be significant, it means that you have violated the assumption of homogeneity of variance. This means that a separate formula must be used to obtain the standard deviation for each group in the t test or, in other words, you may not use the pooled variance formula. Another option is to transform each score in both groups by a mathematical formula and conduct the F test for homogeneity of variance again (see Kirk, 1995, for a discussion of transformations). However, we mentioned earlier that large sample sizes and equal numbers of participants in each group have the effect of minimizing violations of the assumption of homogeneity of variance. In addition, the t test is robust to violations of most of its assumptions.

A Computational Example

Let us imagine that a psychiatrist wishes to test a new drug that is purported to be better than Prozac (a popular antidepressant). In a clinical trial, a sample of depressed patients was randomly assigned to either the experimental group or the control group: The experimental group was given the new drug Kojac, and the other group was given Prozac. None of the groups' members knew whether he or she was taking Kojac or Prozac. After 1 month of drug treatment, each patient completed a depression inventory on which the scores can range from 24 (not depressed) to 96 (very depressed).

As far as the formula is concerned, either group can be labeled Group 1. If you wish to have a positive t value, then label the group with the highest mean Group 1. In this example, the Prozac group appears to have the highest mean, so it will be labeled Group 1, while the Kojac group will be Group 2.

Prozac Group	*Kojac Group*
64	53
68	44
84	61
65	50
50	40
60	39
81	58
76	42

Steps in the *t* Test Formula

Step 1. Calculate $\bar{x}_1$:

$$\bar{x}_1 = \frac{64 + 68 + 84 + 65 + 50 + 60 + 81 + 76}{8}$$
$$\bar{x}_1 = \frac{548}{8}$$
$$\bar{x}_1 = 68.500$$

Step 2. Calculate $\bar{x}_2$:

$$\bar{x}_2 = \frac{53 + 44 + 61 + 50 + 40 + 39 + 58 + 42}{8}$$
$$\bar{x}_2 = \frac{387}{8}$$
$$\bar{x}_2 = 48.375$$

Step 3. Calculate $\Sigma\, x_1^2$:

$$\sum x_1^2 = 64^2 + 68^2 + 84^2 + 65^2 + 50^2 + 60^2 + 81^2 + 76^2$$
$$\sum x_1^2 = 4096 + 4624 + 7056 + 4225 + 2500 + 3600 + 6561 + 5776$$
$$\sum x_1^2 = 38{,}438$$

Step 4. Calculate $\Sigma\, x_2^2$:

$$\sum x_2^2 = 53^2 + 44^2 + 61^2 + 50^2 + 40^2 + 39^2 + 58^2 + 42^2$$
$$\sum x_2^2 = 2809 + 1936 + 3721 + 2500 + 1600 + 1521 + 3364 + 1764$$
$$\sum x_2^2 = 19{,}215$$

Step 5. Calculate $(\Sigma\, x_1)^2$:

$$\left(\sum x_1\right)^2 = (64 + 73 + 84 + 65 + 50 + 60 + 81 + 76)^2$$
$$\left(\sum x_1\right)^2 = (548)^2$$
$$\left(\sum x_1\right)^2 = 300{,}304$$

Step 6. Calculate $(\Sigma\, x_2)^2$:

$$\left(\sum x_2\right)^2 = (53 + 44 + 61 + 50 + 40 + 39 + 58 + 42)^2$$
$$\left(\sum x_2\right)^2 = (387)^2$$
$$\left(\sum x_2\right)^2 = 149{,}769$$

Step 7. Obtain the number of scores in each group.

$N_1 = 8$ (the number of scores in Group 1)

$N_2 = 8$ (the number of scores in Group 2)

Step 8. Enter the values obtained in Steps 1 through 7 into the formula for the t test.

$$t = \frac{\bar{x}_1 - \bar{x}_2}{\sqrt{\left[\dfrac{\sum x_1^2 - \dfrac{\left(\sum x_1\right)^2}{N_2} + \sum x_2^2 - \dfrac{\left(\sum x_2\right)^2}{N_2}}{N_1 + N_2 - 2}\right]\left[\dfrac{1}{N_1} + \dfrac{1}{N_2}\right]}}$$

$$t = \frac{68.500 - 48.375}{\sqrt{\left[\dfrac{38,438 - \dfrac{(548)^2}{8} + 19{,}215 - \dfrac{(387)^2}{8}}{8 + 8 - 2}\right]\left[\dfrac{1}{8} + \dfrac{1}{8}\right]}}$$

$$t = \frac{20.1250}{\sqrt{\left[\frac{38{,}438 - 37{,}538 + 19{,}215 - 18{,}721.125}{14}\right]\left[\frac{1}{4}\right]}}$$

$$t = \frac{20.1250}{\sqrt{\left[\frac{1393.875}{14}\right][.25]}}$$

$$t = \frac{20.1250}{\sqrt{24.890625}}$$

$$t = \frac{20.1250}{4.9890505}$$

$$t = 4.034$$

Testing the Null Hypothesis

To test the null hypothesis, you can compare the derived *t* value to critical values of *t* on the *t* distribution. Refer to the table of critical *t* test values in Appendix B.

Steps in Determining Significance

Step 1. To use the table of *t* values, you need the degrees of freedom (*df*). The *df* formula is $N_1 + N_2 - 2$ or, in the previous example, $8 + 8 - 2 = 14$. The concept of *df* is a complex one, but one simple aspect is that the *df* for a *t* test is roughly related to the total number of participants but is always less.

Step 2. Determine whether you will conduct a one-tailed or two-tailed test of significance. If you have chosen a nondirectional alternative hypothesis (as has been suggested throughout the book), then you have already decided on a two-tailed test of significance. In our example, we had a nondirectional alternative hypothesis; therefore, we have chosen a two-tailed test of significance.

Step 3. Determine the level of significance at which you will conduct the test of the null hypothesis. Statisticians are in nearly complete agreement that the starting level of significance should be $p = .05$ (this starting level is also called the conventional level of significance). Therefore, the null hypothesis will be tested against the conventional level of $p = .05$.

Step 4. Determine the critical values of *t* at $p = .05$ with $df = N_1 + N_2 - 2$. In our example, for a two-tailed test with $p = .05$ and $df = 14$, Appendix B reveals that the critical values of $t = \pm 2.145$.

Step 5. Finally, compare the formula-derived value of t to the critical values of t. If the t value does exceed this tabled value, then the null hypothesis is rejected, and it is concluded that the means are significantly different from each other. If the t value does not exceed the tabled value at the .05 level of significance, then the null hypothesis is retained, and it is concluded that there is no significant difference between the means (this does not imply that the means are equal, only that they are not significantly different from each other).

Step 6. Report the findings. The discipline of psychology, as well as more than 200 journals, has adopted the American Psychological Association (2001) format for the reporting of statistical tests. Their recommendations are to report the test statistic (r or t), its derived value, the degrees of freedom, and the significance level (p level).

In our example, the null and alternative hypotheses are as follows:

$$H_0: \mu_1 = \mu_2$$
$$H_a: \mu_1 \neq \mu_2$$

where

μ_1 = Prozac group, and

μ_2 = Kojac group.

Our informal write-up might look like this:

The derived $t = 4.034$ exceeds the critical value of $t = +2.145$ at $p = .05$ with $df = 14$. Therefore, H_0 is rejected, and it is concluded that the mean Depression score for the Kojac group (48.38) was significantly lower than the mean of the Prozac group (68.50), $t(14) = 4.034$, $p < .05$. In terms of the research question, it appears that the new drug Kojac relieves depression better than Prozac.

In a research article, the t test results might be reported as follows:

As predicted, depressed patients who receive the new drug Kojac had a significantly lower depression score ($M = 48.38$, $SD = 8.40$) than the patients who received Prozac ($M = 68.50$, $SD = 11.34$), $t(14) = 4.034$, $p < .05$.

When H_0 Has Been Rejected

When H_0 has been rejected, the standard statistical procedure is to report the lowest p level (also known as the alpha level). In the present case, H_0 was rejected at the .05 level, but the derived t value also exceeded the critical value of t at $p = .01$, but it did not exceed the critical value at $p = .001$. Therefore, the t will be reported as statistically significant at $p < .01$.

Note that this significance procedure is counterintuitive. This means that the derived t value must *exceed* the tabled critical value of t in order to report it significant at *less than* .05. Remember, however, what this testing process represents: If there is no effect of the independent variable between the two groups (the treatment does not work), then theoretically, the derived t value should be zero or very close to zero. If the treatment does work, then the two groups' means should be far apart, creating a large t value. Thus, if the independent variable does have an effect on the two groups' means, the null hypothesis will be rejected because the derived t value is likely to be very large.

The Power of a Statistical Test

The **power** of an inferential statistical test refers to its ability to detect a false null hypothesis or its ability to detect a real difference between two groups' means. In general, the independent t test is a powerful test because it does have a solid ability to detect true treatment differences between two means. However, the power of a test can be experimentally manipulated, and the power of a statistical test can be underused or abused. The power of a statistical test can be underused if we fail to use enough participants. In this case, although there may be a real difference between two groups' means (the treatment really works), we may fail to reject the null hypothesis (committing the Type II error) if we do not have enough participants in each group. Remember, an experiment is psychometrically large if the total N is about 30 or greater. Thus, to have sufficient power, it is recommended that each group have at least 15 participants. Another way of experimentally increasing the power of a test would be to try to make the participants as homogeneous as possible (e.g., restricting the ages of the participants). While restricting the ages of the participants may reduce the effect of within-groups variance or noise in the experiment, increasing the power of the statistical test, this procedure might limit the generalizability of the results to other age groups.

It is also common in the sciences to witness an abuse of power with statistical tests. In this case, the abuse comes from having too many participants in the study. While it may appear that "the larger, the better" or "the more, the merrier" may be true in surveys, in an inferential statistical test such as the independent t test, too many participants may result in an abuse of statistical power. Using more participants than needed does increase the chance of detecting a real difference between two means. Thus, using a lot of participants does reduce the chance of committing the Type II error (retaining H_0 when it is false). However, there is a cost for this increased detection ability: We are much more likely to commit the Type I error (rejecting H_0 when it is true) when we use too many participants. This occurs because the increased sensitivity to real differences also makes us vulnerable to claiming chance differences as real differences. The solution to the power dilemma is to conduct a power analysis. A power analysis determines how many participants would be appropriate for a study. Power analysis tends to be an advanced topic in statistics, so consult an advanced statistical text if you want to learn more about it.

Effect Size

Remember when we ran into the curious artifact in correlation where a correlation might be weak (even as low as $r = .06$) and still be statistically significant or a correlation could be strong (as high as $r = .56$) yet not be statistically significant? This curiosity often occurs as an interaction between the power of a statistical test and the effect size. **Effect size** refers to the strength of the influence of the independent variable on the dependent variable. Another way to consider effect size is how well the treatment works. If the treatment really works (as detected by a large difference between the two groups' means), then there is said to be a large effect size. If the difference between the two groups' means is small, then there is said to be a small effect size. Notice the same curious dilemma as we witnessed with correlation: We can have a strong effect size but fail to reject H_0. This may occur because we failed to have sufficient statistical power (such as too few participants in each group or too much variance around a group's mean). We can also have a very weak effect size, and yet we may still reject H_0. This may occur in cases where we have abused statistical power by using too many participants. Once again, the solution to the effect size dilemma is to conduct a power analysis.

The Correlation Coefficient of Effect Size

To determine the effect size in the independent *t* test, a correlation coefficient of effect size can be derived. The formula is the following:

$$r = \sqrt{\frac{t^2}{t^2 + df}}$$

The correlation coefficient of effect size will always be positive and range from 0 to 1.00. Use the following scale to interpret the magnitude of the effect size:

Effect Size	*Minimum r Value*
Small	.100
Medium	.243
Large	.371

Notice that these r values are considerably smaller than we had for the same labels when we used the Pearson product-moment correlation coefficient r. Thus, keep in mind that these are two separate statistical procedures, although both use the coefficient r.

In the previous example, we obtained a t value of 4.034. If we enter this value into the correlation coefficient of effect size formula, we obtain the following:

$$r = \sqrt{\frac{t^2}{t^2 + df}}$$

$$r = \sqrt{\frac{4.034^2}{4.034^2 + 14}}$$

$$r = .733$$

Thus, we can see that a very large effect size was observed in the Kojac-Prozac experiment, or in other words, there was a major difference in the way the two drugs affected depression in these clinically depressed patients.

Confidence Intervals

It is rare that researchers gather information from an entire population. If we did, many inferential statistics would be unnecessary. Furthermore, even if there was an extremely rare instance where we could work with an entire population, we might treat it as a sample because we would want our results to extend to future cases that were not in the present population. Yet, error is inherent in any experimental procedure or survey. It has been argued (e.g., Schmidt, 1996) that traditional significance tests such as the classic t test for independent groups fall short as measures of the true reality of an outcome of an experiment because (a) when a null hypothesis is rejected, p levels tend to be misinterpreted, and (b) a retained null hypothesis tells us very little.

Schmidt (1996), among others, argues for the use of confidence intervals. Confidence intervals give us an estimate of the amount of error involved in our data. They tell us about the precision of the statistical estimates (e.g., means, standard deviations, correlations) we have computed. Confidence intervals are related to the concept of the power. Although it may sound counterintuitive, a larger confidence interval means the study has less power to detect genuine differences between treatment conditions in experiments or between groups of respondents in survey research.

A confidence interval is based on three elements:

1. a value of a parameter around which the confidence interval will be built (mean, correlation coefficient, etc.),
2. the standard error (SE) of the measure, and

3. the desired width of the confidence interval (e.g., the 95% confidence interval or the 99% confidence interval).

In this section, we will construct a confidence interval around a single mean. The standard error of the mean is the standard deviation of all means taken from repeated samples from a population. For example, suppose we randomly selected 200 samples of 30 cases ($n = 30$) from a population and found the mean for each of the 200 samples. Also, suppose that the population mean is zero. We can ask, "Is the mean of the distribution of sampled means different from zero?" To answer this question, we will build a confidence interval around the mean of the distribution of means. A confidence interval (CI) is defined by the following formula:

$$95\%\ \text{CI} = M \pm (z \cdot SE) = M - (z \cdot SE) \text{ to } M + (z \cdot SE)$$

where

M = the mean of the distribution of means,

SE = the standard error of the mean, and

z = the z score for the particular confidence interval of interest.

For example, if we want a 95% confidence interval, the value of z would be 1.96. The value 1.96 comes from our understanding of the normal curve. The areas between ±1.96 standard deviations of the mean of the z distribution cover 95% of the cases, if the means are normally distributed. Substituting 1.96 for z the formula for the 95% CI would be

$$95\%\ \text{CI} = M \pm (1.96 \cdot SE) = M - (1.96 \cdot SE) \text{ to } M + (1.96 \cdot SE)$$

The standard deviation of the distribution of means is called the *standard error*. The standard deviation of this distribution of sample means is 1.00, and hence the standard error is also 1.00. The 95% confidence interval for this distribution is indicated by the gray portion in Figure 7.1. The 95% confidence interval for this set of data ranges from −1.96 to 1.96 because $M = 0$ and $SE = 1.00$. Thus, 95% of the 200 means will fall within the 95% confidence interval. The darker areas at either end of the distribution represent the areas that are outside the 95% confidence interval. The 95% CI includes zero, so the mean of the scores is not significantly different from zero.

Another random sample of 200 sets of 30 cases each is drawn. The distribution of the means is shown in Figure 7.2. In this case, the mean of the distribution is 2.50. The standard deviation of the distribution of means is 1.00, and hence the standard error is 1.00. Applying the formula for the 95% CI:

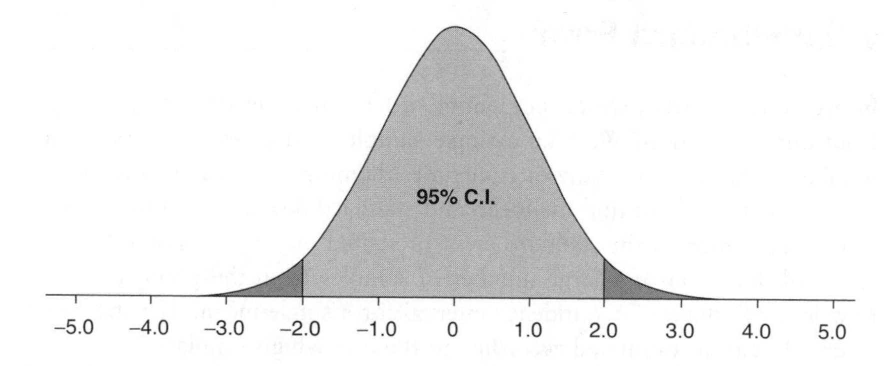

Figure 7.1 95% Confidence Interval: CI Includes Zero

$$95\%\ \text{CI} = 2.50 \pm (1.96 \cdot 1.00)$$
$$95\%\ \text{CI} = 2.50 \pm 1.96$$
$$95\%\ \text{CI} = (2.50 - 1.96) \text{ to } (2.50 + 1.96)$$
$$95\%\ \text{CI} = 0.54 \text{ to } 4.46$$

Thus, we find that the 95% CI ranges from 0.54 to 4.46. The confidence interval does not include zero, so the mean of this distribution of means is different from zero at $p < .05$.

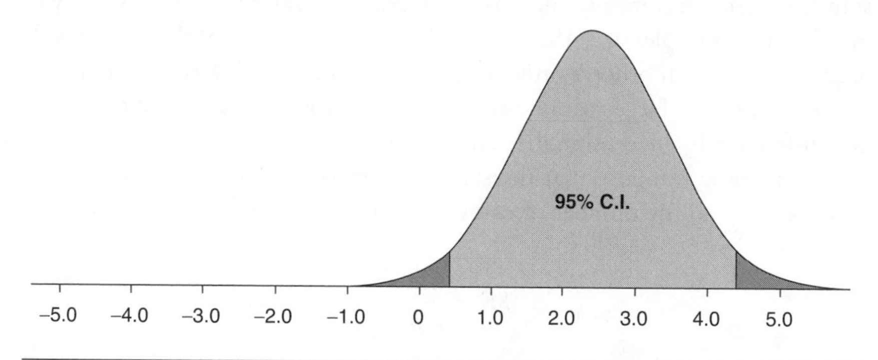

Figure 7.2 95% Confidence Interval: CI Does Not Include Zero

The distribution could be toward the negative end of the scale. Suppose that the mean of a distribution of 200 randomly sampled sets of 30 cases was −2.20, and the standard deviation of the distribution of means was 1.00. Then the 95% confidence interval would range from −4.16 to −0.24. The 95% CI does not include zero, so the mean of −2.20 is significantly lower than zero at $p < .05$.

$$95\%\ \text{CI} = -2.20 \pm (1.96 \cdot 1.00)$$
$$95\%\ \text{CI} = -2.20 \pm 1.96$$
$$95\%\ \text{CI} = (-2.20 - 1.96) \text{ to } (-2.20 + 1.96)$$
$$95\%\ CI = -4.16 \text{ to } -0.24$$

Estimating the Standard Error

In everyday research, we do not sample a large number of sets of n cases from our population. We take a single sample of n cases, where n is the number of cases in the study or sometimes the number of cases in each cell of the design. We can find the mean and standard deviation of our sample. But the definition of the standard error presented previously is that it is the standard deviation of a large number of samples from the population. So now let us find a 95% confidence interval for a single mean. The standard error (*SE*) can be estimated according to the following formula:

$$SE = \frac{SD}{\sqrt{n}}$$

where

SD = the standard deviation of our sample, and

n = the number of cases.

That is, the standard error is estimated by dividing the obtained standard deviation by the square root of the number of cases. Notice that the standard error becomes smaller as the size of the sample increases. As we increase our sample size, the standard error and the confidence interval become smaller. In other words, we can detect smaller differences between means if we have larger sample sizes. We can increase the power to detect any difference by increasing the sample size.

For example, suppose that the sample mean is 4.52, the standard deviation is 6.28, and the number of cases is 15. Then the standard error is

$$SE = \frac{SD}{\sqrt{n}}$$
$$SE = \frac{6.28}{\sqrt{15}}$$
$$SE = \frac{6.28}{3.873}$$
$$SE = 1.62$$

Applying the formula for the 95% CI,

$$95\%\ \text{CI} = M \pm (1.96 \cdot SE)$$
$$95\%\ \text{CI} = 4.52 \pm (1.96 \cdot 1.62)$$
$$95\%\ \text{CI} = 4.52 \pm 3.175$$
$$95\%\ \text{CI} = (4.52 - 3.175) \text{ to } (4.52 + 3.175)$$
$$95\%\ \text{CI} = 1.345 \text{ to } 7.695$$

The 95% CI ranges from 1.345 to 7.695. It does not include zero, so the mean, 4.52, is different from zero at $p < .05$. A graphical representation of the 95% confidence interval using the estimated standard error is shown in Figure 7.3. The area under the 95% CI bar represents the 95% confidence interval around the mean of 4.52.

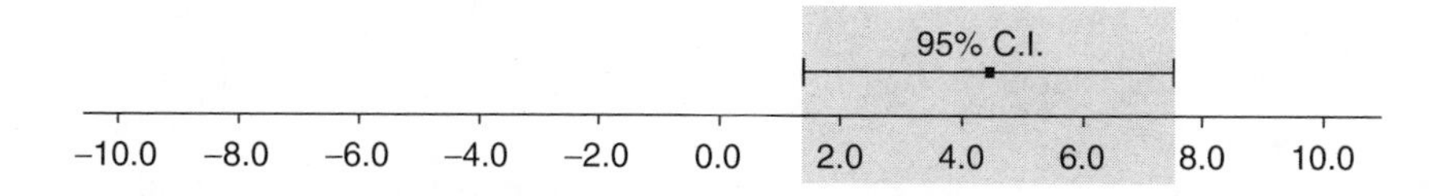

Figure 7.3 95% Confidence Interval Using the Estimated Standard Error

In these examples, we asked, "Is the sample mean different from zero?" Usually we are more interested in the question, "Is Mean A different from Mean B?" Confidence intervals are also useful for that question. Does the 95% confidence interval for one of the means include the other mean? If so, then the means are not different from each other. If not, then the means are different from each other at $p < .05$.

History Trivia

Gosset and Guinness Brewery

William Gosset (1876–1937) is attributed the discovery of the *t* distribution. He worked part-time for the Guinness Brewery in Ireland while pursuing his graduate studies. Later, he was employed by the brewery to do research, and he came to discover that if he used small samples in sampling procedures, it led to errors when trying to estimate population values. This led him to the discovery of the family of *t* distributions based on varying small sample sizes. In 1908, he published an article about his findings, but Guinness Brewery had strict regulations about publications by its employees (apparently because a previous employee had published Guinness's recipe for beer), so Gosset chose a pen name, "Student." It is not exactly known why he chose to label the new distributions *t*, although it may have had something to do with "tea" breaks in the afternoon. It is also not exactly known why he chose the pen name Student, although it is suggested that it in part represented his modest nature.

By 1920, Gosset, under the pen name of Student, had published tables of the *t* distribution and the *t* statistic. In 1925, Ronald Fisher (1890–1962) published one of his two major statistics books, titled *Statistical Methods for Research Workers*. One major contribution of this work was the idea of using Student's t statistic to test a hypothesis about a single mean and to test the difference between two means. Thus, the **t test** was actually created by Fisher based on Gosset's family of t distributions.

Ironically, the most important concept that Fisher *did not directly address* was the interpretation of the p level after a test of significance. He did, however, leave a context for its interpretation, and that is the replicability of a finding. He wrote that a researcher might judge the results of an experiment significant if the results were of such a magnitude that it would not have been produced by pure chance not more frequently than once in 20 trials. However, Fisher cautioned experimenters that they should not allow themselves to be deceived once in every 20 experiments. Fisher wrote that researchers should only claim that a phenomenon is real when they could design an experiment so that it would rarely fail to give a significant result. Fisher was highly skeptical of isolated significant results. Thus, Fisher appeared to be a strong advocate of a sequence of experiments to establish the replicability of a finding.

Key Terms and Definitions

Controlled experiment—The most powerful experimental design because it allows for the inference of causation. The participants are randomly chosen from a population and randomly assigned to one of two (or more) groups, usually an experimental group that receives some treatment and a control group that receives a placebo.

Effect size—How strongly the independent variable affects the dependent variable. Effect size ranges from small to large.

Heterogeneity of variance—A violation of the assumption that two groups have equal or approximately equal variances in the independent groups t test; in this case, the variances are not equal.

Homogeneity of variance—An assumption in the independent groups t test that the two groups have equal or approximately equal variances.

In situ design—Latin for "in its original place." An often used t test design where the participants are not randomly assigned to the two groups but have been preassigned by nature or God, such as dyslexic children and nondyslexic children, brain-injured patients and controls, and so on.

Independent groups t test—A statistical procedure designed to evaluate the difference between two groups' means on the same continuous variable.

Levene's test of homogeneity of variance—A statistical test used to assess the homogeneity of variances assumption in the independent t test. A significant value of F in Levene's test indicates that the two groups' variances are significantly different from each other, and the pooled estimate of variance may not be used.

Pooled variance—A combination of the separate variances for two groups into a single estimate. The denominator of the independent t test formula contains the pooled variance estimate. If the assumption of homogeneity of variance has been met, then the two groups' variances may be pooled together to make a single estimate. If the assumption of homogeneity has not been met, then separate estimates of variance must be conducted for the two groups.

Power—The probability that the null hypothesis will be rejected correctly. In significance testing, the power of a test is estimated by $1 - \beta$, where β is the probability of committing a Type II error (retaining H_0 when it is false).

Robustness—A characteristic of statistical tests that indicates the ability to detect a real difference between the two groups' means or the ability to detect a false null hypothesis despite violations of the assumptions of the statistical test.

***t* test**—A test to determine whether there are significant differences between two independent groups' means on the same dependent variable.

Chapter 7 Practice Problems

1. What is the primary focus of the independent *t* test?
2. Explain the assumption of homogeneity of variance.
3. What can be done to reduce the effects of heterogeneity of variance?
4. Describe the three factors that can make a *t* value larger.
5. If the derived *t* value exceeds the tabled critical *t* value, are the results statistically significant?

Chapter 7 Test Questions

Problems 1–5. A business statistician wishes to determine whether current unemployment rates differ between former communist countries and European/Asian nations that have not been under communist control. A sample of 10 countries each was randomly chosen from all eligible countries. The data are as follows:

Former Communist (%)	*Not Former Communist (%)*
23	12
13	8
33	9
21	14
17	7
24	8
12	10
18	11
27	12
16	7

1. Are the mean percentages significantly different?
   a. yes, at $p < .05$
   b. yes, at $p < .01$
   c. yes, at $p < .001$
   d. no, at $p > .05$

2. The degrees of freedom for this problem are
   a. 20
   b. 18
   c. 10
   d. 9

3. The derived t value for this problem is
   a. 7.338
   b. 4.816
   c. 11.373
   d. 12.000

4. The correct statistical decision would be
   a. retain the null hypothesis
   b. reject the null hypothesis
   c. retain the alternative hypothesis
   d. reject the directional null hypothesis

5. The appropriate conclusion of this statistical analysis would be
   a. The mean unemployment rate (20.4%) in former communist countries is significantly greater ($p < .001$) than the mean unemployment rate (9.8%) in countries not formerly under communist control.
   b. The mean unemployment rate (9.8%) in former communist countries is significantly greater ($p < .001$) than the mean unemployment rate (20.4%) in countries not formerly under communist control.
   c. The mean unemployment rate (20.4%) in former communist countries is significantly greater ($p < .05$) than the mean unemployment rate (9.8%) in countries not formerly under communist control.
   d. The mean unemployment rate (20.4%) in former communist countries is not significantly greater ($p > .05$) than the mean unemployment rate (9.8%) in countries not formerly under communist control.

Problems 6–10. A psychiatrist is interested in the effects of a new drug for the control of depression. In a clinical trial, 18 depressed patients were randomly assigned to one of two groups, either the drug group or the placebo group. One half hour after the administration of the drug or placebo, their depression was measured by the Coolidge Axis II Inventory's Depression scale and reported as T scores ($M = 50$, $SD = 10$).

6. The value of the numerator in the t formula for this problem (either + or –) is
   a. 3.058
   b. 55.33
   c. 62.67
   d. 7.33

Drug	*Placebo*
52	65
53	59
58	68
50	53
53	59
58	67
55	61
66	70
53	62

7. The degrees of freedom for this problem are
   a. 18
   b. 16
   c. 8
   d. 20

8. The derived t value for this problem (either + or −) is
   a. 3.058
   b. 7.33
   c. .008
   d. 2.120

9. The correct statistical decision would be to
   a. reject the null hypothesis at $p < .05$ (although it would be reported significant at $p < .01$)
   b. retain the null hypothesis ($p > .05$)
   c. reject the null hypothesis at $p < .05$ (although it would be reported significant at $p < .001$)

10. The overall results would mean that
    a. the new drug significantly lowers anxiety levels compared to a placebo
    b. the new drug lowers anxiety levels by a factor of .01
    c. the new drug is significant
    d. the new drug lowers anxiety levels compared to a placebo but not significantly

Problems 11-15. The Food and Drug Administration wishes to determine whether the claim that Vitamin C prevents colds has any truth. In a clinical drug trial, 30 subjects were randomly assigned to either the Vitamin C group (receiving a daily supplement of the minimum daily requirement) or the placebo group (who received no supplement but did get a placebo pill). The number of self-reported colds is recorded after 3 years.

11. The value of the numerator (either + or –) in the t test formula for this problem is
    a. 0.378
    b. 6.67
    c. 7.00
    d. .051

Number of Colds in 3 Years	
Vitamin C	*Placebo*
3	8
8	8
6	7
7	10
4	11
9	4
2	3
5	7
7	6
11	8
10	4
8	10
7	6
6	8
7	5

12. The degrees of freedom for this problem are
    a. 30
    b. 28
    c. 15
    d. 14

13. The derived t value (either + or −) for this problem is
    a. 0.708
    b. 0.378
    c. 0.333
    d. 0.05

14. What can be concluded from the statistical analysis of this problem?
    a. the null hypothesis cannot be rejected
    b. Vitamin C doesn't seem to prevent colds any better than a placebo
    c. the mean number of colds is not significantly different between the two groups
    d. all of the above are true

15. What significance level should be reported?
    a. $p < .05$
    b. $p < .01$
    c. $p < .001$
    d. $p > .05$

8 The *t* Test for Dependent Groups

Chapter 8 Goals

- Learn to recognize variations on the design of the controlled experiment
- Learn to statistically analyze the controlled experiment with the *t* test for dependent groups

Variations on the Controlled Experiment

Imagine if a psychologist wishes to test a group of patients before and after their treatment with a new type of psychotherapy. In this case, the scores in the two groups would be related to each other and not independent because a pair of scores would represent the same patient before and after treatment. The appropriate statistical analysis for this design would be the *t* test for dependent groups (or **dependent *t* test**). The *t* test for dependent groups is used to analyze the difference between two groups' means in experimental designs where the participants in both groups are related to each other in some way. The two most common designs are where the same participant serves in both groups or where pairs of matched participants are split between the two groups. The dependent *t* test has an interesting number of names, despite a limited number of designs. For example, this test has also been called the **repeated-measures *t* test,** paired-samples *t* test, dependent-samples *t* test, within-subjects *t* test, paired *t* test, **matched *t* test,** paired difference *t* test, and correlated *t* test. Although the independent *t* test is far more common than the dependent *t,* the dependent *t* test can, in many circumstances, be a more powerful test than the independent *t* test.

Remember that the power of a statistical test is its ability to detect a genuine difference between two means. Before we examine the reasons for the increased power of the dependent t test, let us review the three essential experimental designs of the dependent t test in order of their prevalence.

Design 1

The same participant is used in both groups, but the presentation of the independent variable is fixed by God or nature.

Example of Design 1

This design is the most common of the three dependent t test designs. It occurs when any participants are measured before and after some treatment, and their "before" scores are compared to their "after" scores. It is also referred to as the "pre- and posttreatment" design. For example, we might compare patients' level of distress before psychotherapy and then measure their distress after psychotherapy. If the distress scores go down after psychotherapy, we may tentatively and cautiously conclude that psychotherapy may have been the cause of the decline in distress scores. However, this is a very tenuous assumption. Since we were not able to randomly present the independent variable, a slew of other factors may have actually been responsible, such as the simple passage of time. Another possibility is the **regression effect,** where extreme scores when measured again tend to "regress" toward the mean. Because there may have been factors other than the treatment effects causing the difference in before-and-after experimental t designs, we must be very cautious about any implications of causation.

Design 2

Different participants are used in both groups, but the participants are paired or matched on some critical variables (to make them as alike as possible). The matching critical variables must be known to be related to the dependent variable.

Example of Design 2

This design is not as popular as the previous design. For example, imagine a study where the researcher is interested in long-term memory. The researcher knows that long-term memory varies widely among people, so to control for this variability, the researcher employs the matched t test design because the variability across the same participant over time will be less than the variability of long-term memory between two different participants. The researcher wishes to know whether visual imagery leads to better long-term memory or whether simple auditory rehearsal of the material leads to better long-term memory. If the researcher decides to match the participants, it had better be on variables critical to long-term memory. One variable critical to long-term memory is age, and another critical variable may be IQ. Therefore, the researcher should match the participants on age and IQ as carefully as possible to control for these effects on the dependent variable. Thus, for each matched pair, one will receive the imagery condition, and one will be in the rehearsal condition. The dependent *t* test will compare the means of the two types of memory enhancement for the amounts of memory retained after a set period of time.

One major drawback to this design is that the researcher often may not know or be able to control for all of the critical variables related to the dependent variable. Also remember that for all three of these designs, there must be an equal number of scores in each group since the scores represent either the same participants in each group or matched pairs of participants.

Design 3

The same participant is used in both groups. The independent variable is balanced in presentation across all of the participants (for half of the participants, they receive Level 1 of the independent variable first, and for the other half of the participants, they receive Level 2 of the independent variable first).

Example of Design 3

This design is far less popular than either of the previous two designs. In this design, one concern is the order of presentation of the independent variable. The **order of presentation** of the two levels of an independent variable may affect the results of an experiment since the participants may suffer from fatigue or boredom after a period of time. Thus, if we can balance the presentation of the two levels of an independent variable in an experiment, we can experimentally control for the order of presentation effect. With one of the original tests of intelligence, it was noticed that most people scored lower on the five performance subtests than on the six verbal subtests. In part, this finding may have been due to order of presentation since all six verbal subtests were administered first (often taking an hour) before the five performance subtests. When the order of presentation was balanced across the verbal and performance subtests, it was found that the magnitude of the discrepancy decreased, but the difference did persist.

Assumptions of the Dependent *t* Test

The assumptions of the dependent *t* test are similar to the independent *t* test. The dependent variable is assumed to come from a population of scores that is normally distributed. As noted earlier, there must also be an equal number of scores in each group, although this is an experimental design requirement as well as a statistical requirement. It is also assumed that the scores within a group are independent of one another. As an example of a violation of the latter assumption, imagine if as one tested participants, one's laboratory became hotter and smellier. The participants tested later might have scores that were affected by the previous participants, thus violating the assumption of independence of scores within a group.

Why the Dependent *t* Test May Be More Powerful Than the Independent *t* Test

As we mentioned earlier, the variability of the same participant over time is less than the variability between two unrelated participants. For example, imagine pairing your IQ this year and next year to two unrelated people's IQs. There will be greater variability, of course, between two unrelated people's IQ scores. Also, imagine pairing the IQs of two identical twins (matched pairs) compared to the variability of two unrelated people's IQs. Once again, there will be less variability between the matched pairs of scores than two unrelated scores. Thus, the dependent *t* test design is more powerful because it can reduce the noise in an experiment (better known as the within-subject variance or within-subject error).

How to Increase the Power of a *t* Test

There are at least three ways to increase the power of an independent *t* test or dependent *t* test. One way is to increase the difference between the two means, perhaps by creating a stronger treatment effect. In the example of the memory enhancement experiment, the treatment may be strengthened by training the participants in the visual imagery and rehearsal conditions for a longer period of time, such as instead of just one session of training, train the participants for five sessions. In the pre- and posttreatment example with psychotherapy, the length of psychotherapy might be increased from 1 month of stress reduction to 3 months. Another way to increase the power of a *t* test is to increase the number of participants in a study. Remember that a statistically large sample begins at $N = 30$. Yet also remember that in inferential statistics, we can abuse power by using far too many participants. A power analysis can determine the appropriate number of participants in a study. A third way of increasing power is to decrease the within-subject variance in each group. The dependent *t* test automatically decreases the within-subject variance because either the participants are used twice (under both conditions of the independent variable), or variance is reduced by matching the participants on critical variables. In addition, within-subject variance may be reduced by trying to homogenize the participants, that is, by using similar groups such as only ages 18 to 21, middle-class, and so on. Homogenization does have the negative effect of restricting generalizations from the results to other populations, but the power of the experiment to detect genuine differences in the data is enhanced.

Drawbacks of the Dependent *t* Test Designs

A word of caution is in order. The dependent *t* test does have some drawbacks. One issue deals with the concept of degrees of freedom. Remember that the formula for degrees of freedom in the independent *t* test is $N_1 + N_2 - 2$. In a dependent *t* test, the dependence between the pairs of scores means that knowing a score under one condition fixes (to some extent) the score in the other condition. For example, if a researcher has used twins in an IQ study, knowing one twin's IQ means that the other twin's IQ in the second condition is not really free to vary but will be highly correlated to his or her matching twin's IQ score. Thus, the formula for the degrees of freedom in the dependent *t* test is half of the degrees of freedom in the independent *t* test, and it is $N - 1$, where *N* is the *number of pairs* of scores. This reduction in the number of degrees of freedom actually reduces the statistical power of the dependent *t* test since the reduction produces a higher critical value of *t* in the family of *t* distributions (see Appendix B). The higher critical value of *t* makes it more difficult to reject the null hypothesis.

One-Tailed or Two-Tailed Tests of Significance

The way in which the alternative hypothesis is stated determines whether a test of significance is one-tailed or two-tailed. It is one-tailed if the alternative hypothesis has been stated directionally (for example, the pretest mean will be less than the posttest mean). It is two-tailed if the alternative hypothesis is stated nondirectionally (for example, there will be a difference between the pretest and posttest means, but we are not sure which mean will be higher). Ideally, the formal statement of the alternative hypothesis is made before the experiment has been conducted but certainly before the data analysis begins. Each test of significance has its advantages and disadvantages. The one-tailed test of significance is more sensitive to genuine differences between the two means, and it is also more sensitive to smaller real differences between the two means. However, the cost of this sensitivity is that it is also more sensitive to chance differences and to interpreting chance differences as real differences (the Type I error).

The two-tailed test of significance protects against the Type I error better than the one-tailed test of significance. It is less likely to have us interpret chance differences between two means as being real differences. However, this protection comes at a cost: It will be less sensitive to smaller but genuine differences between two means. Many statisticians feel that the nondirectional alternative hypothesis provides sufficient sensitivity to real differences and provides better protection against interpreting chance differences as real differences than the one-tailed test. Also, I have two other thoughts about directional alternative hypotheses. First, it is not uncommon in scientific research to have a finding the exact opposite of what was predicted. The one-tailed test of significance puts the entire rejection region (.05) in a single tail of the *t* distribution. If the results are the exact opposite of what has been predicted, we are forced to retain the null hypothesis since we can only reject the null hypothesis if the direction of the finding is as we predicted. If it is the opposite of what we predicted, we have no alternative but to retain the null hypothesis in the one-tailed test. My second thought is that if one is so certain of the outcome of an experiment, why is one doing the experiment in the first place? In fact, two-tailed tests are so prevalent in research that it has been often wondered whether, in experiments that report one-tailed tests of significance, the experimenters failed to attain significance with a two-tailed test and then switched to a one-tailed test to attain significance.

Hypothesis Testing and the Dependent *t* Test: Design 1

The null and nondirectional alternative hypotheses for Design 1 of the dependent *t* test are as follows:

$$H_0: \mu_1 = \mu_2 \text{ or } \mu_1 - \mu_2 = 0$$
$$H_a: \mu_1 \neq \mu_2 \text{ or } \mu_1 - \mu_2 \neq 0$$

The formula for the dependent *t* test is

$$t = \frac{\bar{x}_1 - \bar{x}_2}{\sqrt{\dfrac{\sum D^2 - \frac{(\sum D)^2}{N}}{N(N-1)}}}$$

where

$\bar{x}_1$ = the mean of the pretest scores,

$\bar{x}_2$ = the mean of the posttest scores,

ΣD^2 = the sum of the squares of the differences between the pretest scores and the posttest scores,

$(\Sigma D)^2$ = the square of the sum of the differences between the pretest scores and the posttest scores,

N = the number of *pairs* of scores.

Design 1 (Same Participants or Repeated Measures): A Computational Example

A police department is interested in reducing stress in its employees, who have been exposed to extreme levels of stress in the past 6 months. Those officers who have been exposed to extremely stressful situations are encouraged to enter into free psychotherapy (paid for by the department) and were assessed for their stress levels before and after three psychotherapy sessions. The purpose of the study was to determine whether a new type of brief psychotherapy was successful after just three sessions. The employees' stress was measured by standardized stress scores (reported as T scores, where higher values indicate higher stress; remember, for T scores, mean $T = 50$ and $SD = 10$).

	Stress Scores (as t Scores)	
	Before Therapy	*After Therapy*
Employee 1	68	63
Employee 2	58	60
Employee 3	74	65
Employee 4	55	62
Employee 5	81	54
Employee 6	59	73
Employee 7	47	45
Employee 8	75	73

Step 1. Subtract the pairs of scores from each other in the following manner.

	Before		*After*		*Difference Scores (D Scores)*
Employee 1	68	–	63	=	5
Employee 2	58	–	60	=	–2
Employee 3	74	–	65	=	9
Employee 4	55	–	62	=	–7
Employee 5	81	–	54	=	27
Employee 6	59	–	73	=	–14
Employee 7	47	–	45	=	2
Employee 8	75	–	73	=	2

Step 2. Calculate the mean of the pretest scores ($\bar{x}_1$):

$$\bar{x}_1 = \frac{68 + 58 + 74 + 55 + 81 + 59 + 47 + 75}{8}$$

$$\bar{x}_1 = \frac{517}{8}$$

$$\bar{x}_1 = 64.625$$

Step 3. Calculate the mean of the posttest scores ($\bar{x}_2$):

$$\bar{x}_2 = \frac{63 + 60 + 65 + 62 + 54 + 73 + 45 + 73}{8}$$
$$\bar{x}_2 = \frac{495}{8}$$
$$\bar{x}_2 = 61.875$$

Step 4. Calculate the sum of the squares of the differences between the pretest scores and the posttest scores ΣD^2:

$$\sum D^2 = 5^2 + (-2)^2 + 9^2 + (-7)^2 + 27^2 + (-14)^2 + 2^2 + 2^2$$
$$\sum D^2 = 25 + 4 + 81 + 49 + 729 + 196 + 4 + 4$$
$$\sum D^2 = 1092$$

Step 5. Obtain the square of the sum of the differences between the pretest scores and the posttest scores $(\Sigma D)^2$:

$$\left(\sum D\right)^2 = (5 - 2 + 9 - 7 + 27 - 14 + 2 + 2)^2$$
$$\left(\sum D\right)^2 = (22)^2$$
$$\left(\sum D\right)^2 = 484$$

Step 6. Determine the number of pairs of scores. In the formula, N refers to the number of pairs of scores.

$$N = 8$$

Step 7. Enter the values obtained from Steps 1 through 6 into the formula for the dependent test.

$$t = \frac{\bar{x}_1 - \bar{x}_2}{\sqrt{\dfrac{\sum D^2 - \dfrac{(\sum D)^2}{N}}{N(N-1)}}}$$
$$t = \frac{64.625 - 61.875}{\sqrt{\dfrac{1092 - \frac{22^2}{8}}{8(8-1)}}}$$
$$t = \frac{2.75}{\sqrt{18.4196}}$$
$$t = 0.641$$

Step 8. Compare the derived *t* value to the critical values of *t* in Appendix B of this book. The formula for the *df* in a dependent *t* test is $df = N - 1$, where N = the number of pairs of scores. In the present example, there were eight pairs of scores, so $df = 8 - 1$, or $df = 7$.

Step 9. Determine whether the H_0 should be retained or rejected by comparing the derived *t* value to the tabled critical values of *t* at $p = .05$, with $df = 7$. If the derived *t* value exceeds the tabled critical value of *t* (or if it is less than the negative *t* critical value), then H_0 is rejected. If not, then H_0 is retained.

The null and alternative hypotheses are as follows:

$$H_0: \mu_{\text{pretest}} = \mu_{\text{posttest}}$$
$$H_a: \mu_{\text{pretest}} \neq \mu_{\text{posttest}}$$

Step 10. Write up your findings and make a conclusion. Our informal write-up might appear as follows:

> The derived $t = 0.641$ does not exceed the tabled critical value of $t = \pm 2.365$ at $p = .05$ with $df = 7$. Therefore, H_0 is retained, and it is concluded that the mean Stress scale *T* score for the employees after therapy (61.88; $SD = 9.33$) was not significantly lower than the *T* scores before psychotherapy (64.63; $SD = 11.67$), $t(7) = 0.641$, $p > .05$. In terms of the research question, it appears that this type of psychotherapy did not appear to be effective in reducing stress in this sample of police officers.

In a research article, the dependent *t* test results might be reported as follows:

> There was no significant reduction in stress for the police officers who received the novel psychotherapy. The mean stress level before treatment ($M = 64.63$, $SD = 11.67$) was not significantly different from the mean stress level after treatment ($M = 61.88$, $SD = 9.33$), $t(7) = 0.641$, $p > .05$.

Determination of Effect Size

As noted earlier, the significance of a statistical test presents only part of the picture. Effect size estimation is useful for determining the strength of the independent variable, regardless of whether H_0 was retained or rejected. In the present example, the effect size determination would be

$$r = \sqrt{\frac{t^2}{t^2 + df}}$$

$$r = \sqrt{\frac{.641^2}{.641^2 + 7}}$$

$$r = .235$$

By comparing this value to the table of effect sizes in Chapter 7, we see that it translates into slightly less than a medium-size effect. In essence, this determination of effect size indicates that despite a lack of statistical significance, there appears to be a nearly medium-strength effect of psychotherapy in reducing the distress of this sample of patients. This discrepancy preliminarily suggests that the lack of significance was due to a lack of power (in this case, not enough participants in the study). Another way to judge the practical value of this nearly medium effect size is to take the difference between the means and divide by the original stress level. This procedure reveals a 4.3% reduction in stress, perhaps as a result of the new therapy. At this point, practical matters may intervene in determining whether the police department decides to adopt this stress reduction therapy. For example, if the therapy can be delivered relatively inexpensively, then a 4.3% reduction in stress may be acceptable. Also, if no other type of psychotherapy has even approached this reduction figure, then the department may adopt it, regardless of cost.

Design 2 (Matched Pairs): A Computational Example

A researcher wishes to determine which of two methods of memory training works best with the elderly. Twins older than age 65 were selected for the study. Half were trained to memorize a list of words using visual imagery, and the other half spent the same amount of time simply memorizing the words. The number of correct words (maximum 20 correct) recalled was measured in both groups 24 hours later. The results are as follows:

Correct Number of Words			
Visual Imagery		*Simple Rehearsal*	
Twin 1a	15	Twin 1b	12
Twin 2a	10	Twin 2b	11
Twin 3a	12	Twin 3b	9
Twin 4a	9	Twin 4b	7
Twin 5a	16	Twin 5b	13
Twin 6a	18	Twin 6b	15
Twin 7a	11	Twin 7b	12
Twin 8a	14	Twin 8b	12

Step 1. Subtract the pairs of scores from each other in the following manner.

Visual Imagery		Simple Rehearsal		Difference Scores (D Scores)
Twin 1a	15	Twin 1b	12	15 – 12 = 3
Twin 2a	10	Twin 2b	11	10 – 11 = –1
Twin 3a	12	Twin 3b	9	12 – 9 = 3
Twin 4a	9	Twin 4b	7	9 – 7 = 2
Twin 5a	16	Twin 5b	13	16 – 13 = 3
Twin 6a	18	Twin 6b	15	18 – 15 = 3
Twin 7a	11	Twin 7b	12	11 – 12 = –1
Twin 8a	14	Twin 8b	12	14 – 12 = 2

Step 2. Calculate the mean of the visual imagery condition ($\bar{x}_1$):

$$\bar{x}_1 = \frac{15 + 10 + 12 + 9 + 16 + 18 + 11 + 14}{8}$$

$$\bar{x}_1 = \frac{105}{8}$$

$$\bar{x}_1 = 13.125$$

Step 3. Calculate the mean of the simple rehearsal condition ($\bar{x}_2$):

$$\bar{x}_2 = \frac{12 + 11 + 9 + 7 + 13 + 15 + 12 + 12}{8}$$

$$\bar{x}_2 = \frac{91}{8}$$

$$\bar{x}_2 = 11.375$$

Step 4. Calculate the sum of the squares of the differences between the pretest scores and the posttest scores ΣD^2:

$$\sum D^2 = 3^2 + (-1)^2 + 3^2 + 2^2 + 3^2 + 3^2 + (-1)^2 + 2^2$$

$$\sum D^2 = 9 + 1 + 9 + 4 + 9 + 9 + 1 + 4$$

$$\sum D^2 = 46$$

Step 5. Obtain the square of the sum of the differences between the pretest scores and the posttest scores $(\Sigma D)^2$:

$$\left(\sum D\right)^2 = (3 - 1 + 3 + 2 + 3 + 3 - 1 + 2)^2$$

$$\left(\sum D\right)^2 = (14)^2$$

$$\left(\sum D\right)^2 = 196$$

Step 6. Determine the number of pairs of scores. In the formula, N refers to the number of pairs of scores.

$$N = 8$$

Step 7. Enter the values obtained from Steps 1 through 6 into the formula for the dependent t test.

$$t = \frac{\bar{x}_1 - \bar{x}_2}{\sqrt{\dfrac{\sum D^2 - \frac{(\sum D)^2}{N}}{N\,(N-1)}}}$$

$$t = \frac{13.125 - 11.375}{\sqrt{\dfrac{46 - \frac{14^2}{8}}{8(8-1)}}}$$

$$t = \frac{1.75}{\sqrt{\dfrac{21.5}{56}}}$$

$$t = \frac{1.75}{.61962}$$

$$t = 2.824$$

Step 8. Compare the derived t value to the critical values of t in Appendix B. The formula for the df in a dependent t test is $df = N - 1$, where $N =$ the number of pairs of scores. In the present example, there were eight pairs, so $df = 8 - 1$, or $df = 7$.

Step 9. Determine whether H_0 should be retained or rejected by comparing the derived t value to the tabled critical values of t at $p = .05$ with $df = 7$. If the derived t value exceeds the tabled critical value of t (or if it is less than the negative t critical value), then H_0 is rejected. If not, then H_0 is retained.

The null and alternative hypotheses are as follows:

$$H_0: \mu_{\text{visual}} = \mu_{\text{rehearsal}}$$
$$H_a: \mu_{\text{visual}} \neq \mu_{\text{rehearsal}}$$

Step 10. Write up your findings and make a conclusion. Our informal write-up might look like this:

> The derived $t = 2.824$ exceeds the tabled critical value of $t = \pm 2.365$ at $p = .05$ with $df = 7$. Therefore, H_0 is rejected, and it is concluded that the mean number of words correct under the visual imagery

> condition (13.13; $SD = 3.14$) was significantly greater than the mean number of words correct in the simple rehearsal condition (11.38; $SD = 2.45$), $t(7) = 2.824$, $p < .05$. In terms of the research question, it appears that visual imagery appears to enhance memory recall better than simple rehearsal in this sample of twins.

In a research article, the dependent *t* test results might be reported as follows:

> Twins in the visual imagery condition obtained significantly more correct words ($M = 13.13$, $SD = 3.14$) than their co-twins in the rehearsal condition ($M = 11.38$, $SD = 2.45$), $t(7) = 2.824$, $p < .05$.

Determination of Effect Size

The effect size determination would be

$$r = \sqrt{\frac{t^2}{t^2 + df}}$$

$$r = \sqrt{\frac{2.824^2}{2.824^2 + 7}}$$

$$r = .730$$

By comparing this value to the table of effect sizes in Chapter 7, we see that it translates into a very large effect size. Thus, it appears that not only is visual imagery statistically significantly superior to simple rehearsal, but the strength of this effect is very large.

Design 3 (Same Participants and Balanced Presentation): A Computational Example

A researcher wished to determine whether consistently higher verbal intelligence (VIQ) scores than performance intelligence (PIQ) scores are due to order of presentation effects on an intelligence test because the six subtests of the verbal portion are all administered before the five subtests on the performance section. The researcher counterbalances the presentation of the verbal and performance subtests, assigning half of the participants to take the verbal portion first and half of the participants to take the performance subtests first. The data are tabulated as follows:

Participant 1	VIQ: 112	PIQ: 109
Participant 2	VIQ: 117	PIQ: 110
Participant 3	VIQ: 99	PIQ: 105
Participant 4	VIQ: 128	PIQ: 114
Participant 5	PIQ: 103	VIQ: 120
Participant 6	PIQ: 108	VIQ: 117
Participant 7	PIQ: 111	VIQ: 135
Participant 8	PIQ: 88	VIQ: 96

Step 1. Subtract the pairs of scores from each other in the following manner.

	VIQ		*PIQ*		*Difference Scores (D Scores)*
Participant 1	112	–	109	=	3
Participant 2	117	–	110	=	7
Participant 3	99	–	105	=	–6
Participant 4	128	–	114	=	14
Participant 5[a]	120	–	103	=	17
Participant 6[a]	117	–	108	=	9
Participant 7[a]	135	–	111	=	24
Participant 8[a]	96	–	88	=	8

a. Note that Participants 5 through 8 have had their scores reversed from the initial table so that all of the VIQ scores can be compared to all of the PIQ scores now that counterbalancing of the order of presentation has taken place. This means that order of presentation will no longer be a factor in determining whether VIQ is usually higher than PIQ.

Step 2. Calculate the mean of the VIQ scores ($\bar{x}_1$):

$$\bar{x}_1 = \frac{112 + 117 + 99 + 128 + 120 + 117 + 135 + 96}{8}$$

$$\bar{x}_1 = \frac{924}{8}$$

$$\bar{x}_1 = 115.5$$

Step 3. Calculate the mean of the PIQ scores ($\bar{x}_2$):

$$\bar{x}_2 = \frac{109 + 110 + 105 + 114 + 103 + 108 + 111 + 88}{8}$$

$$\bar{x}_2 = \frac{848}{8}$$

$$\bar{x}_2 = 106.0$$

Step 4. Calculate the sum of the squares of the differences between the pretest scores and the posttest scores:

$$\sum D^2 = 3^2 + 7^2 + (-6)^2 + 14^2 + 17^2 + 9^2 + 24^2 + 8^2$$
$$\sum D^2 = 9 + 49 + 36 + 196 + 289 + 81 + 576 + 64$$
$$\sum D^2 = 1300$$

Step 5. Obtain the square of the sum of the differences between the pretest scores and the posttest scores $(\Sigma D)^2$:

$$\left(\sum D\right)^2 = (3 + 7 - 6 + 14 + 17 + 9 + 24 + 8)^2$$
$$\left(\sum D\right)^2 = (76)^2$$
$$\left(\sum D\right)^2 = 5776$$

Step 6. Determine the number of pairs of scores. In the formula, N refers to the number of pairs of scores.

$$N = 8$$

Step 7. Enter the values obtained from Steps 1 through 6 into the formula for the dependent t test.

$$t = \frac{\bar{x}_1 - \bar{x}_2}{\sqrt{\dfrac{\sum D^2 - \frac{(\sum D)^2}{N}}{N(N-1)}}}$$
$$t = \frac{115.500 - 106.000}{\sqrt{\dfrac{1300 - \frac{76^2}{8}}{8(8-1)}}}$$
$$t = 2.957$$

Step 8. Compare the derived t value to the critical values of t in Appendix B. The formula for the df in a dependent t test is $df = N - 1$, where N = the number of pairs of scores. In the present example, there were eight pairs, so $df = 8 - 1$, or $df = 7$.

Step 9. Determine whether the H_0 should be retained or rejected by comparing the derived t value to the tabled critical values of t at $p = .05$ with $df = 7$. If the derived t value exceeds the tabled critical value of t (or if it is less than the negative t critical value), then H_0 is rejected. If not, then H_0 is retained.

The null and alternative hypotheses are as follows:

$$H_0\text{: } \mu_{VIQ} = \mu_{PIQ}$$
$$H_a\text{: } \mu_{VIQ} \neq \mu_{PIQ}$$

Step 10. Write up your findings and make a conclusion. Our informal write-up might appear as follows:

> The derived $t = 2.957$ does exceed the tabled critical value of $t = \pm 2.365$ at $p = .05$ with $df = 7$. Therefore, H_0 is rejected, and it is concluded that the mean VIQ score (115.5; $SD = 13.23$) was significantly greater than the mean PIQ score (106.0; $SD = 8.04$), $t(7) = 2.957$, $p < .05$. In terms of the research question, it appears that in this sample of participants, the VIQ scores do exceed their PIQ scores, and the finding does not appear to be accounted for by order of presentation.

In a research article, the dependent t test results might be reported as follows:

> The results appear to indicate that VIQ scores ($M = 115.50$, $SD = 13.23$) are significantly greater than PIQ scores ($M = 106.00$, $SD = 8.04$), $t(7) = 2.957$, $p < .05$, and the difference does not appear to be accounted for by the order of presentation.

Determination of Effect Size

Once again, the significance of a statistical test presents only part of the picture. Effect size estimation is useful for determining the strength of the independent variable, regardless of whether H_0 was retained or rejected. In the present example, the effect size determination would be

$$r = \sqrt{\frac{t^2}{t^2 + df}}$$

$$r = \sqrt{\frac{2.957^2}{2.957^2 + 7}}$$

$$r = .745$$

By comparing this value to the table of effect sizes in Chapter 7, we see that it translates into a very large effect size.

History Trivia

Fisher to Pearson

Ronald Fisher grew up as a sickly child with severe eye impairments. To protect his limited eyesight, his doctors forbade him to read by electric light. He excelled in mathematics, but his mathematics teachers would help him in the evening without light, pencil, paper, or any other aid. Perhaps as a result, Fisher developed a highly intuitive and geometric sense for mathematical and statistical problem solving. While at Cambridge as an undergraduate, he managed to solve difficult mathematical problems, often in multidimensional geometric space. He managed to publish one of these solutions while still an undergraduate in the premiere statistical journal of its time, Karl Pearson's *Biometrika*. Subsequently, Fisher was able to meet Pearson, who gave him a difficult problem of determining the statistical distribution of the correlation coefficient produced by Galton. Fisher solved the problem in a week and submitted it for publication to Pearson. Pearson, however, could not understand the mathematics and asked William Gosset to explain it. However, Gosset also had a problem understanding it. Pearson apparently was jealous of the young genius and effectively held off publication for over a year. Fisher's paper was finally published, but it appeared almost as a footnote to a larger paper by Pearson, and Fisher was never to publish another paper in *Biometrika,* or in many other premiere statistical journals, due to Pearson's influence. Fisher nonetheless managed to remain controversial throughout his lifetime by his own efforts and beliefs. We mentioned earlier that he argued strongly in the 1950s against the validity of studies that found smoking to be dangerous. He also joined the eugenics movement and called for national policies to discourage reproduction of the lower classes and to encourage reproduction among professional classes and artists.

Key Terms and Definitions

Dependent *t* test—A test designed to determine the statistical difference between two means where the participants in each group are either the same or matched pairs.

Matched *t* test—A dependent *t* test design where the participants are paired or matched on some critical variables that are known to be related to the dependent variable in the experiment.

Order of presentation—The order in which levels of the independent variable are presented to the participants. If the conditions are not counterbalanced, the participants may become fatigued or bored and do more poorly in the later conditions.

Regression effect—A historically old finding in statistics where it was first noticed that the tallest people in families tended to have children shorter than they were and closer to the mean of the entire family's height (called regression to the mean). Also, people who first measure at the extreme end on a variable tend to have more moderate scores later.

Repeated-measures *t* test—A dependent *t* test design where the participants are the same in both groups and usually are measured pretreatment and posttreatment.

Chapter 8 Practice Problems

A psychologist wants to determine whether a controversial therapy, eye movement desensitization reprocessing (EMDR), works because of the placebo effect. An actor is hired to mimic the enthusiasm, belief, and excitement of EMDR therapists but not mimic the eye movement techniques. Ten patients with traumatic memories are exposed to EMDR or the actor's placebo therapy; half get EMDR first and then the placebo, and half get the placebo first and then EMDR. After therapy, the patients rate their trauma on a 1 (*feel great*) to 10 (*feel terrible*) scale.

EMDR	*Placebo Therapy*
5	4
3	2
4	7
2	1
1	1
3	2
3	1
7	3
6	2
2	4

1. Conduct a dependent t test and test of significance.
2. What can be concluded?

Chapter 8 Test Questions

Problems 1–6. A psychiatrist wishes to determine whether a new medication controls seizures better than the patients' previous medications. The number of seizures is recorded on previous and new medications for a period of 1 year for each patient. The research question is focused on determining whether there are differences between the two means of the numbers of seizures between the previous and new medications.

	Number of Seizures	
	Previous Medication	*New Medication*
Patient 1	4	7
Patient 2	5	6
Patient 3	4	3
Patient 4	3	3
Patient 5	4	5
Patient 6	3	2
Patient 7	4	5
Patient 8	6	7
Patient 9	4	4
Patient 10	3	2
Patient 11	4	5
Patient 12	5	4

1. The value of the numerator in the dependent *t* test formula for this problem is approximately
   a. 0.08
   b. 4.50
   c. 4.42
   d. .333
2. The *df* for this problem are
   a. 24
   b. 12
   c. 11
   d. 10
3. The derived *t* value from the formula is
   a. 0.938
   b. 4.500
   c. 4.417
   d. 0.200
4. Assume this *t* test had a nondirectional alternative hypothesis, and we will use the conventional level of statistical significance. The critical value (plus and minus) of the *t* test for this problem (obtained from Appendix B: *t* distribution) is
   a. 2.179
   b. 2.201
   c. 3.106
   d. 4.437
5. What statistical decision is appropriate?
   a. retain the null hypothesis
   b. reject the null hypothesis
   c. reject the alternative hypothesis
   d. the results are significant at $p < .05$
6. In nonstatistical language, what can be concluded?
   a. the new drug works better
   b. the new drug works no better than the old drug
   c. because the mean number of seizures was lower for the new drug, this indicates that it works better than the old drug
   d. because the mean number of seizures was higher for the old drug, this indicates that the old drug works better than the new drug

Problems 7–13. A psychologist wishes to determine whether drinking coffee improves one's mood. The psychologist hangs out at the Café Coffay and asks customers to fill out a 10-point good-mood bad-mood questionnaire (where 10 = *great mood* and 1 = *bad mood*) before and after their first cup of coffee that day. The research question is whether there is any evidence that drinking coffee improves mood.

7. The value of the numerator in the dependent *t* test formula for this problem is approximately
   a. 0.08
   b. 2.87
   c. 5.00
   d. 7.87

	Before Mood Rating	*After Mood Rating*
Participant 1	5	9
Participant 2	3	8
Participant 3	9	9
Participant 4	3	8
Participant 5	2	7
Participant 6	4	9
Participant 7	1	10
Participant 8	8	8
Participant 9	6	5
Participant 10	7	7
Participant 11	4	3
Participant 12	3	10
Participant 13	8	7
Participant 14	7	9
Participant 15	5	9

8. The *df* for this problem are
   a. 0
   b. 8
   c. 5
   d. 4

9. The derived t value (either + or −) from the formula is
   a. 5.00
   b. 7.87
   c. 3.44
   d. .05

10. Assume this t test had a nondirectional alternative hypothesis, and we will use the conventional level of statistical significance. The critical value (plus and minus) of the t test for this problem (obtained from Appendix B: t distribution) is
    a. 2.145
    b. 2.977
    c. 4.140
    d. 2.131

11. What statistical decision is appropriate?
    a. retain the null hypothesis
    b. reject the null hypothesis
    c. reject the alternative hypothesis
    d. the results are significant at $p < .001$

12. In nonstatistical language, what can be concluded?
    a. it appears that coffee drinking is associated with improved mood
    b. coffee drinking does not cause mood improvement
    c. coffee drinking is correlated with people's moods
    d. the null hypothesis is retained

13. In statistical language, the results are significant at (and remember statisticians report the lowest p level possible)

    a. $p < .05$
    b. $p < .01$
    c. $p < .001$

14. The ______________ effect is said to occur when extreme scores, which are measured again, tend to be closer to the mean.

    a. correlation
    b. regression
    c. dependent
    d. diminishing

15. Design 3 of the dependent *t* test controls for the

    a. order of presentation
    b. regression effect
    c. diminishing effect
    d. abuse of power

16. Which of the following is not a way to increase the power of a *t* test?

    a. increase the difference between two means
    b. increase the number of participants
    c. decrease within-subject variance
    d. increase the number of independent variables

17. One drawback of dependent *t* test designs versus independent *t* test designs is the degrees of freedom. Given the same number of participants in both designs, which has the greater degrees of freedom?

    a. dependent
    b. independent
    c. they are equal

18. The way in which the ______________ hypothesis is stated determines whether a test of significance is one-tailed or two-tailed.

    a. null
    b. alternative
    c. neutral
    d. pejorative

19. The two-tailed test of significance protects against the Type ____ error better than the one-tailed test of significance.

    a. I
    b. II
    c. III
    d. IV

20. Which of the following is not a name for the dependent *t* test?

    a. repeated-measures *t* test
    b. paired-samples *t* test
    c. correlated *t* test
    d. multiple *t* test

9 Analysis of Variance

One-Factor Completely Randomized Design

Chapter 9 Goals

- Learn the limitations of multiple *t* tests
- Learn about analysis of variance (ANOVA) and the *F* distribution
- Learn how to conduct a multiple comparison test
- Learn how to estimate effect sizes in ANOVA

In this chapter, you will learn about a variation on the controlled experiment design, analysis of variance (ANOVA). It is essentially a multigroup *t* test design, usually consisting of three or more groups, thus giving much more information than the typical two-group controlled experiment. By having three groups, two different levels of a treatment can be tested along with a control group that receives a placebo. An experimenter might find that a lower level of a treatment might not be any better than the placebo, but the greater level of treatment does work.

A Limitation of Multiple *t* Tests and a Solution

While *t* tests are highly useful when we are comparing just two groups' means, a problem occurs when we try to compare three or more groups' means. For example, if we had an experiment where we were comparing reading comprehension levels for three different types of reading teaching methods (Types A, B, and C), it would require three different t tests to make all possible mean comparisons (A vs. B, A vs. C, and B vs. C). This experimental design is a common one, and its analysis through t tests is called **multiple *ts*.** A problem occurs, however, because by doing each t test at the conventional level of statistical significance (p or $\alpha = .05$), we inflate the probability of committing the Type I error. In multiple t test designs, the overall α level is obtained by the following formula: overall $\alpha = 1 - (1 - \alpha)^n$, where $n =$ the number of t tests performed and α is the conventional level of significance (.05). Thus, in the previous example for three t tests, overall $\alpha = 1 - (.95)^3$ or $\alpha = .14$, and as you are probably now aware, this is an unacceptable overall Type I error rate. In the reading comprehension experiment, there would be about a one in seven chance of claiming that a real difference occurred between two of the methods when the difference might actually be due to chance.

The Equally Unacceptable Bonferroni Solution

A solution to this dilemma was proposed, known as the **Bonferroni correction.** It takes the number of t tests to be performed (k) and divides this value into the conventional level of significance ($\alpha = .05$). In our previous example, this would be $p = \frac{\alpha}{k}$ or $\frac{.05}{3} = .017$. Thus, $p = .017$ becomes the error rate for each t test to be performed. This would mean that each t test must meet at least the $p = .017$ level of significance instead of $p = .05$ to be labeled statistically significant. The t value from the t distribution would be higher at $p = .017$ than at $p = .05$; thus, it would be harder to attain statistical significance. The problem with the Bonferroni correction is that by protecting against the Type I error so strenuously, it increases the probability of Type II error and reduces the power of our statistical analysis. In other words, it would mean that now we are more likely to conclude that there are no real differences among the three teaching methods when, in reality, they do really differ.

The Acceptable Solution: An Analysis of Variance

The theoretical foundations of **analysis of variance (ANOVA)** were developed by Ronald Fisher. This is the same person who helped William Gosset refine the development of the t test and its distribution. One original purpose of ANOVA was to control for the automatic increase in the probability of the Type I error when more than two groups' means are being compared. As its name suggests, ANOVA is concerned with analyzing the variance produced in multiple mean comparisons to determine whether genuine differences exist among the means of a response variable (or dependent variable) as the result of some independent variable.

The Null and Alternative Hypotheses in Analysis of Variance

In ANOVA, the focus is on different types of variance inherent in a multigroup design, yet ANOVA is very much an extension of the *t* test for independent groups. A null hypothesis will be tested that states there is no difference among a number of group means on a response variable. We will still obtain a single derived value and compare it to a distribution of values, the *F* distribution (named in honor of Fisher). If the derived *F* value exceeds the tabled critical value of *F*, then the null hypothesis will be rejected. The null hypothesis in ANOVA is as follows:

$$H_0: \mu_1 = \mu_2 = \cdots = \mu_k$$

where

μ_1 = the mean for Group 1 on the response variable,

μ_k = the last group's mean.

Although we will always analyze sample means, notice that the null hypothesis states that the population means are not different from each other. Thus, if there is no real effect of the independent variable on the dependent variable, then it is as if each of these samples was taken from the same population. If the samples were all taken from the same population, then the sample means should not vary from each other based on reasons stated earlier in the book about the sampling distribution of means and the standard error of the mean. Only chance or random differences will differentiate among the group means.

Imagine if the dependent variable is significantly affected by the independent variable. In this case, one sample mean or all of them will be different from each other, and it will be as if each mean was sampled from a different population. Thus, the null hypothesis would be false. It is said, however, that in ANOVA, the null hypothesis is an **omnibus hypothesis,** and the *F* statistic in ANOVA is an **omnibus test.** *Omnibus* means "covering many different situations at the same time." For ANOVA, this means that if the null hypothesis is really false, we will not know which means are different from each other among the various groups. It could be that Group 1's mean is significantly lower than Group 2's mean, but Group 3's mean is not different from Group 2's mean or a plethora of other outcomes. Thus, in ANOVA, the alternative hypothesis is simply stated "the null hypothesis is not true," and the alternative hypothesis cannot be represented symbolically.

The Beauty and Elegance of the *F* Test Statistic

The derived *F* value will be a ratio of variances. At its heart, this ratio (with a variance estimate in the numerator and a variance estimate in the denominator) is an elegant and ingenious comparison of two different forms of variance, **between-groups variance** and **within-groups variance**. The *F* ratio is as follows:

$$F = \frac{\text{Between-Groups Variance}}{\text{Within-Groups Variance}}$$

Let us first examine the denominator's variance. Within-group variance is a measure of each score in a group from that group's mean. It is called variance because we will square the difference between any individual score and the group's mean and sum the resulting squares. For example, we could measure reading comprehension scores in the Type A teaching method group, but not all scores in the group will be the same. Even if Type A works much better than Type B or Type C, the reading comprehension scores in that group will differ from each other. The participants' scores will vary from their group's mean mostly because of these "individual" differences. As noted earlier, statisticians have the rather bizarre name of **subject error** for this difference between any participant's score and his or her group mean. The notion of *error* comes from the archaic scientific view that all the sciences were governed by firm rules and laws that rarely, if ever, varied. In budding sciences where the subject of interest could be described by a probability distribution, it was initially thought that any differing scores were errors.

Subject error or within-subject variance is not systematically applied by the experimenter but randomly appears in the experiment by nature. There is also another unsystematic and random form of variance, called **experimental error**, which means that testing conditions may vary from score to score or participant to participant: Temperature, lighting conditions, noise levels, and so on may vary. Because experimental error appears unsystematically and randomly, it is usually assumed that it will not affect one testing condition more than another.

The other major form of variance in an experiment, and the numerator of the *F* statistic, is the between-groups variance. This form of variance is systematic and applied by the experimenter. When the experimenter tests three different reading teaching methods, the experimenter is generally assuming that there will be mean differences among the groups. It is called between-group variance because it compares each group's mean with the grand mean of all the scores or participants' scores in the study and takes the sum of these squares. If there are really no differences among any of the means (the null hypothesis is true), then the scores or participants will probably vary slightly from the grand mean because of nothing more than the individual differences (within-subject variance). If there are really differences among the means of the groups (H_0 is false), then there will be a large between-groups variance because the group means will vary more widely from the grand mean. Remember, the between-groups variance will contain not only the variation of each group's mean from the grand mean but also the variation of the individuals about their group mean (within-group variance).

The *F* Ratio

The derived *F* ratio will consist of the between-groups variance in the numerator and the within-groups variance in the denominator. What would happen in an experiment where the null hypothesis is true? As we stated earlier, the between-groups variance actually has two sources, the variance from the differences among groups (if they really are different from one another) and within-groups variance (from individual differences). If the null hypothesis is true, then there would be little or just tiny chance contributions in the numerator for the between-groups variance. The numerator would mostly contain only within-groups variance. The denominator of the *F* ratio is another estimate of within-groups variance. If the null hypothesis is true, then we would have only within-groups variance in the numerator and within-groups variance in the denominator; thus, the resulting *F* value would be 1.00 because both within-groups variance estimates are trying to assess the same value. To repeat: In ANOVA, if the null hypothesis is true, then the derived *F* value should be 1.00. To the extent that the null hypothesis is false—that is, that there are real differences among the groups' means—then the numerator of the derived *F* value will contain an estimate of the between-groups variance *and* within-groups variance. This will mean that the resulting *F* value will become greater than 1.00, and if the groups' mean is very far apart (the various levels of the independent variable really are different), then the resulting *F* value will be much larger than 1.00. A summary of these variance discussions appears in Table 9.1.

How Can There Be Two Different Estimates of Within-Groups Variance?

You probably would not be surprised if you found out that a mathematician claimed to have proved that day is night and yin is yang, but for the curious, here is an explanation of how there can be two different mathematical estimates of the same entity. The numerator's estimate of

Table 9.1

When the Null Hypothesis Is True	*When the Null Hypothesis Is False*
$F = \frac{\text{Between-Groups Variance} + \text{Within-Groups Variance}}{\text{Within-Groups Variance}}$	$F = \frac{\text{Between-Groups Variance} + \text{Within-Groups Variance}}{\text{Within-Groups Variance}}$
yet between-groups variance is nil,	and the between-groups variance is substantial,
$F = \frac{\text{Within-Groups Variance}}{\text{Within-Groups Variance}}$	$F = \frac{\text{Between-Groups Variance} + \text{Within-Groups Variance}}{\text{Within-Groups Variance}}$
Thus, $F = 1.00$	Thus, $F > 1.00$

within-groups variance comes from a concept known as the sampling distribution. The **sampling distribution,** as noted earlier in the book, is a theoretical distribution made up of all possible random samples from the same population. Any finite sample mean from a population of scores will probably vary slightly and sometimes more than just slightly from any other sample mean from this same population. A standard deviation derived for all of these sample means from the population mean is known as the **standard error of the mean.** If the square of the standard error of the mean (making it variance) is multiplied by the number of samples taken, the resulting value will be an estimate of the population's variance, s^2, which also just happens to be an estimate of the within-groups variance.

The denominator's estimate of within-groups variance is a bit more straightforward. Within-groups variance in the denominator is obtained by simply taking each group's variance (the square of the standard deviation), adding them together, and dividing by the number of groups. The result (a mean of the groups' variances) is a very good estimate of the population's variance, s^2. Interestingly, and here is the really ingenious part of ANOVA, the numerator's estimate of within-groups variance is only accurate if the null hypothesis is true and there are no differences among the groups' means. If the null hypothesis is false, then the within-groups variance also contains between-groups variance because remember, it was derived from the standard error of the means. Thus, when the null hypothesis is false, and there really are differences between the groups' means, then the numerator's estimate of within-group variance is inflated. Because the derived F value is a ratio of these two variances, then the F value will become increasingly greater than 1.00 as the null hypothesis is increasingly more false (i.e., there are greater and greater differences among the groups' means).

A friend of mine from college took an advanced statistics class with statistics and mathematics majors. The professor assigned an ANOVA problem, and the result was an *F* value greater than 0 but less than 1.00. "Impossible," shouted the students. "Possible," shouted my friend. He knew from empirical research that *F* values less than 1.00 would occasionally occur because the two estimates of within-group variance are independent of each other, and, although they are measuring the same value, they may vary slightly. This situation may occur if the null hypothesis is true and if the numerator estimate is slightly less than the denominator estimate. An *F* value < 1.00 may also be indicative of violations of the assumptions of ANOVA, one of which is that there is homogeneity of variance about each group's means.

ANOVA Designs

The single most popular and powerful ANOVA design is called the **completely randomized ANOVA.** A completely randomized ANOVA has a single independent variable but two or more levels or conditions. These conditions will form the basis for the groups. For example, imagine an experiment to measure student satisfaction with an extremely long, boring, but simple mathematical task for three different payment options ($1, $5, and $20). Let's say 30 students volunteer to participate. The independent variable is the type of payment, and there are three levels or conditions: $1, $5, and $20. The design is said to be completely randomized because the students are randomly assigned to one of the three conditions. The resulting three groups will be independent of each other, and no participant will be in any condition other than the one to which he or she was assigned.

The completely randomized ANOVA is a powerful one because, unlike correlation, we can imply a causative relationship between the independent and dependent variables. Another ANOVA design, the in situ design, is analyzed by the same formula as the completely randomized design, but the implication of causation is weaker. As you may recall from the in situ *t* test design, the participants in the in situ ANOVA are not randomly assigned to the levels of the independent variable, but they are preassigned by God or nature. For example, in one strange study, self-esteem (the dependent variable) was assessed between college students and death row inmates. In this case, the participants were not randomly assigned to the two conditions of the independent variable (diagnosis or group membership), yet the completely randomized ANOVA formula will derive the *F* value the same way for this design, only the implication of causation will be weaker.

ANOVA Assumptions

The assumptions of an ANOVA are similar to the independent *t* test. First, it is assumed that the dependent variable is drawn from a population of values that is normally distributed (also called the normality assumption). Second, it is assumed that the participants have been randomly assigned to each group and the scores in each group are independent of each other. Third, it is assumed that the variances about each group's means are not substantially different from each other (also called the assumption of homogeneity of variance).

It is important to note, however, that ANOVA is a robust statistical test, and violations of the assumptions may still result in a correct statistical decision (to reject or retain the null hypothesis). Also, large samples (minimum of 10 to 15 participants per group) help minimize violations of the assumption of normality, and using an equal number of participants in each group helps to minimize violations of the assumption of homogeneity of variance.

Pragmatic Overview

The whole purpose of an ANOVA is to determine whether there are significant differences between two or more groups' means. All of the data in the experiment will be combined to form a single *F* value. This value will be compared to a distribution of critical *F* values. If the obtained *F* value exceeds the tabled critical *F* value, then the null hypothesis will be rejected, and it will be concluded that there is at least one mean significantly different from one other mean. If the obtained *F* value does not exceed the tabled critical value, then H_0 will not be rejected, and it will be concluded that there are no significant differences between any of the means (see Figure 9.1).

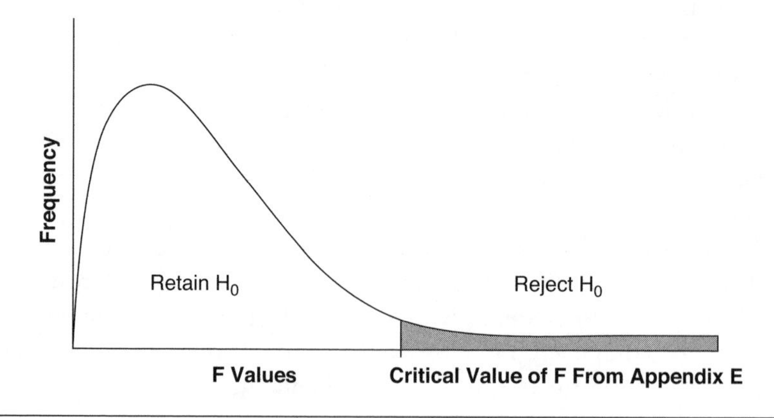

Figure 9.1

The F distribution is positively skewed with the rejection region only in the positive tail. Remember, if the null hypothesis is true, the resulting F value should be close to 1.00. If the null hypothesis is false, the resulting F value should be greater than 1.00. The critical F value will have two different types of degrees of freedom, one associated with the numerator (between-groups variance) of the F ratio and one associated with the denominator (within-groups variance).

What a Significant ANOVA Indicates

The interest in ANOVA is whether there are significant differences between two or more groups' means. Remember, ANOVA is an omnibus test, and a significant F value does not reveal which groups' means are different from each other. There are a group of tests that are to be used after the null hypothesis has been rejected in ANOVA. These tests are called **multiple comparison tests**, a posteriori tests (Latin for "what comes later"), or post hoc tests (Latin for "after this"). These tests are designed to tell which means are significantly different from each other. A significant ANOVA simply reveals whether there is at least one mean different from one other mean. A multiple comparison test reveals the pattern of significant differences between the means.

A Computational Example

Let us imagine you are a forensic psychologist at your state's mental hospital. It has been brought to the attention of your state's department of transportation that two thirds of all people who have been admitted to your state's mental hospital are subsequently released, medicated, and driving automobiles. How would you determine whether there is any evidence that one of the most common forms of antipsychotic medication interferes with driving ability? A one-factor ANOVA would be an ideal experimental design because it could assess more than one level of medication at once. So let us set up an experiment with volunteers from the community and test to see whether Thorazine, a major tranquilizer for the treatment of psychoses, affects driving ability. Twenty participants have volunteered to participate in the study. Five students are randomly assigned to each of four different groups: (a) placebo, (b) 100 mg, (c) 250 mg, and (d) 500 mg. The scores obtained will represent the results on a 10-point (10 = *excellent,* 1 = *poor*) driving skills test (computer simulated), which was performed 1 hour after ingestion of the drug. The data are tabled as follows:

1. Placebo	*2. 100 mg*	*3. 250 mg*	*4. 500 mg*
8	7	7	4
9	9	6	5
6	8	8	6
9	10	7	5
9	8	9	7

Note that each score represents a different participant and that there are a total of 20 participants.

Step 1. Calculate $\Sigma\, x$ and $\Sigma\, x^2$ for each of the individual groups.

	1	*2*	*3*	*4*
Σx	41	42	37	27
Σx^2	343	358	279	151

Step 2. Obtain the total or overall $\Sigma\, x$ and $\Sigma\, x^2$.

$$\text{Total } \Sigma\, x = 147$$
$$\text{Total } \Sigma\, x^2 = 1131$$

Step 3. Calculate the correction for the mean (CM). This is accomplished by taking the grand sum, squaring it, and dividing by the total number of observations.

$$\text{CM} = \frac{(\text{Total } \sum x)^2}{\text{Total Number of Scores}}$$
$$\text{CM} = \frac{(147)^2}{20}$$
$$\text{CM} = 1080.450$$

Step 4. The sum of squares total (SST) is obtained by taking the total $\Sigma\, x^2$ and subtracting the CM.

$$\text{SST} = \text{Total } \sum x^2 - \text{CM}$$
$$\text{SST} = 1131 - 1080.450$$
$$\text{SST} = 50.550$$

Step 5. The sum of squares between groups (SSB) is obtained by squaring the sum of each group total and dividing by the number of observations within each group, summing the quotients, and subtracting the CM from the resultant value.

$$\text{SSB} = \left[\frac{(\sum x\ for\ Gp\ 1)^2}{\#\ Scores\ in\ Gp\ 1} + \frac{(\sum x\ for\ Gp\ 2)^2}{\#\ Scores\ in\ Gp\ 2} + \frac{(\sum x\ for\ Gp\ 3)^2}{\#\ Scores\ in\ Gp\ 3} + \frac{(\sum x\ for\ Gp\ 4)^2}{\#\ Scores\ in\ Gp\ 4}\right] - \text{CM}$$

$$\text{SSB} = \left[\frac{(41)^2}{5} + \frac{(42)^2}{5} + \frac{(37)^2}{5} + \frac{(27)^2}{5}\right] - \text{CM}$$

$$\text{SSB} = 1108.600 - 1080.450$$

$$\text{SSB} = 28.150$$

Step 6. The error term or sum of squares within groups (SSW) is obtained by subtracting SSB from SST.

$$\text{SSW} = \text{SST} - \text{SSB}$$

$$\text{SSW} = 50.550 - 28.150$$

$$\text{SSW} = 22.400$$

Step 7. The *df* are now calculated:

$$df_{\text{SST}} = \text{the total number of scores} - 1$$
$$df_{\text{SST}} = 20 - 1 = 19$$
$$df_{\text{SSB}} = \text{the total number of groups} - 1$$
$$df_{\text{SSB}} = 4 - 1 = 3$$
$$df_{\text{SSW}} = df_{\text{SST}} - df_{\text{SSB}}$$
$$df_{\text{SSW}} = 19 - 3 = 16$$

Step 8. The mean squares (MS) are computed by dividing the sum of squares by the appropriate *df*.

$$\text{MSB} = \frac{\text{SSB}}{df_{\text{SSB}}} = \frac{28.150}{3} = 9.383$$

$$\text{MSW} = \frac{\text{SSW}}{df_{\text{SSW}}} = \frac{22.400}{16} = 1.400$$

Step 9. The F value is obtained by dividing MSB by MSW.

$$F = \frac{\text{MSB}}{\text{MSW}} = \frac{9.383}{1.400} = 6.70$$

Step 10. Obtain the degrees of freedom. To obtain the critical value of F from the F distribution table in Appendix E, the df for SSB and SSW are needed.

Degrees of Freedom for the Numerator

The numerator degrees of freedom or df_1 refers to the df_{SSB}. This value was obtained in Step 7. For educational purposes, it will be repeated:

df_1 = the number of groups − 1. Since there were four groups:
$df_1 = 4 - 1$
$df_1 = 3$

Degrees of Freedom for the Denominator

The denominator degrees of freedom, or df_2, refer to the df_{SSW}. This value was obtained in Step 7 but will be repeated:

$$df_2 \text{ or } df_{\text{SSW}} = df_{\text{SST}} - df_{\text{SSB}} = 19 - 3$$
$$df_2 = 16$$

Step 11. Check for significance. Compare the derived F value = 6.70 to the tabled critical value of F at $p = .05$ with $df_1 = 3$ and $df_2 = 16$. If the derived F value exceeds the tabled critical value, then H_0 is rejected. If not, then H_0 is retained.

The null and alternative hypotheses are as follows:

$$H_0: \mu_1 = \mu_2 = \mu_3 = \mu_4$$

where μ_1 = the mean for the first group on the driving skills test, and so on.

$$H_a: H_0 \text{ is not true.}$$

Step 12. An informal write-up and conclusion would look like this:

> The derived $F = 6.70$ exceeds the tabled critical value of $F = 3.24$ at $p = .05$ with $df_1 = 3$ and $df_2 = 16$. Therefore, H_0 is rejected, and it is concluded that at least one mean is significantly different from one other mean, $F(3, 16) = 6.70$, $p < .01$. To determine the pattern of mean differences, a multiple comparison test is needed. In terms of the research question, it does appear that Thorazine does significantly affect driving ability.

In the first sentence of the previous paragraph, note that we will test the derived F and reject or retain the null hypothesis at the conventional level of significance, $p = .05$. In the second sentence, we will report the F value to the lowest p level possible. Also, because the F test is an omnibus test, we do not yet know which groups' means are different until we perform a multiple comparison test. In the sentence that begins, "In terms of the research question," notice that our conclusion is just that the independent variable (Thorazine) has an effect on the dependent variable (score on the driving skills test).

In a research article, the results of the ANOVA might be reported as follows:

> Thorazine levels had a significant effect on driving ability, $F(3, 16) = 6.70$, $p < .01$.

This sentence would then be followed by a summary of the multiple comparison test results. See Chapter 10 for a discussion of multiple comparison tests. If the ANOVA had yielded a nonsignificant F, then there is the option of stating $p > .05$ or stating the exact p level (e.g., $p = .13$) if the computation of the ANOVA was performed through statistical software.

Step 13. The following is known as a "source table" in ANOVA:

Source	*Sum of Squares*	*df*	*Mean Squares*	*F*	*p*
Between groups	28.15 (SSB)	3	9.38 (MSB)	6.70	< .01
Within groups	22.40 (SSW)	16	1.40 (MSW)		
Total	50.55 (SST)	19			

Determining Effect Size in ANOVA

Although we found significance in this example, we do not know the strength of the relationship between the independent variable (Thorazine levels) and the dependent variable (driving skills test scores). Just as in the t test, significance is only part of the picture. For ANOVA, one popular measure of the magnitude of the effect of the independent variable on the dependent variable is omega-squared (ω^2). Omega-squared is obtained through the following formula:

$$\omega^2 = \frac{\text{SSB} - (k-1)\text{MSW}}{\text{SST} + \text{MSW}}$$

where k = the number of levels of the independent variable.

In the previous example,

$$\omega^2 = \frac{28.15 - (4-1)1.40}{50.55 + 1.40}$$

$$\omega^2 = \frac{23.95}{51.95}$$

$$\omega^2 = .46$$

According to omega-squared interpretation guidelines,

$$\omega^2 > .15 = \text{large effect}$$

$$\omega^2 > .06 = \text{medium effect}$$

$$\omega^2 > .01 = \text{small effect}$$

Thus, in our example, $\omega^2 = .46$, which is indicative of a large effect size, may also be interpreted as the percentage of variance in the dependent variable (driving skills test scores) accounted for by the independent variable (Thorazine levels). ω^2 is a measure of how much variation is accounted for by the treatment and, by default, how much of the total variation in scores is accounted for by unknown or random factors. In this example, 46% of the total variance in the scores can be accounted for by the treatment with Thorazine, and 54% of the total variance is accounted for by unknown or random factors.

History Trivia

Gosset to Fisher

In 1908, William Gosset (1876–1937) published a table of ratios of differences between sample means and population means. Gosset called the ratios of differences *zs* and found that they did not fit the normal distribution. He entitled the paper, "The Probable Error of a Mean." He also published a second paper in *Biometrika* that same year called "Probable Error of a Correlation Coefficient."

Ronald Fisher (1890–1962) worked for an experimental agricultural station for 14 years analyzing, among other data, wheat crop yields and weather patterns. In 1912, Gosset was 36 years old and Fisher was 22. Fisher had just graduated from Cambridge University, and in his first paper, he derived an estimate for the variance of a normal sample, which differed from Gosset's. Fisher used a divisor of n where Gosset used $n - 1$. Fisher's college tutor encouraged him to write Gosset, and he did so. Thus began a friendship and an exchange of letters until Gosset's death in 1937.

Fisher's first letter to Gosset apparently gave Fisher another idea: to represent geometrically the configuration of a sample in n-dimensional space. This, in turn, led him to the notion of degrees of freedom and to the correct divisor for the formula to estimate variance ($n - 1$). Finally, it led Fisher to a mathematical proof of Gosset's z distribution. Fisher sent the proof to Gosset, and Gosset sent it to Pearson with a recommendation to publish it. About the paper, Gosset wrote, "I couldn't understand his stuff . . . I don't feel at home in more than three dimensions even if I could understand it otherwise." But Gosset concluded, "It's so nice and mathematical that it might appeal to some people" (Salsburg, 2001, pp. 31–32). Thus began their collaboration, which would lead to a retabulation of Gosset's *z*s into *t*s, and it was to be known as Student's t distribution. Fisher went on to solve the problem posed in Gosset's second 1908 paper by deriving the general sampling distribution of the correlation coefficient. Upon receipt of Fisher's paper, Gosset, with his typical humility and generosity, wrote, "When I first saw it, I nearly wrote to thank you for the kind way in which you referred to my unscientific efforts."

When Gosset became aware that Fisher was looking for a job and that Fisher had been doing statistical work on orchards, Gosset told Fisher that John Russell of the Rothamstead Agricultural Experiment Station was looking to hire a statistician. Eight months later, Fisher notified Gosset that he had gotten the appointment. In Fisher's letter, he asked for advice on what calculator (adding machine) he should buy and advice on home brewing. With respect to the latter request, Gosset wrote, "Less trouble to buy Guinness . . . let us do it for you." Fisher ended up with a calculating machine known as the "Millionaire." It was hand cranked and primitive but got its name because it could handle numbers in the millions.

At Rothamstead, Fisher would create two of his greatest works, *Statistical Methods for Research Workers,* published in 1925, and *The Design of Experiments,* published in 1935. Fisher came to America in 1931 and taught at Iowa and Minnesota. At Iowa, he was to influence George W. Snedecor. Snedecor published his own influential statistical text in 1937, and analysis of variance would become common for statistical analyses in agriculture. Beyond Fisher's creative contributions to hypothesis testing with small samples and their test distributions, he would ultimately contribute two of statistics' most popular tests, the t test and analysis of variance.

It is interesting to note that Fisher dealt with probability in significance testing in a very brief way. It is a problem that still partially plagues modern statistics. The problem is, "What does $p < .05$ really mean?" Fisher was relatively clear on this matter, at least at a practical level. For Fisher, the significance of a test would make sense only in the context of repeated experiments. These experiments would all be aimed at determining the various effects of specific treatments. If a p value was very small (e.g., less than .01), he would claim that an effect has been shown but only if the experiment was repeatable (replicated). If the p value was large (e.g., $p > .20$), he argued that there was an effect, but it was so small that no experiment of that size would be able to detect it. If a p value fell somewhere between .01 and .20, then he argued that a subsequent experiment should be designed and improved to get a better idea of the treatment effect. Jerzy Neyman later advanced the concept of probability and significance levels, and statisticians now say that $p < .05$ means the probability that we are rejecting the null hypothesis when the null hypothesis is actually true is less than 5 chances out of 100. However, in its original conception, significance was not the probability of any kind of error. It was originally conceived as the probability that something other than what we think affected the data actually affected the data.

An instructor went to teach at University College during the years of Karl Pearson. What was her reaction to these famous figures? Of Karl Pearson, she wrote, "Endured K. P." Of Jerzy Neyman, she wrote, "I was baby-sitting for Neyman, explaining to the students what the hell he was up to." She wrote, "Went fly fishing with Gosset. A nice man . . . [he] didn't have a jealous bone in his body."

Key Terms and Definitions

Analysis of variance (ANOVA)—A popular, powerful, and robust statistical test that assesses the differences between two or more groups' means by analyzing a ratio of variances between groups and within groups.

Between-groups variance—A measure of the sum of the squared differences between a series of groups' means and the population mean (or grand mean). Between-groups variance is a measure of how much the independent variable affects the dependent variable in ANOVA.

Bonferroni correction—A solution created to correct for a high overall α level when using multiple t tests that divides the overall α level by the number of t tests to be performed. The Bonferroni correction does reduce the probability of the Type I error by tremendously increasing the probability of the Type II error.

Completely randomized ANOVA—The most powerful ANOVA design where participants are randomly assigned to one of the two or more groups. Causation may be implied from a completely randomized ANOVA design.

Experimental error—The statistical name for the random and unsystematic effects that influence an experiment but are not due to the treatment or independent variable. The largest source of experimental error is subject error and may also include variations in testing conditions such as temperature, lighting, experimenter's changing enthusiasm for the experiment, participant's variation in interest, boredom, and so on.

Multiple comparison tests—Also known as a posteriori or post hoc tests. Multiple comparison tests determine the pattern of mean differences after the null hypothesis has been rejected in an ANOVA design.

Multiple *ts*—A design where mean differences among three or more groups are assessed by more than one *t* test. Using multiple *ts* has the effect of inflating the overall α level (the probability of committing the Type I error) to unacceptably high levels ($p > .05$).

Omnibus hypothesis and omnibus test—The word *omnibus* means covering many situations at once. ANOVA has omnibus hypotheses because the actual pattern of mean differences is not specified. ANOVA is an omnibus test because any pattern of mean differences can cause the rejection of the null hypothesis.

Sampling distribution—A theoretical distribution made up of all possible finite random samples from the same population.

Standard error of the mean—The standard deviation of the sampling distribution's means.

Subject error—A curious name for the difference between a participant's score in a group and the group's mean.

Within-groups variance—A measure of the sum of the differences of each participant's score from the group's mean. Within-groups variance is a measure of how much individual differences affect the dependent variable even when all of the participants are treated alike.

Chapter 9 Practice Problems

1. What is the general purpose of an ANOVA?
2. What are the assumptions of an ANOVA?
3. Why is ANOVA an omnibus test?
4. Under what conditions do multiple comparison tests follow an ANOVA?
5. What do the words *completely randomized* mean in an ANOVA?
6. A cognitive psychologist wishes to determine how short-term verbal memory capacity varies under three distraction conditions: no distraction, music, and music plus video. Thirty college students are randomly assigned to one of the three conditions. The scores represent how many digits they could successfully recall (where larger numbers indicate better memory).

No Distraction	*Music*	*Music + Video*
7	7	4
9	8	5
6	7	3
8	6	6
7	8	3
8	7	2
9	7	4
6	9	3
7	5	4
7	9	2

Conduct an ANOVA, create a source table, and write up your results.

Chapter 9 Test Questions

1. If a three-group experiment is performed and the means for the three groups are analyzed with three *t* tests (Group 1 vs. Group 2, Group 2 vs. Group 3, Group 1 vs. Group 3), what is the probability of committing the Type I error?
   a. .05
   b. .10
   c. .14
   d. .15

2. The original Bonferroni correction for three multiple *t* tests would have the conventional level of significance change to
   a. .05
   b. .01
   c. .001
   d. .017

3. The problem with the original Bonferroni correction is that
   a. one is more likely to commit the Type I error
   b. the probability of the Type II error increases dramatically
   c. Type I and Type II errors increase dramatically
   d. the power of the test is increased

4. The procedures of ANOVA were primarily developed by
   a. Galton
   b. Pearson
   c. Fisher
   d. Gosset

5. An omnibus hypothesis means that
   a. there are many different possible outcomes
   b. there is a single outcome
   c. there are two outcomes but three alternatives
   d. there are two null hypotheses

6. The derived *F* value in ANOVA is a _______ of variances.
   a. collection
   b. group
   c. ratio
   d. matrix

7. A standard deviation derived from a finite sample of means from a population of scores is known as the
   a. standard deviation
   b. sampling deviation
   c. standard error of the mean
   d. power of a test

8. Theoretically, in ANOVA, if there were virtually no group mean differences, then the value of the derived *F* would be

   a. 0.00
   b. 0.05
   c. 1.00
   d. 0.01

9. Which of the following is not an assumption of ANOVA?

   a. the dependent variable is drawn from a normally distributed population of scores
   b. the participants are randomly assigned to each group, and the scores are independent
   c. the variances about each group mean are not too different from each other
   d. the independent variable has an infinite number of levels

10. When an ANOVA is significant,

    a. a multiple comparison test must be performed
    b. an a posteriori test must be performed
    c. a post hoc test must be performed
    d. all of the above

Problems 11–20. A psychologist wishes to determine which of four diet plans works best. The psychologist randomly assigns 32 overweight clients to one of four diet plans: Life on Herbs (LOH), E-mail-A-Meal (EAM), Weight-Lookers (WLK), and Low-Carbs (LOC). The numbers represent weight loss in pounds for 3 months for each client.

LOH	*EAM*	*WLK*	*LOC*
14	11	10	25
19	15	5	18
12	9	6	24
9	19	11	29
14	12	4	17
15	10	5	20
18	20	4	23
8	8	3	27

11. The sum of squares between groups (SSB) for this problem is

    a. 1150.75
    b. 436.75
    c. 1587.50
    d. 24.591

12. The sum of squares within groups (SSW) for this problem is

    a. 1150.75
    b. 436.75
    c. 1587.50
    d. 24.591

13. The sum of squares total (SST) for this problem is
    a. 1150.75
    b. 436.75
    c. 1587.50
    d. 24.591

14. The *F* value for this problem is
    a. 1150.75
    b. 436.75
    c. 1587.50
    d. 24.591

15. The degrees of freedom for the numerator (df_1) is
    a. 3
    b. 28
    c. 31
    d. 1

16. The degrees of freedom for the denominator (df_2) is
    a. 3
    b. 28
    c. 31
    d. 1

17. The mean squares within groups (MSW) is
    a. 436.750
    b. 15.598
    c. 383.583
    d. 24.591

18. The null hypothesis for this problem should be
    a. rejected
    b. not rejected
    c. retained
    d. none of the above

19. Which of the following can be concluded about this experiment?
    a. There are differences in weight loss among the groups.
    b. There are no differences in weight loss among the groups.
    c. One group was superior to the other groups in weight loss.
    d. Two of the groups were better than the other groups in weight loss.

20. One way to determine the effect size in ANOVA is
    a. omega-squared
    b. *F* squared
    c. alpha-squared
    d. beta-squared

Problems 21–30. A consumer psychologist wishes to determine whether playing music increases sales. The owner plays no music, soft rock, classical, or jazz and notes the sales for 28 people under these conditions. The following numbers are in dollar amounts per shopper.

No Music	*Soft Rock*	*Classical*	*Jazz*
128	57	87	55
101	32	40	24
157	93	21	95
196	35	92	19
163	58	55	60
144	61	59	85
170	82	39	54

21. The sum of squares between groups (SSB) for this problem is
    a. 17,451.714
    b. 63,902.714
    c. 21.293
    d. 46,451.0
22. The sum of squares within groups (SSW) for this problem is
    a. 17,451.714
    b. 63,902.714
    c. 21.293
    d. 46,451.0
23. The sum of squares total (SST) for this problem is
    a. 17,451.714
    b. 63,902.714
    c. 21.293
    d. 46,451.0
24. The *F* value for this problem is
    a. 17,451.714
    b. 63,902.714
    c. 21.293
    d. 46,451.0
25. The degrees of freedom for the numerator (df_1) is
    a. 3
    b. 24
    c. 27
    d. 28
26. The degrees of freedom for the denominator (df_2) is
    a. 3
    b. 24
    c. 27
    d. 28
27. The mean squares within groups (MSW) is
    a. 727.155
    b. 15,483.667

c. 21.293
d. 46,451.0

28. The null hypothesis for this problem should be
    a. rejected
    b. not rejected
    c. retained
    d. neither rejected nor not rejected

29. Which of the following can be concluded about this experiment?
    a. There are differences in sales among the groups.
    b. There are no differences in sales among the groups.
    c. One group was superior to the other groups in sales.
    d. Two of the groups were better than the other groups in sales.

30. The significance level for this problem is
    a. $p < .01$
    b. $p < .10$
    c. $p > .05$
    d. $p > .001$

10 After a Significant Analysis of Variance

Multiple Comparison Tests

Chapter 10 Goals

- Learn about multiple comparison tests
- Learn to conduct Tukey's HSD test

The purpose of ANOVA was to determine whether significant differences existed among two or more groups' means without inflating the probability of the Type I error. One limitation of ANOVA is that, if the null hypothesis is rejected, ANOVA is an omnibus test with a vague alternative hypothesis (H_0 is not true). ANOVA does not tell us what the exact pattern of mean differences is. **Multiple comparison tests,** also called **a posteriori** (Latin for "what comes after") **tests** or **post hoc** (Latin for "after this") **tests,** were designed to be used after a significant ANOVA where the null has been rejected. Multiple comparison tests help to determine the pattern of significant differences among the means, and they also keep the Type I error rate at an acceptable level ($p = .05$ or less) despite whether there are two or more comparisons to be made among the groups' means.

There are many different multiple comparison tests such as Tukey's, Newman-Keuls, Scheffé's, Duncan's, and so on. Some of these are considered to be more conservative in controlling the Type I error, and others are considered more liberal. This book will present only one: **Tukey's HSD test.** It is a popular multiple comparison test, and it is considered to be neither too conservative nor too liberal.

Historically, one of the first multiple comparison tests to be used after a significant ANOVA was the **least significant difference test** (LSD test). While college statistics students in the late 1960s snickered at the initials, it quickly came to be known that the LSD test was far too liberal; that is, the probability of the Type I error was increased beyond an acceptable level (e.g., $p > .05$). Princeton professor John Tukey created a better test that came to be known as the honestly significant difference test or Tukey's HSD test or now simply as Tukey's test.

The basic assumptions made for t tests are also required for Tukey's test (e.g., normal distributions, homogeneity of variance, and approximately equal numbers of participants).

Conceptual Overview of Tukey's Test

Each mean in the rejected null hypothesis will be compared to every other mean. If there are three group means, then there will be three comparisons (Group 1 vs. Group 2, Group 1 vs. Group 3, and Group 2 vs. Group 3). If there are four group means, then there will be 6 comparisons, and if there are five group means, then there will be 10 comparisons. The value of the difference between two means will be compared to a critical value, known as Tukey's HSD value. The absolute value of the difference between the two means must exceed Tukey's HSD value to be statistically significant.

Computation of Tukey's HSD Test

Step 1. Determine all possible mean comparisons that can be made from the rejected null hypothesis. For example, if the rejected null hypothesis had three means, then there are three possible mean comparisons: Group 1 mean versus Group 2 mean, Group 1 mean versus Group 3 mean, and Group 2 mean versus Group 3 mean. If the rejected null hypothesis contained four means, then there are six mean difference comparisons.

For example, in the previous chapter (one-factor completely randomized design), a significant F value was obtained. The null hypothesis contained four means (placebo group, 100-mg group, 250-mg group, and 500-mg group). There are six possible mean comparisons, which are as follows:

$$\bar{x}_{\text{placebo}} \text{ vs. } \bar{x}_{100\text{mg}}$$

$$\bar{x}_{\text{placebo}} \text{ vs. } \bar{x}_{250\text{mg}}$$

$$\bar{x}_{\text{placebo}} \text{ vs. } \bar{x}_{500\text{mg}}$$

$$\bar{x}_{100\text{mg}} \text{ vs. } \bar{x}_{250\text{mg}}$$

$$\bar{x}_{100\text{mg}} \text{ vs. } \bar{x}_{500\text{mg}}$$

$$\bar{x}_{250\text{mg}} \text{ vs. } \bar{x}_{500\text{mg}}$$

Step 2. Determine the absolute value of the difference between each comparison.

Comparisons		*Means*		*Absolute Value of Difference*
$\bar{X}_{placebo} - \bar{X}_{100mg}$	=	8.20–8.40	=	0.20
$\bar{X}_{placebo} - \bar{X}_{250mg}$	=	8.20–7.40	=	0.80
$\bar{X}_{placebo} - \bar{X}_{500mg}$	=	8.20–5.40	=	2.80
$\bar{X}_{100mg} - \bar{X}_{250mg}$	=	8.40–7.40	=	1.00
$\bar{X}_{100mg} - \bar{X}_{500mg}$	=	8.40–5.40	=	3.00
$\bar{X}_{250mg} - \bar{X}_{500mg}$	=	7.40–5.40	=	2.00

Step 3. Determine Tukey's HSD critical value according to the following formula

$$\text{HSD} = q\sqrt{\frac{\text{MSW}}{N}}$$

First, obtain the value of q from Appendix F (Tukey's HSD q Table of Critical Values). To enter this table, we will be required to know three values:

1. α level or p level = .05 or .01 (we will use .05).
2. r = the number of different means that will be compared (hint: look at H_0 and count the means).
3. df_{MS} or df_{error} = the degrees of freedom associated with the MS error term or the significant F value.

Our Tukey's test will be conducted at $p = .05$. Regardless of the significance level of F, statisticians have the option of performing Tukey's test at either the .05 or the .01 level or both. In the latter case, the results would be reported at the level that made the most conceptual sense. Note also that Tukey's post hoc test should not be performed if the ANOVA F value was not significant.

In our previous example, we had four unique means in the null hypothesis, so $r = 4$, and $df_{\text{error}} = 16$ (df for MSW). Thus, the q value is 4.05.

Next, determine the value of MSW (or MS within groups). This value is also called the MS error term. The value of MSW was tabled in Step 13 of Chapter 9 as the mean squares for the within-groups source. It was also the value of the denominator in the F value. In the present example, MSW = 1.400.

Finally, N = number of scores on which each mean was based (if Ns are not equal, use the harmonic mean described later in this chapter).

Thus,

$$\text{HSD} = 4.05\sqrt{\frac{1.400}{5}}$$

$$\text{HSD} = (4.05)(.5292)$$

$$\text{HSD} = 2.14$$

Step 4. The obtained HSD value is a critical value. The differences between the means in Step 2 must exceed this critical value to be significantly different at $p < .05$. Thus, to be significantly different, the absolute value of a mean pair difference must exceed 2.14.

Now, attach an asterisk to those mean pairs that exceed the critical HSD value.

Comparisons		*Means*		*Absolute Value of Difference*
$\bar{X}_{placeb} - \bar{X}_{100mg}$	=	8.20−8.40	=	0.20
$\bar{X}_{placebo} - \bar{X}_{250mg}$	=	8.20−7.40	=	0.80
$\bar{X}_{placebo} - \bar{X}_{500mg}$	=	8.20−5.40	=	2.80*
$\bar{X}_{100mg} - \bar{X}_{250mg}$	=	8.40−7.40	=	1.00
$\bar{X}_{100mg} - \bar{X}_{500mg}$	=	8.40−5.40	=	3.00*
$\bar{X}_{250mg} - \bar{X}_{500mg}$	=	7.40−5.40	=	2.00

*$p < .05$.

Note that an asterisk in statistics has no standard meaning. It can refer to any level of significance as designated by the statistician, or it can refer to degrees of freedom or other statistical parameter. However, if an asterisk is used, it must be defined near the place or on the same page as it first appears.

What to Do If the Error Degrees of Freedom Are Not Listed in the Table of Tukey's *q* Values

Suppose that the error degrees of freedom were 28 for an ANOVA. Tukey's table of *q* values lists 24 and then jumps to 30. If there are 28 degrees of freedom for the error term, are there at least 24 degrees of freedom? Yes. If you have 28 degrees of freedom, then you have at least 24 degrees of freedom. If you have 28 degrees of freedom for the error term, are there at least 30 degrees of freedom? No. Therefore, if there are 28 degrees of freedom, use the Tukey's *q* value for 24 degrees of freedom for the error term. Here is another example. If an ANOVA has 59 degrees of freedom for the error term, would you use the Tukey's *q* value for the error degrees of freedom at 40 or at 60? Even though 59 is closer to 60 than to 40, 59 degrees of freedom contains at least 40 degrees of freedom, but it does not contain at least 60 degrees of freedom. Therefore, if an ANOVA had 59 degrees of freedom for the error term, you should use the *q* value at 40 degrees of freedom.

Determining What It All Means

More often than not, the overall pattern of significant and nonsignificant differences among the means will be understandable. It is also important to remember *that nonsignificant differences may have just as much meaning as significant ones*. Let us start with the first pair of means. From that pair (the placebo and the 100-mg group), it can be determined that the mean driving score for the placebo group (8.20) is not significantly different from the mean score for the 100-mg group (8.40). What does that indicate? Well, our preliminary conclusion would be that 100 mg of Thorazine does not significantly impair (or improve) driving ability since there was no significant difference between those two means. You may have also noticed that *mathematically*, the mean of the 100-mg group (8.40) was higher than the placebo group's mean (8.20). However, although the means are mathematically different, they are *not* to be interpreted as statistically different. Therefore, the interpretation would be that 100 mg of Thorazine does not impair driving ability! It is extremely important to remember that Tukey's test has shown that this difference is a nonsignificant one, so ignore any mathematical differences. If there is no statistically significant difference between two means, then any mathematical differences between them are attributed to chance, and these chance differences should be completely ignored.

The next pair of means (placebo and 250-mg groups) is also not significantly different from one another. This indicates that the mean driving score for the placebo group (8.20) is not significantly higher or lower than the 250-mg group's mean (7.40). Again, the interpretation is the same as before. It appears that 250 mg of Thorazine also does not impair driving ability. Our preliminary conclusion would be that up to 250 mg of Thorazine does not impair driving ability.

Once again, it is important to warn you that a mathematical difference between the two means is apparent. However, do not interpret this mathematical difference. Because it is not statistically significant according to Tukey's test, then the differences between the two means are attributed to chance.

The third pair of means (placebo and 500-mg groups) is significantly different from each other. This indicates that the placebo mean driving score (8.20) is significantly higher than the 500-mg group's driving score mean (5.40). Thus, our conclusion would be that 500 mg of Thorazine does appear to impair driving ability, and while up to 250 mg of Thorazine does not impair driving ability, 500 mg does impair driving ability.

We still have three more pairs of means to interpret, yet notice that a meaningful overall pattern has already emerged: 100 mg and 250 mg of Thorazine do not affect driving ability, but 500 mg has a negative effect on driving ability. We discovered this pattern by comparing all three Thorazine groups (100 mg, 250 mg, and 500 mg) to the placebo group's mean. The last three pairs of differences do not provide any more appreciable meaning to our overall pattern. The 100-mg group's driving ability is not different from the 250-mg group's, the 100-mg group's driving ability is significantly less than the 500-mg group's driving ability, and the 250-mg group's driving ability is not significantly different from the 500-mg group's driving ability.

On the Importance of Nonsignificant Mean Differences

It is very important to note that most researchers, casual readers, journal editors, and just about everybody else in the world are biased toward significant differences. It is important to remember that *nonsignificant differences may be just as important as significant differences.* In the previous example, the first and second mean pairs are not significantly different, but they contain very important information. The first pair reveals that the mean driving score for the placebo group (8.20) is not significantly different from the mean score for the 100-mg group (8.40). This means that Thorazine does not seem to impair driving ability when given in a 100-mg dose, and that may be very important information to many people. In addition, there was no significant difference between the mean driving score for the placebo group (8.20) and the mean score for the 250-mg group (7.40). This means that Thorazine may not impair driving ability in a 250-mg dose. Again, this may be important information, although it was not statistically significant. However, remember that the word *significance* or *nonsignificance* in a statistical context means *not likely due to chance* or *likely due to chance,* and it does not represent a value judgment about the value of the results. So remember, when interpreting the results of a multiple comparison test, pay close attention to the significant *and* the nonsignificant mean pairs.

Final Results of ANOVA

In a research article, the results of the ANOVA and Tukey's post hoc test might be reported as follows:

> Thorazine levels had a significant effect on driving ability, $F(3, 16) = 6.70$, $p < .01$. A Tukey's post hoc test ($p < .05$) revealed that there was no significant difference between the mean driving skills test scores for the placebo (8.20), 100-mg (8.40), and 250-mg (7.40) groups, which indicates that Thorazine does not appear to affect driving ability at 250-mg and lower dosages. The mean driving skills test score for the group that received 500 mg of Thorazine (5.40) was significantly lower than the mean scores for the placebo and 100-mg groups, indicating that Thorazine does significantly impair driving ability at the 500-mg dosage.

Tukey's With Unequal *N*s

If the means are based on unequal *N*s, the *N* in Tukey's formula is based on the **harmonic mean** ($\tilde{N}$). The harmonic mean is obtained by the following formula:

$$\tilde{N} = \frac{k}{\frac{1}{N_1} + \frac{1}{N_2} + \ldots + \frac{1}{N_k}}$$

where

k = the number of means,

N_1 = the number of scores in Group 1,

N_k = the number of scores in the last group,

$\tilde{N}$ = the harmonic mean.

Key Terms, Symbols, and Definitions

A posteriori test—An alternative name for a multiple comparison test. It is Latin for "what comes later."

Harmonic mean ($\tilde{N}$)—An alternative measure of central tendency to the arithmetic mean. It is defined as the reciprocal of the arithmetic mean of the reciprocals of the scores. It is used in Tukey's HSD test when there is an unequal number of scores in each group.

Least significant difference test—Historically, one of the first multiple comparison tests that is presently considered too liberal (i.e., *p* level may exceed .05).

Multiple comparison test—A test to be used only after a significant ANOVA. It determines which pairs of means are significantly different from each other.

Post hoc test—An alternative name for a multiple comparison test. It is Latin for "after this."

Tukey's HSD test—A popular multiple comparison test, considered neither too liberal nor too conservative, that maintains the Type I error rate regardless of the number of means to be compared.

Chapter 10 Practice Problems

1. If the rejected null hypothesis had five means, what are the possible mean comparisons?
2. What are the basic assumptions required for Tukey's post hoc test?
3. What should be done if the means are based on unequal *N*s?
4. What is the difference between the LSD test and HSD test?

Chapter 10 Test Questions

Problems 1–7 (same as Problems 11–20 in Chapter 9). A psychologist wishes to determine which of four diet plans works best. The psychologist randomly assigns 32 overweight clients to one of four diet plans: Life on Herbs (LOH), E-mail-A-Meal (EAM), Weight-Lookers (WLK), and Low-Carbs (LOC). The numbers represent weight loss in pounds for 3 months for each client.

LOH	*EAM*	*WLK*	*LOC*
14	11	10	25
19	15	5	18
12	9	6	24
9	19	11	29
14	12	4	17
15	10	5	20
18	20	4	23
8	8	3	27

1. The value of N in the Tukey's HSD formula for the above problem is
   a. 8
   b. 16
   c. 32
   d. 40
2. The value of r in Appendix F for Tukey's q values is
   a. 2
   b. 4
   c. 8
   d. 32
3. What error degrees of freedom would be chosen in the appendix for Tukey's q value, because the actual error degrees of freedom for Problem 1 is not listed in the appendix?
   a. 3
   b. 24
   c. 30
   d. 8
4. The value of q in the appendix for Tukey's q values is
   a. 3.90
   b. 4.91
   c. 3.85
   d. 4.80
5. Tukey's HSD value for Problem 1 would be
   a. 5.45
   b. 3.90
   c. 4.91
   d. 4.80
6. The results of Tukey's HSD test for Problem 1 revealed
   a. the Weight-Lookers group was the best diet
   b. the Weight-Lookers group lost the least weight

c. there were no differences in mean weight loss between the Weight-Lookers group and the Low-Carbs group
d. the E-mail-A-Meal group lost more weight than the Life on Herbs group

7. Based on Tukey's HSD results, all but one of the following is true. Which is not a true statement?

a. The Low-Carbs diet produced the most weight loss.
b. There was no significant difference in weight loss between the E-mail-A-Meal and the Life on Herbs diets.
c. The Weight-Lookers diet produced the least weight loss.
d. The Life on Herbs diet produced a greater weight loss than the E-mail-A-Meal diet.

Problems 8–14 (same as Problems 21–30 in Chapter 9). A consumer psychologist wishes to determine whether playing music increases sales. The owner plays either no music, soft rock, classical, or jazz and notes the sales for 28 people under these conditions. The following numbers are in dollar amounts per shopper.

No Music	*Soft Rock*	*Classical*	*Jazz*
128	57	87	55
101	32	40	24
157	93	21	95
196	35	92	19
163	58	55	60
144	61	59	85
170	82	39	54

8. The value of N in the Tukey's HSD formula for the above problem is

a. 7
b. 14
c. 28
d. 40

9. The value of r in the appendix for Tukey's q values is

a. 2
b. 4
c. 8
d. 32

10. What error degrees of freedom would be chosen in the appendix for Tukey's q value?

a. 3
b. 24
c. 30
d. 8

11. The value of q in the appendix for Tukey's q values is

a. 3.90
b. 4.91

c. 3.85
d. 4.80

12. Tukey's HSD value for this problem would be
    a. 39.7
    b. 91.5
    c. 45.3
    d. 91.2

13. Remember for this question, you are answering based on the results of Tukey's test, not on the mathematical results. The results of Tukey's HSD test for this problem revealed that
    a. the soft rock condition produced the second greatest sales, that is, better than the classical and jazz conditions
    b. the soft rock, classical, and jazz conditions all were tied as second best, next to the no-music condition
    c. the jazz condition produced the least amount of sales
    d. the classical condition produced the third best amount of sales

14. Based on Tukey's HSD results, all but one of the following is true. Which is not a true statement?
    a. The no-music condition produced the greatest sales.
    b. The soft rock condition produced the second best sales.
    c. There were no differences in sales between soft rock, classical, and jazz.

15. A harmonic mean is used instead of the arithmetic mean when
    a. there are an unequal number of scores in each group
    b. the scores are not harmonic
    c. the standard deviations are unequal
    d. the homogeneity of variance assumption has been violated

16. In the sample problem in Chapter 10, which of the following is true?
    a. The mean driving score for the 100-mg group was significantly greater than the placebo group.
    b. The mean driving score for the 100-mg group was significantly less than the placebo group.
    c. The mean driving score of the 100-mg group was not significantly different from the placebo group.
    d. The 100-mg group statistically had the highest mean driving score.

11 Analysis of Variance

One-Factor Repeated-Measures Design

Chapter 11 Goals

- Learn the statistical analysis of a repeated-measures design
- Learn how to evaluate effect size in ANOVA

The Repeated-Measures ANOVA

A one-factor **ANOVA repeated-measures design** is similar to a repeated-measures *t* test. The same participant serves under all levels of the single factor. In the repeated-measures *t* test, the same participant serves in both groups. In the one-factor repeated measures ANOVA, the same participant serves under all levels of the single factor. The interpretation of the repeated-measures ANOVA is the same as the completely randomized ANOVA. The interest in ANOVA is whether there are significant differences between two or more groups' means. Remember that a significant ANOVA does not reveal which groups' means are different from each other. A multiple comparison test or post hoc test is designed to tell which means are significantly different from each other. A significant ANOVA simply tells whether there is at least one mean different from one other mean. The multiple comparison test reveals the pattern of significant differences between the means.

One variant of the repeated-measures *t* test is that there could be pairs of participants who were matched on some critical variable or variables. The same is true of the repeated-measures ANOVA. The variant of the repeated-measures ANOVA is called the **randomized block design.** Instead of the same participant serving under all levels of a factor, a block of participants who are all alike is randomly assigned to each level of the single factor. The randomized block design is probably more common in animal experiments where the various blocks may be litters of rats or armadillos (a litter of armadillos consists of identical quadruplets, which makes them ideal participants for a one-factor randomized block ANOVA).

In some experiments, the repeated-measures design may not be used. Experiments where the participants must be naive in each condition will preclude the use of a repeated-measures ANOVA. Other examples are in drug studies where one level of a drug will affect other levels or in designs where the levels of a factor are fixed, such as an analysis of three different types of personality disorders or a cross-sectional study of four different age groups.

One advantage of the repeated-measures design is that it uses fewer participants than the completely randomized design. Another advantage is that the statistical power of the test is increased because the variability of one participant across all levels of the factor is usually less than the variability between different participants across levels of the factor.

Assumptions of the One-Factor Repeated-Measures ANOVA

The repeated-measures ANOVA assumptions are the same as in the one-factor completely randomized ANOVA with one addition. Thus, it is assumed that the distributions of the populations of scores are normal, the scores are independent of one another, and the variances of the populations are homogeneous. The additional assumption concerns the homogeneity of variance of the difference scores between each level of the independent variable for the participants. A difference score is derived for each participant by subtracting his or her score on one level from his or her score on another level. These difference scores across all participants are assumed to have equal (or homogeneous) variances in the population of difference scores.

Computational Example

Eight patients with a diagnosis of Alzheimer's disease were tested by a neuropsychologist at the end of each of 4 consecutive years. The scores were examined to determine whether the cognitive deficits increased over time. The patients were rated on a scale from 0, which represented no deficits, to 12, which indicated maximum deficits. The data are tabled as follows:

	Year			
Patients	*First*	*Second*	*Third*	*Fourth*
P_1	7	8	8	7
P_2	9	8	10	11
P_3	5	6	8	7
P_4	10	9	10	11
P_5	4	5	7	8
P_6	5	4	5	5
P_7	6	5	6	7
P_8	8	9	9	10

Step 1. Obtain Σx and Σx^2 for the four different years and their total.

	Years				
	First	*Second*	*Third*	*Fourth*	*Overall*
Σx	54	54	63	66	$\Sigma x = 237$
Σx^2	396	392	519	578	$\Sigma x^2 = 1885$

Step 2. The correction for the mean (CM) is obtained by squaring the total Σx and dividing by N (the total number of observations).

$$\text{CM} = \frac{(\text{total} \sum x)^2}{N} = \frac{237^2}{32} = 1755.281$$

Step 3. The sum of squares total (SST) is obtained by taking the total Σx^2 and subtracting the CM.

$$\text{SST} = \text{total} \sum x^2 - \text{CM}$$
$$\text{SST} = 1885 - 1755.281$$
$$\text{SST} = 129.179$$

Step 4. Obtain Σx for each patient across the four years.

$$P_1 = (7 + 8 + 8 + 7) = 30$$
$$P_2 = (9 + 8 + 10 + 11) = 38$$
$$P_3 = 26$$
$$P_4 = 40$$
$$P_5 = 24$$
$$P_6 = 19$$
$$P_7 = 24$$
$$P_8 = 36$$

Next, square each of these values, add these results together, and divide by the number of scores upon which each sum was based.

$$\frac{30^2 + 38^2 + 26^2 + 40^2 + 24^2 + 19^2 + 24^2 + 36^2}{4} = 1857.250$$

Next, the sum of squares for subjects (SSS) is obtained by subtracting the CM from the previously obtained value.

$$\text{SSS} = 1857.250 - \text{CM}$$
$$\text{SSS} = 1857.250 - 1755.28$$
$$\text{SSS} = 101.969$$

Step 5. To obtain the sum of squares for treatments (SSB), obtain $\Sigma\, x$ for each level of the factor across all the participants in a group (which was done in Step 1), square each of these four values and add them together, and divide by the number of scores upon which each sum was based.

$$\frac{(\sum x_1)^2 + (\sum x_2)^2 + (\sum x_3)^2 + (\sum x_4)^2}{N} = \frac{54^2 + 54^2 + 63^2 + 66^2}{8}$$
$$= 1769.625$$

To obtain SSB, take the resulting value and subtract the CM.

$$\text{SSB} = 1769.625 - \text{CM}$$
$$\text{SSB} = 1769.625 - 1755.281$$
$$\text{SSB} = 14.344$$

Step 6. In order to obtain the error term sum of squares (SSW), take SST and subtract SSS and SSB.

$$\text{SSW} = \text{SST} - \text{SSS} - \text{SSB}$$
$$\text{SSW} = 129.719 - 101.969 - 14.344$$
$$\text{SSW} = 13.406$$

Step 7. The degrees of freedom (*df*) are now calculated.

$$df_{\text{SST}} = \text{total number of scores} - 1$$
$$df_{\text{SST}} = 32 - 1 = 31$$
$$df_{\text{SSS}} = \text{total number of participants} - 1$$
$$df_{\text{SSS}} = 8 - 1 = 7$$
$$df_{\text{SSB}} = \text{total number of years} - 1$$
$$df_{\text{SSB}} = 4 - 1 = 3$$
$$df_{\text{SSW}} = df_{\text{SST}} - df_{\text{SSS}} - df_{\text{SSB}}$$
$$df_{\text{SSW}} = 31 - 7 - 3 = 21$$

Step 8. The mean squares (MS) are now calculated and are obtained by dividing the sum of squares by the appropriate *df*.

$$\text{MSB} = \frac{\text{SSB}}{df_{\text{SSB}}} = \frac{14.344}{3} = 4.781$$

MSS = This value is not needed. The interest in this design is not in the difference among subjects but in the stability of cognitive deficits across time.

$$\text{MSW} = \frac{\text{SSW}}{df_{\text{SSW}}} = \frac{13.406}{21} = .638$$

Step 9. The *F* value for testing significance is obtained by dividing MSB by MSW.

$$F = \frac{\text{MSB}}{\text{MSW}} = \frac{4.781}{.638} = 7.49$$

Step 10. To check the obtained *F* value for significance, refer to the *F* distribution in Appendix E. To use the table, df_{SSB} and df_{SSW} are needed. The *df* for the numerator, or df_1, refers to df_{SSB} (or number of levels of the main factor minus 1). The *df* for the denominator, or df_2, refers to df_{SSW}. In this example, $df_1 = 3$ and $df_2 = 21$. The critical value at $p = .05$ is 3.07. The obtained *F* value exceeds this value; therefore, it is significant, $F(3, 21) = 7.49$, $p < .05$. Note that the obtained *F* value also exceeds the critical value at $p < .01$ ($F = 4.87$). Therefore, remember to report the lowest significance level possible, and in this case, it would be $p < .01$.

The null and alternative hypotheses are as follows:

$$\text{H}_0\text{: } \mu_1 = \mu_2 = \mu_3 = \mu_4$$

where μ_1 = the mean cognitive deficit score for all participants after the first year, and so on.

$$\text{H}_a\text{: H}_0 \text{ is not true.}$$

Because the *F* value was significant, the null hypothesis is not retained.

Step 11. The analysis is tabled as follows:

Source	*Sum of Squares*	*df*	*Mean Squares*	*F*	*p*
Between groups	14.34	3	4.78	7.49	<.01
Subjects	101.97	7			
Within groups (error term)	13.41	21	0.64		
Total	129.72	31			

Step 12. The final write-up would look like the following:

> The derived value of $F = 7.49$ exceeds the tabled critical value of $F = 3.07$ at $p = .05$ with $df_1 = 3$ and $df_2 = 21$. Therefore, H_0 is rejected, and it is concluded that at least one mean is significantly different from one other mean, $F(3, 21) = 7.49, p < .01$. To determine the pattern of mean differences, a post hoc test is necessary. In terms of the research question, it appears that Alzheimer's patients' cognitive deficits do change over time.

Remember, a significant analysis of variance does not reveal which means are significantly different from each other. To determine the pattern of mean differences, a post hoc or multiple comparison test must be employed. In terms of this research question, it appears that cognitive deficits do change significantly over the four years.

Determining Effect Size in ANOVA

Just as in the one-factor completely randomized ANOVA, we still do not know the strength of the relationship between the independent variable and the dependent variable after an ANOVA. If there is a significant F value, we know that at least one mean is significantly different from one other mean, but only an effect size analysis will give an estimate of how much the independent variable affects the dependent variable. Once again, the magnitude of the effect of the independent variable on the dependent variable will be calculated by the omega-squared (ω^2) statistic. Omega-squared is obtained through the same formula presented in Chapter 9 and is as follows:

$$\omega^2 = \frac{\text{SSB} - (k - 1)\,\text{MSW}}{\text{SST} + \text{MSW}}$$

where k = number of levels of the independent variable.

In the previous example,

$$\omega^2 = \frac{14.34 - (4 - 1)0.64}{129.72 + 0.64}$$

$$\omega^2 = \frac{12.42}{130.36}$$

$$\omega^2 = .095$$

According to omega-squared interpretation guidelines,

$$\omega^2 > .15 = \text{large effect}$$

$$\omega^2 > .06 = \text{medium effect}$$

$$\omega^2 > .01 = \text{small effect}$$

Thus, in this example, $\omega^2 = .10$, which is indicative of somewhere between a medium and large effect size. ω^2 may also be interpreted as the percentage of variance in the scores accounted for by the independent variable. ω^2 is a measure of how much variation is accounted for by the treatment and how much of the total variation in scores is accounted for by random factors. In this example, 10% of the total variance in the cognitive deficits can be accounted for by the four years, and 90% of the total variance is accounted for by unknown or random factors.

Key Terms and Definitions

ANOVA repeated-measures design—Similar to the dependent *t* test design in that the same participants are used for each level of the independent variable. The research interest is whether the means for each level of the independent variable are significantly different from each other.

Randomized block design—A form of repeated-measures design in ANOVA that uses groups of similar participants (e.g., triplets or quadruplets) so that, although the participants are different for each level of the independent variable, the repeated-measures ANOVA formula is still used to analyze the data.

Chapter 11 Practice Problems

1. Describe how the one-factor repeated-measures ANOVA is an extension of the dependent *t* test.

2. Be able to explain a randomized block design.

3. Be able to explicate the assumptions of the one-factor repeated-measures ANOVA.

4. A school psychologist wishes to determine whether a new charter school's curriculum helps to improve reading scores over six 6-week periods. A sample of nine students' reading scores is measured across the six periods. Perform an ANOVA one-factor repeated-measures design and write up the results. The reading scores are measured as *T* scores (mean = 50, standard deviation = 10) according to national standards.

	Period 1	*Period 2*	*Period 3*	*Period 4*	*Period 5*	*Period 6*
Student 1	49	50	51	52	52	54
Student 2	52	54	54	54	55	58
Student 3	50	51	50	53	52	53
Student 4	42	45	48	51	50	52
Student 5	57	56	59	62	63	65
Student 6	56	59	60	64	66	69
Student 7	52	52	55	55	58	59
Student 8	46	47	47	47	47	47
Student 9	48	51	55	54	55	57

Chapter 11 Test Questions

Problems 1–10. A psychologist wishes to determine whether hypnosis could be used on morbidly obese patients (more than 100 lbs. over ideal weight) to help them lose weight. The hypnosis sessions were conducted by a psychologist-hypnotist once a month for 4 months. The patients' weights were measured at the end of each month for the 4-month period.

	Month 1	*Month 2*	*Month 3*	*Month 4*
Patient 1	245	225	219	202
Patient 2	340	311	305	284
Patient 3	298	280	277	244
Patient 4	355	328	309	285
Patient 5	321	322	330	325
Patient 6	414	387	375	340
Patient 7	309	285	277	249

1. The correction for the mean (CM) for this problem is
   a. 4,662,720.4
   b. 12,956
   c. 507,963.6
   d. 5,152,897
2. The sum of squares total (SST) for this problem is
   a. 4,662,720.4
   b. 12,956
   c. 507,963.6
   d. 5,152,897
3. The sum of squares for subjects (SSS) for this problem is
   a. 14,426.0
   b. 490,176.6
   c. 3361
   d. 3475
4. The sum of squares for treatments (SSB) for this problem is
   a. 14,426.0
   b. 490,176.6
   c. 5,152,897
   d. 24
5. The error term sum of squares (SSW) for this problem is
   a. 14,426.0
   b. 490,176.6
   c. 507,963.6
   d. 3361
6. The degrees of freedom for SST are
   a. 3
   b. 8

c. 24
d. 35

7. The degrees of freedom for SSS are
   a. 3
   b. 8
   c. 24
   d. 35

8. The degrees of freedom for SSB are
   a. 3
   b. 8
   c. 24
   d. 35

9. The degrees of freedom for SSW are
   a. 3
   b. 8
   c. 24
   d. 35

10. The obtained *F* value for this problem is
    a. 31.385
    b. 34.337
    c. 37.766
    d. 76.099

12 Analysis of Variance

Two-Factor Completely Randomized Design

Chapter 12 Goals

- Learn how to analyze a two-factor ANOVA design
- Learn about main and simple effects in ANOVA
- Learn about fixed and random effects in ANOVA
- Learn about effect size analyses in two-factor ANOVA designs

Factorial Designs

While one-factor ANOVA designs are powerful and elegant, the world is a complex place. It is rare that we can attribute behavior to a single variable. For example, the roots of war, poverty, aggression, crime, suicide, or a plethora of other societal ills can never be attributed to a single causal agent. In recognition of this fact, statisticians have devised multifactor experimental designs and analyses to account for the confluence of more than one variable on behavior. In ANOVA, these designs are referred to as factorial designs. A **factorial design** evaluates the effects of two or more independent variables (called factors) simultaneously on a single dependent variable. The simplest factorial ANOVA design is the **completely randomized factorial design,** where there are two or more factors (or treatments) that each has two or more levels. The phrase "completely randomized" indicates that the participants have been randomly assigned to one of the unique levels of the factors. If there are x levels of one factor and y levels of another factor, then there are x times y number of factor combinations, and each participant can serve in only one unique condition (also called a cell). Factorial designs are sometimes referred to as a 2 × 3 (pronounced "two by three") factorial or 2 × 3 × 3 factorial design. The individual numbers refer to the number of levels of each factor, and the number of numbers refers to the number of factors in the experiment. Thus, there are three independent variables in a 2 × 3 × 3 factorial ANOVA, and there are 18 different treatment conditions. In my master's thesis, I had a 2 × 2 × 2 × 2 × 3 ANOVA design. Therefore, I had a five-factor ANOVA design where I was testing five independent variables, and four of them had two levels and one had three levels.

The Most Important Feature of a Factorial Design: The Interaction

The most important feature of a factorial design is the **interaction** between the two independent variables on the dependent variable. An interaction between the two factors allows the experimenter to determine what the effect is of both independent variables simultaneously on the dependent variable. If we performed an experiment with one independent variable and analyzed it with a one-factor ANOVA and then performed a second one-factor ANOVA on a second independent variable, we would still not know the effects of the two variables together. This interaction between the two independent variables is one of the powerful advantages of factorial designs in ANOVA, and it explains their popularity in the scientific literature. Each of the two independent variables is called a **main effect.** Thus, there are two main effects and one interaction in a two-factor ANOVA.

There are also said to be **simple effects** in a factorial ANOVA. In this case, the experimenter is interested in all levels of one of the independent variables under only one level of the other independent variable. The interaction should be significant for the simple effects to be of interest to an experimenter since, if there was no significant interaction, the main effects alone could have predicted the outcome of the experiment.

There is a cost of a factorial ANOVA design, and it is that we will sacrifice experimental simplicity, ease of interpretation, and ease of computation. A factorial design is also usually a large experiment, and if many treatment levels are involved, the number of participants can become cumbersomely large. Overall, however, the benefits of the two-factor ANOVA and its interaction far outweigh these costs.

Fixed and Random Effects and In Situ Designs

The independent variable in ANOVA designs may vary in nature. It is said to be a *fixed effect* if the experimenter is only interested in those particular levels of the independent variable and is not interested in generalizing to other levels not included in the experiment. For example, if gender was chosen as the independent variable, then it would be a fixed effect because there are only (typically) two levels of gender. In this case, it would also be considered an in situ independent variable because the participants are not randomly assigned to the two levels of the independent variable; thus, any implications of causation will be limited because of the in situ nature of the design.

A random effects ANOVA design is not often used in psychological research, or the random effects factor is rarely the focus of the experiment. In theory, a *random effects* factor is one where the experimenter is interested in generalizing to other levels of the independent variable, and the levels of the independent variable are randomly chosen from all possible levels of the independent variable. The participants in a repeated-measures ANOVA could be considered a random effects factor, but they are very rarely the focus of an experiment. The random effects are of concern in advanced topics in ANOVA because random effects require a different error term in the denominator of the F value than the typical within-subjects mean square error term.

The Null Hypotheses in a Two-Factor ANOVA

Three null hypotheses will be tested in a two-factor ANOVA: one for each of the main effects and one for the interaction. Each null hypothesis will have a corresponding alternative hypothesis. The final computational analysis will yield three F values, one for each of the two main effects and one for the interaction.

Assumptions and Unequal Numbers of Participants

The assumptions are the same for a factorial ANOVA as for the one-factor ANOVA. The dependent variable should come from a population that is normally distributed (normality assumption). The variances of the levels of the independent variables should come from populations whose variances do not differ (homogeneity of variance assumption). Finally, the scores should be independent of one another throughout all levels of the independent variables.

A factorial ANOVA is robust with regards to violations of the first two assumptions, but the latter assumption may never be violated. The formulas in this book have been written to handle unequal numbers of participants in each group; however, the first two assumptions become more important when there are unequal numbers of participants in the different conditions. Large numbers of participants in each condition (minimum of 10) and equal numbers of participants in each condition make the first two assumptions less important.

Computational Example

A psychologist wishes to determine how inattention varies as a function of gender and age in a group of adult patients with attention-deficit hyperactivity disorder (ADHD). The three age groups were selected as follows: young = 16 to 29 years old, middle = 30 to 49 years old, and older = 50 or more years old. The scores represent the number of inattention incidents during a 10-minute behavioral observation. Scores ranged from 1 incident (*relatively little inattention*) to 16 (*a lot of inattention*). The data are tabled as follows:

Men			Women		
Young	Middle	Older	Young	Middle	Older
12	8	3	5	3	8
14	6	2	2	8	5
16	2	1	9	3	1
10	1	3	10	2	4
12	9	4	6	5	7
13	3	2	7	8	2
15	4	1	1	4	3
14	5	2	5	7	4

Step 1. The first step is to obtain the Σx and Σx^2 for each of the six groups, as well as the grand total Σx and Σx^2.

	Men			Women			
	Young	Middle	Older	Young	Middle	Older	Grand Total
Σx	106	38	18	45	40	34	281
Σx^2	1430	236	48	321	240	184	2459

$$N = 48$$

Step 2. The correction for the mean (CM) is obtained by squaring the grand total Σx and dividing by N (the total number of observations).

$$\text{CM} = \frac{(\text{total} \sum x)^2}{N}$$

$$\text{CM} = \frac{(281)^2}{48} = 1645.021$$

Step 3. The sum of squares total (SST) is obtained by taking the total Σx^2 and subtracting the CM.

$$\text{SST} = \text{total} \sum x^2 - \text{CM}$$
$$\text{SST} = 2459 - 1645.021$$
$$\text{SST} = 813.979$$

Computation of the First Main Effect

In this example, there are two main effects: *gender* (with its two levels, males or females) and *age* (with its three levels, young, middle, and older age). In Step 4, choose either of the two main effects. It does not matter which one is chosen first. In Step 6, the other main effect will be computed.

Step 4: First Main Effect. To compute the effects of the gender factor (males or females), add the sums of the male conditions together (regardless of age group), and add the sums of the female conditions together (regardless of age group).

Sum of male conditions: $106 + 38 + 18 = 162$
Sum of female conditions: $45 + 40 + 34 = 119$

Next, square these sums and divide by the number of observations upon which these sums were based and add the quotients.

$$\frac{(162)^2}{24} + \frac{(119)^2}{24} = 1093.500 + 590.042 = 1683.542$$

Step 5. To compute the sum of squares for gender (SSG), take the number resulting from Step 4 and subtract the CM. Note that I use *G* for gender. You may use any letter except *T*, *W*, or *B*, because they are used later in the calculations.

$$\text{SSG} = 1683.542 - \text{CM}$$
$$\text{SSG} = 1683.542 - 1645.021$$
$$\text{SSG} = 38.521$$

Computation of the Second Main Effect

Because we arbitrarily chose gender as the first main effect, now age (with its three levels) will be computed as the second main effect.

Step 6: Second Main Effect. To compute the effects of age (young, middle, or older), add the sums of the young conditions together (regardless of gender), add the sums of the middle conditions together (regardless of gender), and add the sums of the older age conditions (regardless of gender).

Sum of young age group conditions: $106 + 45 = 151$
Sum of middle age group conditions: $38 + 40 = 78$
Sum of older age group conditions: $18 + 34 = 52$

Next, square these sums, divide by the number of observations upon which these sums were based, and add the quotients.

$$\frac{(151)^2}{16} + \frac{(78)^2}{16} + \frac{(52)^2}{16} = 1425.063 + 380.250 + 169.000 = 1974.313$$

Step 7. To compute the sum of squares for age (SSA), take the number resulting from Step 6 and subtract the CM.

$$\begin{aligned} SSA &= 1974.313 - CM \\ SSA &= 1974.313 - 1645.021 \\ SSA &= 329.292 \end{aligned}$$

Computation of the Interaction Between the Two Main Effects

In Step 8, the sums of squares for the interaction will be computed. In a 2 × 3 design, there will be six sums (in other words, two times three).

Step 8. To compute the interaction effects between gender and age, take the sum of each of the six conditions or groups, square them, divide them each by the number of observations upon which each was based, and add the quotients.

$$\begin{aligned} &\frac{(106)^2}{8} + \frac{(38)^2}{8} + \frac{(18)^2}{8} + \frac{(45)^2}{8} + \frac{(40)^2}{8} + \frac{(34)^2}{8} \\ &= 1404.500 + 180.500 + 40.500 + 253.125 \\ &\quad + 200.000 + 144.500 = 2223.125 \end{aligned}$$

Step 9. To obtain the sum of squares for the interaction effects ($SS_{G \times A}$), take the value obtained from Step 8 and subtract the CM, SSG, and SSA.

$$\begin{aligned} SS_{G \times A} &= 2223.125 - CM - SSG - SSA \\ SS_{G \times A} &= 2223.125 - 1645.021 - 38.521 - 329.292 \\ SS_{G \times A} &= 210.291 \end{aligned}$$

Step 10. To obtain the error term sum of squares (SSW), take SST and subtract SSG, SSA, and $SS_{G\times A}$.

$$SSW = SST - SSG - SSA - SS_{G\times A}$$
$$SSW = 813.979 - 38.521 - 329.292 - 210.291$$
$$SSW = 235.875$$

Step 11. The degrees of freedom (*df*) are now calculated:

$$df_{SST} = \text{total number of scores} - 1$$
$$df_{SST} = 48 - 1 = 47$$
$$df_{SSG} = \text{total number of gender levels} - 1$$
$$df_{SSG} = 2 - 1 = 1$$
$$df_{SSA} = \text{total number of age levels} - 1$$
$$df_{SSA} = 3 - 1 = 2$$
$$df_{SS_{G\times A}} = df_{SSG} \times df_{SSA}$$
$$df_{SS_{G\times A}} = 1 \times 2 = 2$$
$$df_{SSW} = df_{SST} - df_{SSG} - df_{SSA} - df_{SS_{G\times A}}$$
$$df_{SSW} = 47 - 1 - 2 - 2 = 42$$

Step 12. The mean squares are computed by dividing the sum of squares by the appropriate *df.*

$$MSG = \frac{SSG}{df_{SSG}} = \frac{38.521}{1} = 38.521$$
$$MSA = \frac{SSA}{df_{SSA}} = \frac{329.292}{2} = 164.646$$
$$MS_{G\times A} = \frac{SS_{G\times A}}{df_{SS_{G\times A}}} = \frac{210.291}{2} = 105.146$$
$$MSW = \frac{SSW}{df_{SSW}} = \frac{235.875}{42} = 5.616$$

Step 13. The F values for testing the significance of the two main effects (gender and age) and for the interaction are obtained by dividing the appropriate mean squares by MSW.

$$F \text{ value for gender} = \frac{MSG}{MSW} = \frac{38.521}{5.616} = 6.859$$
$$F \text{ value for age} = \frac{MSA}{MSW} = \frac{164.646}{5.616} = 29.317$$
$$F \text{ value for gender} \times \text{age} = \frac{MS_{G\times A}}{MSW} = \frac{105.146}{5.616} = 18.723$$

Step 14. Determine the significance levels of the obtained F values by referring to the F distribution table in Appendix E and table the data. To check for significance, the degrees of freedom for the numerator will be the df for the main effect in question, and the df for the denominator will be the degrees of freedom for the error term. In the present example, to determine significance of the F value for gender, the df for the numerator will be 1, and the df for the denominator will be 42. For the main effect of age, the df for the numerator will be 2, and the df for the denominator will be 42. For the interaction, the df for the numerator will be 2, and the df for the denominator will be 42. Thus, the critical value for gender at $p = .05$ will be 4.07, and the critical value for the age main effect and the interaction will be 3.22. It is standard statistical convention to report the lowest significance level possible. If the obtained value exceeds the critical value, then the F value will be reported as significant at that p level.

Two-Factor ANOVA Source Table

Source	*Sum of Squares*	*df*	*Mean Squares*	*F*
Gender	38.521	1	38.521	6.86*
Age	329.292	2	164.646	29.32**
Gender × Age	210.291	2	105.146	18.72**
Error	235.875	42	5.616	
Total	813.979	47		

*p < .05; **p < .01.

Note that if all three F values had exceeded the critical value of F at $p = .01$, then all three F values could have had a single asterisk, and below the source table, there would appear *$p < .01$. In statistics, an asterisk can mean anything one wants it to mean. There is no conventional interpretation of a single asterisk or a double asterisk. However, it is statistical convention to report the lowest p level possible. Many statistical programs will calculate p levels exactly, and often, large F values will be reported significant at $p = .000$. It is not statistical convention to report significance level equal to .000 or .00. Thus, it would be reported significant at $p < .001$. If the computer program reported the significance at $p = .0000$, then it is most often reported significant at $p < .0001$.

Interpretation of the Results

A factorial ANOVA is a powerful but complicated design. Therefore, the interpretation will be relatively complex. The results cannot be stated in a single statement. In the present example, there are two significant main effects and a significant interaction. In any situation where the interaction is significant, there will be a minimal interpretation of the two main effects because the results could not have been predicted from the main effects. We will deal with the interpretation of the present results of this 2 × 3 ANOVA and other issues in the next chapter.

Here are the results, so far, as might be reported in a research article.

> The main effect for gender was significant, $F(1, 42) = 6.86$, $p < .05$, as was the main effect for age, $F(2, 42) = 29.32$, $p < .01$. The interaction was also significant, $F(2, 42) = 18.72$, $p < .01$. A post hoc analysis of the interaction revealed. . . .

Note that although the main effect for age was significant and there were three levels of the effect, which would normally require a post hoc analysis, it is not performed because the interaction was significant.

Key Terms and Definitions

Completely randomized factorial design—A design in ANOVA where the participants are randomly assigned to the conditions. In situ ANOVA designs may actually be more common, and the former design allows for the implication of causation while the latter does not.

Factorial design—A design in ANOVA where two or more main effects and the interaction between the main effects can be evaluated for a single dependent variable.

Interaction—The unique confluence of two main effects or independent variables on a single dependent variable in factorial designs. The interaction effect may reveal outcomes that could not have been predicted had the experiment consisted of two separate one-factor ANOVAs of the same independent variables.

Main effect—Another name for the effect of a single independent variable in ANOVA designs. Each main effect must have two or more levels.

Simple effects—A situation where the interaction is significant and the experimenter is interested in all levels of one independent variable under just one level of the other independent variable.

Chapter 12 Practice Problems

1. Be able to state the advantages and disadvantages of a factorial ANOVA.
2. In a 2 × 2 × 3 factorial ANOVA, be able to state how many factors and how many levels of each factor there are.
3. Understand the terms *main effect, simple effects,* and *interaction.*
4. Be able to state the difference between fixed and random effects in ANOVA.
5. Understand how in situ designs limit the conclusions after an ANOVA.

6. A statistician for a drug company wishes to determine whether older adults differ from younger adults in a complex reaction time experiment involving three levels of a minor tranquilizer (Valium). Fifteen older adults (mean age 65 years, range = 60–75 years) were compared to 15 young adults (mean age 21 years, range = 18–23) on a computer task measuring reaction time. The reaction times varied on the task from 2 seconds (a very quick reaction time or a good score) to over 10 seconds (a very slow reaction time or bad score). The participants were randomly assigned to receive either 0 milligrams (placebo), 5 mg, or 10 mg, and one half hour later, they were tested for their reaction times in total seconds (rounded to whole numbers). Perform a two-factor ANOVA and summarize the results in a source table.

Young			*Older*		
0 mg	*5 mg*	*10 mg*	*0 mg*	*5 mg*	*10 mg*
7	7	13	5	4	18
4	8	10	7	3	16
5	7	10	5	2	16
5	6	9	6	6	20
4	7	10	4	4	14

Chapter 12 Test Problems

1. The simplest factorial ANOVA is the
   a. completely randomized design
   b. split plot design
   c. repeated-measures design
   d. randomized block design

2. The phrase "completely randomized" indicates that
   a. the dependent variable has been completely randomized
   b. the results are completely randomized
   c. the participants have been randomly assigned to the groups
   d. the blocks have been randomized to the degrees of freedom

3. In a 2 × 2 × 3 ANOVA, there are
   a. 2 independent variables
   b. 7 independent variables
   c. 12 independent variables
   d. 3 independent variables

4. In a 2 × 3 × 4 ANOVA, the interaction will compare how many means?
   a. 9
   b. 24
   c. 3
   d. 4

5. In the ______________ effects ANOVA design, the experimenter is interested in generalizing to other levels of the independent variable, and the levels of the independent variable are randomly chosen.
   a. fixed
   b. random
   c. mixed
   d. repeated-measures

Problems 6–15. A psychologist wishes to determine whether short-term memory scores are affected by the type of medication in hyperactive children. Sixty hyperactive children (30 boys and 30 girls) were randomly assigned to receive either Cylert or Ritalin (two kinds of amphetamines that are central nervous system stimulants). Their short-term memory was tested on a 20-point scale (where 0 = *no memory* and 20 = *perfect memory*).

Boys			*Girls*		
Cylert	*Placebo*	*Ritalin*	*Cylert*	*Placebo*	*Ritalin*
12	5	3	5	4	12
10	6	2	6	6	9
13	7	5	2	7	14
10	6	6	4	6	13
9	4	1	3	5	9
12	5	4	2	4	12
13	6	2	4	5	8
10	7	6	6	6	15
11	5	1	2	7	12
12	7	5	4	7	11

6. The sums of squares for gender are
   a. .42
   b. 40.83
   c. 593.43
   d. 141.90

7. The sums of squares for drug levels are
   a. .42
   b. 40.83
   c. 593.43
   d. 141.90

8. The sums of squares for the interaction are
   a. .42
   b. 40.83
   c. 593.43
   d. 141.90

9. The error term sums of squares are
   a. .42
   b. 40.83
   c. 593.43
   d. 141.90
10. The degrees of freedom for gender are
    a. 1
    b. 2
    c. 54
    d. 59
11. The degrees of freedom for drug levels are
    a. 1
    b. 2
    c. 54
    d. 59
12. The degrees of freedom for the interaction are
    a. 1
    b. 2
    c. 54
    d. 59
13. The F value for gender is
    a. 0.16
    b. 7.77
    c. 112.92
    d. 0.42
14. The F value for drug levels is
    a. 0.16
    b. 7.77
    c. 112.92
    d. 0.42
15. The F value for the interaction is
    a. 0.16
    b. 7.77
    c. 112.92
    d. 0.42

Problems 16–25. The element mercury causes massive brain dysfunction, including hallucinations and delusions. A "silver filling" from your dentist is actually 50% to 66% mercury and only 35% silver. The American Dentistry Association (ADA) says that "silver amalgams" are safe. A psychologist wishes to determine whether mercury fillings affect a child's brain functioning. The Coolidge Personality and Neuropsychological Inventory Neuropsychological Dysfunction (ND) scale has 20 questions, such as "My child has learning problems," "My child is hyperactive," and so on. The overall score on the ND scale ranges from 0 (*no evidence of brain dysfunction*) to 20 (*severe brain dysfunction*). The survey is given to parents as they visit a dentist with their child. The number of "silver" fillings is noted, as is the child's gender. For analysis purposes, the children are divided up into whether they have no fillings, one to three fillings, or more than five fillings, and their ND score is noted.

No Fillings		*One to Three Fillings*		*Five or More Fillings*	
Male	*Female*	*Male*	*Female*	*Male*	*Female*
7	4	11	6	17	13
4	6	12	9	20	15
6	7	10	9	15	11
7	8	9	9	17	12
8	4	13	8	16	10
5	3	14	5	14	12
9	7	15	7	18	14
5	9	10	5	20	14

16. The sums of squares for gender are
    a. 615.13
    b. 935.31
    c. 157.63
    d. 117.19

17. The sums of squares for fillings are
    a. 615.13
    b. 935.31
    c. 157.63
    d. 117.19

18. The sums of squares for the interaction are
    a. 45.38
    b. 615.13
    c. 935.31
    d. 157.63

19. The error term sums of squares are
    a. 157.63
    b. 45.38
    c. 615.13
    d. 935.31

20. The degrees of freedom for gender are
    a. 1
    b. 2
    c. 42
    c. 47

21. The degrees of freedom for fillings are
    a. 1
    b. 2
    c. 42
    d. 47
22. The degrees of freedom for the interaction are
    a. 1
    b. 2
    c. 42
    d. 47
23. The *F* value for gender is
    a. 6.045
    b. 31.225
    c. 81.952
    d. 1.000
24. The *F* value for fillings is
    a. 6.045
    b. 31.225
    c. 81.952
    d. 1.000
25. The *F* value for the interaction is
    a. 6.045
    b. 31.225
    c. 81.952
    d. 1.000

13 Post Hoc Analysis of Factorial ANOVA

Chapter 13 Goals

- Learn the post hoc analysis of the main effects in factorial ANOVA
- Learn the post hoc analysis of an interaction in factorial ANOVA
- Learn how to conduct effect size analyses in factorial ANOVA

In this chapter, you will learn how to interpret the main effects and interaction in a factorial ANOVA. We will continue with the results from the example in Chapter 12.

Main Effect Interpretation: Gender

The null hypothesis for the gender factor is

$$H_0 : \mu_1 = \mu_2$$
$$H_a : \mu_1 \neq \mu_2$$

where

μ_1 = the mean number of inattention incidents for all the males in the three age conditions,

μ_2 = the mean number of inattention incidents for all the females in the three age conditions.

First, we will review how to check for the significance of the F value for the first main effect, gender. Checking for significance requires two types of degrees of freedom. The df_1 or df for the numerator for the gender main effect refers to the number of levels of the gender main effect minus 1. This value was obtained in Step 11 (df_{SSG}): Since there are two levels of gender, $2 - 1 = 1$, thus $df_1 = 1$. The df_2 or df for the denominator refers to the df_{SSW}, also obtained in Step 11: $df_{SSW} = df_{SST} - df_{SSA} - df_{SS_{G\times A}}$, or 42. Note that in a completely randomized two-factor ANOVA, df_2 (or denominator df) stays the same for both main effects and the interaction. The critical value at $p = .05$ with $df_1 = 1$ and $df_2 = 42$ is $F = 4.07$. A complete informal write-up might look like this (but it would not be published in this form):

> The derived $F = 6.86$ for the gender main effect did exceed the tabled critical value $F = 4.07$ at $p = .05$ with $df_1 = 1$ and $df_2 = 42$. Therefore, H_0 is rejected, and it is concluded that the mean number of inattention incidents for the males ($M = 6.75$, $SD = 5.19$) was significantly greater than the mean number of inattention incidents for the females ($M = 4.96$, $SD = 2.60$), $F(1, 42) = 6.86$, $p < .05$. In terms of the research question, it appears that males are more inattentive than females.

Note that the next to last sentence in the previous paragraph contained the line "$F(1, 42) = 6.86$, $p < .05$." The F value would have been reported to the lowest significance level possible in that line. However, the critical value for gender's F value at $p = .01$ was 7.27, and the obtained F value for gender did not exceed this value, so it could not be reported as significant at that level.

Why a Multiple Comparison Test Is Unnecessary for a Two-Level Main Effect, and When Is a Multiple Comparison Test Necessary?

A multiple comparison test is unnecessary for the main effect for gender because it has only two levels. If a main effect has only two levels, and it produces a significant F, then its two means are automatically significantly different from each other. However, when there are more than two levels of a significant main effect, then a multiple comparison test is necessary to interpret the pattern of mean differences. However, a multiple comparison test would not be necessary for a significant main effect with more than two levels if the interaction between the two main effects was significant. Of course, if a main effect was not significant, no matter how many levels, then a multiple comparison test should not be conducted.

Main Effect: Age Levels

The null and alternative hypotheses for the age level factor are as follows:

$$H_0: \mu_1 = \mu_2 = \mu_3$$

H_a: The null hypothesis is not true

where

μ_1 = the mean number of inattention incidents for all subjects in the young age conditions,

μ_2 = the mean number of inattention incidents for all subjects in the middle age conditions,

μ_3 = the mean number of inattention incidents for all subjects in the old age conditions.

To check for significance of the F for age levels also requires two types of degrees of freedom. The df_1 or df for the numerator for the main effect for age levels refers to the number of levels of the age minus 1 (in other words, $3 - 1 = 2$). This value was obtained in Step 11 (df_{SSA}). The df_2 or df for the denominator refers to the df for SSW, also obtained in Step 11 (df_{SSW}). The informal write-up might look like this:

> The derived $F = 29.32$ for the age main effect exceeds the critical $F = 3.22$ at $p = .05$ with $df_1 = 2$ and $df_2 = 42$. Therefore, H_0 is rejected, and it is concluded that at least one inattention incident mean for the three age levels is significantly different from the mean for one other age level, $F(2, 42) = 29.32$, $p < .01$. To determine the pattern of mean differences, a multiple comparison test is necessary. In terms of the research question, it appears that the number of inattention incidents varies significantly among the age levels.

Multiple Comparison Test for the Main Effect for Age

As noted in the previous chapter, when the interaction between two main effects is significant, a multiple comparison is unnecessary for any significant main effect with three or more levels. The presence of a significant interaction indicates that something has occurred that could not have been predicted by the interpretation of the two main effects. However, the post hoc analysis of a significant main effect with three levels will be presented here for future analyses where there is not a significant interaction between the two main effects.

The means for the number of inattention incidents for the three age levels are needed. The sums for the three age levels were obtained in Step 6. Each mean was based on 16 scores (8 males and 8 females), so the means are easily derived from the values in Step 6. As presented in Chapter 10, compare each mean to each other and take the absolute difference between the means for the three comparisons.

Comparison		*Means*		*Absolute Value of Difference*
$\bar{X}_{\text{young}} - \bar{X}_{\text{middle}}$	=	9.44−4.88	=	4.56
$\bar{X}_{\text{young}} - \bar{X}_{\text{older}}$	=	9.44−3.25	=	6.19
$\bar{X}_{\text{middle}} - \bar{X}_{\text{older}}$	=	4.88−3.25	=	1.63

Next, determine Tukey's HSD critical value according to the following formula:

$$\text{HSD} = q\sqrt{\frac{\text{MSW}}{N}}$$

First, obtain the value of q from Appendix F (Tukey's HSD q table of critical values). To enter this table, we will be required to know three values:

1. α level or p level = .05 or .01 (we will use .05). Note that you have a choice of conducting Tukey's test at either .05 or .01, whichever gives you a better understanding of the data and regardless of whether the F was significant at .05 or .01.
2. r = the number of different means that will be compared (hint: look at the null hypothesis and count the means). r is also the number of levels in a main effect.
3. df_{MSW} or df_{error} = the degrees of freedom associated with the error term of the significant F value.

Our Tukey's test will be conducted at p = .05. In the age main effect, there are three levels, so $r = 3$, and $df_{\text{error}} = 42$ (df_{SSW} or df_{MSW}). Thus, the q value is 3.44. Note that Tukey's table did not list an error df at 42, so we use the df value at 40. Again, the reasoning is that we have 42 df, so do we have at least 40 df. Yes! Do we have at least 60 df? No!

Next, determine the value of MSW (MS error or within groups). This value is also called the MS error term. The value of MSW was tabled in Step 14. In the present example, MSW = 5.616.

Finally, N = number of scores on which each mean was based (if Ns are not equal, use the harmonic mean described in Chapter 10). Each age level was based on 16 scores.

Thus,

$$\text{HSD} = 3.44\sqrt{\frac{5.616}{16}}$$

$$\text{HSD} = 2.04$$

The obtained HSD value is a critical value. The differences between the means must exceed this critical value to be significantly different at $p < .05$.

Thus, to be significantly different, the absolute value of a mean pair difference must exceed 2.04.

Now, attach an asterisk to those mean pairs that exceed the critical HSD value.

Comparison		*Means*		*Absolute Value of Difference*
$\bar{X}_{young} - \bar{X}_{middle}$	=	9.44 − 4.88	=	4.56*
$\bar{X}_{young} - \bar{X}_{older}$	=	9.44 − 3.25	=	6.19*
$\bar{X}_{middle} - \bar{X}_{older}$	=	4.88 − 3.25	=	1.63

*$p < .05$.

Now, we can finally interpret the pattern of mean differences. The mean number of inattention incidents for the younger group (9.44) was significantly greater than the mean number of inattention incidents for the middle age group (4.88). The mean number of inattention incidents for the younger group (9.44) was also significantly greater than the mean number of inattention incidents for the older age group (3.25). However, the absolute value of the mean differences between the middle and old groups did not exceed the Tukey's HSD value; therefore, these two means are not significantly different from each other. This finding is conceptually important because it indicates that although the two means were *mathematically* different (4.88 vs. 3.25), we have found that statistically, the difference is too small to be considered a reliable difference. Thus, we will conclude that young people are significantly more inattentive than the two older groups. However, people in the middle age group are not more inattentive than older people or vice versa.

Warning: Limit Your Main Effect Conclusions When the Interaction Is Significant

If the interaction is significant in a two-factor ANOVA, there was some unique effect of the two variables together on the dependent variable that *could not have been predicted* from either of the two main effect conclusions independently. Thus, if the interaction is significant, limit your conclusions severely. In fact, some statisticians recommend making *no conclusions at all* about the main effects if the interaction is significant.

As noted in the previous chapter, the write-up to this point in a published journal article might look like this:

> The main effect for gender was significant, $F(1, 42) = 6.86$, $p < .05$, as was the main effect for age, $F(2, 42) = 29.32$, $p < .01$. The interaction was also significant, $F(2, 42) = 18.72$, $p < .01$. A post hoc analysis of the interaction revealed. . . .

Multiple Comparison Tests

Most multiple comparison tests may be performed at the $p < .05$ or $p < .01$ level, regardless of whether the main effects or interaction were significant at $p < .05$ or $p < .01$. Your guiding principle in conducting multiple comparison tests should be which p level explains your data the best. Thus, it would be a good idea to look at the pattern of mean differences at $p = .05$ and $p = .01$ to see which explains your data the best.

Interpretation of Interaction Effect

The null hypothesis for the interaction between the main effects of gender and age levels is

> H_0: There is no interaction between the main effects of gender and age on the number of inattention incidents.
>
> H_a: There is an interaction between the main effects of gender and age on the number of inattention incidents.

The df_1 or df for the numerator for the interaction is the df for gender main effect times the df_1 for age main effect (in other words, $1 \times 2 = 2$). This value was also obtained in Step 11 ($df_{SS_{G \times A}}$). The df_2 or df for the denominator refers to the df for SSW, also obtained in Step 11 (df_{SSW}). The informal write-up might look like this:

> The derived $F = 18.72$ for the interaction effect exceeds the critical $F = 3.22$ at $p = .05$ with $df_1 = 2$ and $df_2 = 42$. Therefore, H_0 is rejected, and it is concluded that there is a significant interaction between gender and age on the number of inattention incidents, $F(2, 42) = 18.72$, $p < .01$. To determine the pattern of mean differences, a multiple comparison test must be performed. In terms of the research question, there is a significant interaction between gender and age on the number of inattention incidents.

Now, Tukey's post hoc test must be performed on the six means in the interaction (in a 2×3 ANOVA, there will be 6 means in the interaction, and in a 3×4 ANOVA, there will be 12 means in the interaction). The means for the number of inattention incidents for each gender at all three age levels are needed. The six sums were obtained in Step 8, and each mean was based on eight scores. As presented in Chapter 10, compare each mean to each other and take the absolute difference between the means for the three comparisons. Note that it will be conceptually easier if the mean comparisons are grouped like simple main effects; that is, all levels of one main effect (e.g., age) will be compared against only one level of the other main effect (e.g., gender). So, let's make all the male mean comparisons at all three age levels first, then compare all the female means at the three age levels. Finally, we'll compare male versus female means at each age level.

Comparison		*Means*		*Absolute Value of Difference*
For the males				
$\bar{X}_{m\text{-young}} - \bar{X}_{m\text{-middle}}$	=	13.25−4.75	=	8.50
$\bar{X}_{m\text{-young}} - \bar{X}_{m\text{-older}}$	=	13.25−2.25	=	11.00
$\bar{X}_{m\text{-middle}} - \bar{X}_{m\text{-older}}$	=	4.75−2.25	=	2.50
For the females				
$\bar{X}_{f\text{-young}} - \bar{X}_{f\text{-middle}}$	=	5.63−5.00	=	0.63
$\bar{X}_{f\text{-young}} - \bar{X}_{f\text{-older}}$	=	5.63−4.25	=	1.38
$\bar{X}_{f\text{-middle}} - \bar{X}_{f\text{-older}}$	=	5.00−4.25	=	0.75
Males vs. females at each age level				
$\bar{X}_{m\text{-young}} - \bar{X}_{f\text{-young}}$	=	13.25−5.63	=	7.62
$\bar{X}_{m\text{-middle}} - \bar{X}_{f\text{-middle}}$	=	4.75−5.00	=	0.25
$\bar{X}_{m\text{-older}} - \bar{X}_{f\text{-older}}$	=	2.25−4.25	=	2.00

Next, determine Tukey's HSD critical value according to the following formula:

$$\text{HSD} = q\sqrt{\frac{\text{MSW}}{N}}$$

First, obtain the value of q from Appendix F (Tukey's HSD q table of critical values). To enter this table, we will be required to know three values:

1. α level or p level = .05 or .01 (we will use .05).
2. r = the number of unique means in the interaction. In a 2 × 3 ANOVA, there will be 2 times 3 unique means, so $r = 6$ in this example.
3. df_{MSW} or df_{error} = the degrees of freedom associated with the error term of the significant F value.

Tukey's test will be conducted at $p = .05$. Thus, the q value is 4.23. Note that Tukey's table did not list an error df at 42, so we used the df value at 40. Again, the reasoning is that we have 42 df, so do we have at least 40 df? Yes! Do we have at least 60 df? No!

Next, determine the value of MSW (also called MS error or MS within groups). The value of MSW was tabled in Step 14. In the present example, MSW = 5.616.

Finally, N = number of scores on which each mean was based (if Ns are not equal, use the harmonic mean described in Chapter 10). Each age level was based on eight scores.

Thus,

$$\text{HSD} = 4.23\sqrt{\frac{5.616}{8}}$$

$$\text{HSD} = 3.54$$

The obtained HSD value is a critical value. The differences between the means must exceed this critical value to be significantly different at $p < .05$. Thus, to be significantly different, the absolute value of a mean pair difference must exceed 3.54. If the absolute value of the mean differences does not exceed the critical value, then the means will not be considered significantly different from each other.

Now, attach an asterisk to those mean pairs exceeding the critical HSD value to indicate they are significantly different at $p < .05$.

Comparison		*Means*		*Absolute Value of Difference*
For the males				
$\bar{X}_{\text{m-young}} - \bar{X}_{\text{m-middle}}$	=	13.25−4.75	=	8.50*
$\bar{X}_{\text{m-young}} - \bar{X}_{\text{m-older}}$	=	13.25−2.25	=	11.00*
$\bar{X}_{\text{m-middle}} - \bar{X}_{\text{m-older}}$	=	4.75−2.25	=	2.50
For the females				
$\bar{X}_{\text{f-young}} - \bar{X}_{\text{f-middle}}$	=	5.63−5.00	=	0.63
$\bar{X}_{\text{f-young}} - \bar{X}_{\text{f-older}}$	=	5.63−4.25	=	1.38
$\bar{X}_{\text{f-middle}} - \bar{X}_{\text{f-older}}$	=	5.00−4.25	=	0.75
Males vs. females at each age level				
$\bar{X}_{\text{m-young}} - \bar{X}_{\text{f-young}}$	=	13.25−5.63	=	7.62*
$\bar{X}_{\text{m-middle}} - \bar{X}_{\text{f-middle}}$	=	4.75−5.00	=	0.25
$\bar{X}_{\text{m-older}} - \bar{X}_{\text{f-older}}$	=	2.25−4.25	=	2.00

*$p < .05$.

At last, we can interpret the pattern of mean differences in the interaction. There is no simple statement that can be made, so let's discuss the pattern of the results in terms of its simple main effects (i.e., one level of one independent variable under all levels of the other independent variable). The first two analyses will be the simple main effects for one level of gender at all three age levels, and the last analysis will be the simple main effects for each age level for both genders.

For the Males

For the males in this study, the mean number of inattention incidents for the young males (13.25) was significantly higher than the mean number of inattention incidents for the middle age males (4.75) and the older males' mean (2.25). There was no significant difference between the latter two groups' means.

For the Females

For the females in this study, there were no significant differences between the mean number of inattention incidents for the young group (5.63), middle age group (5.00), or the old age group (4.25).

Males Versus Females

When comparing males versus females, the younger males had a significantly higher mean number of inattention incidents (13.25) than the younger females (5.63). The middle age male and female groups' mean number of inattention incidents were not significantly different (4.75 and 5.00, respectively), nor were the older age male and female groups' means (2.25 and 4.25, respectively).

Final Summary

Now we can summarize the overall results. In terms of the research question, it appears that younger males are significantly more inattentive than any other age males. Inattention, however, does not appear to continue to decline as men age, as there was no difference in inattention between the middle and older male groups. Females' inattention appears stable across the life span. Finally, there are no significant differences in inattention incidents between middle age and older age males and females. In summary, young males are the most inattentive. Females (in all age groups) and middle age and older males all have similar levels of inattention.

Writing Up the Results Journal Style

Journal article versions of a two-factor ANOVA may actually be rather brief. For our previous example, it might look like this:

> The main effect for gender, $F(1, 42) = 6.86$, $p < .05$; the main effect for age level, $F(2, 42) = 29.32$, $p < .01$; and the interaction, $F(2, 42) = 18.72$, $p < .01$, were all significant. Tukey's post hoc test of the interaction revealed that young males are the most inattentive. Females and middle age and older males all have similar rates but lower inattention rates than the young males.

Once again, the most important feature of a two-factor ANOVA is the interaction. A significant interaction means that the results could not have been predicted by interpreting the main effects alone. In fact, if a significant interaction is present, the interpretation of the two main effects could be *misleading*. If there is no significant interaction, then the main effects can be interpreted without any problems.

Language to Avoid

Statisticians are surprisingly touchy about certain words and phrases. Some people use the phrase **"the null hypothesis was accepted,"** which seems to send most statisticians into a tizzy (a tizzy looks like a huff, only less serious). "The null hypothesis was accepted" has the implication that the null hypothesis is true or that we are endorsing its veracity when this is not really the case. We may not reject the null hypothesis for a variety of reasons *not* related to its truth or falsity. For example, the experiment may simply lack power (too few subjects), the independent variable may have too weak an effect, or the dependent variable may be inappropriate or insensitive. Some statisticians do not even like the phrase "retain or reject the null hypothesis." They feel that even the word *retain* has the connotation of endorsing the truth of the null hypothesis. These statisticians prefer the phrase "reject or fail to reject the null hypothesis."

Another phrase that annoys some statisticians is **"highly significant."** They argue that the null hypothesis is either rejected or not at the .05 level. Because the conventional level of significance is $p = .05$, the null hypothesis is rejected if it exceeds the critical value at this p level. They argue that we do not reject the null hypothesis at $p < .01$ or $p < .001$. It has already been rejected at $p = .05$. So, for example, when we report a finding significant at $p < .001$, it is not "highly significant." It is simply significant (the null hypothesis has been rejected at $p = .05$), and the probability of the Type I error is very low (i.e., less than one chance in 1000 for $p < .001$). The phrase "highly significant" may also confuse the difference between the use of the word *significant* as a value judgment (whether something is worthwhile or valuable) or as a statistical decision (findings are significant if the null hypothesis has been rejected). The phrase "highly significant" has the connotation of a value judgment combined with a statistical decision, which would be inappropriate. It would be best if we learn to report "a significant finding with a low probability of the Type I error."

Exploring the Possible Outcomes in a Two-Factor ANOVA

Because there are two main effects and an interaction effect in a two-factor ANOVA, and there are only two possible hypothesis testing outcomes (significant or not), then there are $2 \times 2 \times 2 = 8$ total possible outcomes. See Figure 13.1 for a summary of these outcomes. Note that the interaction is significant in every case where the lines for the main effects are not parallel.

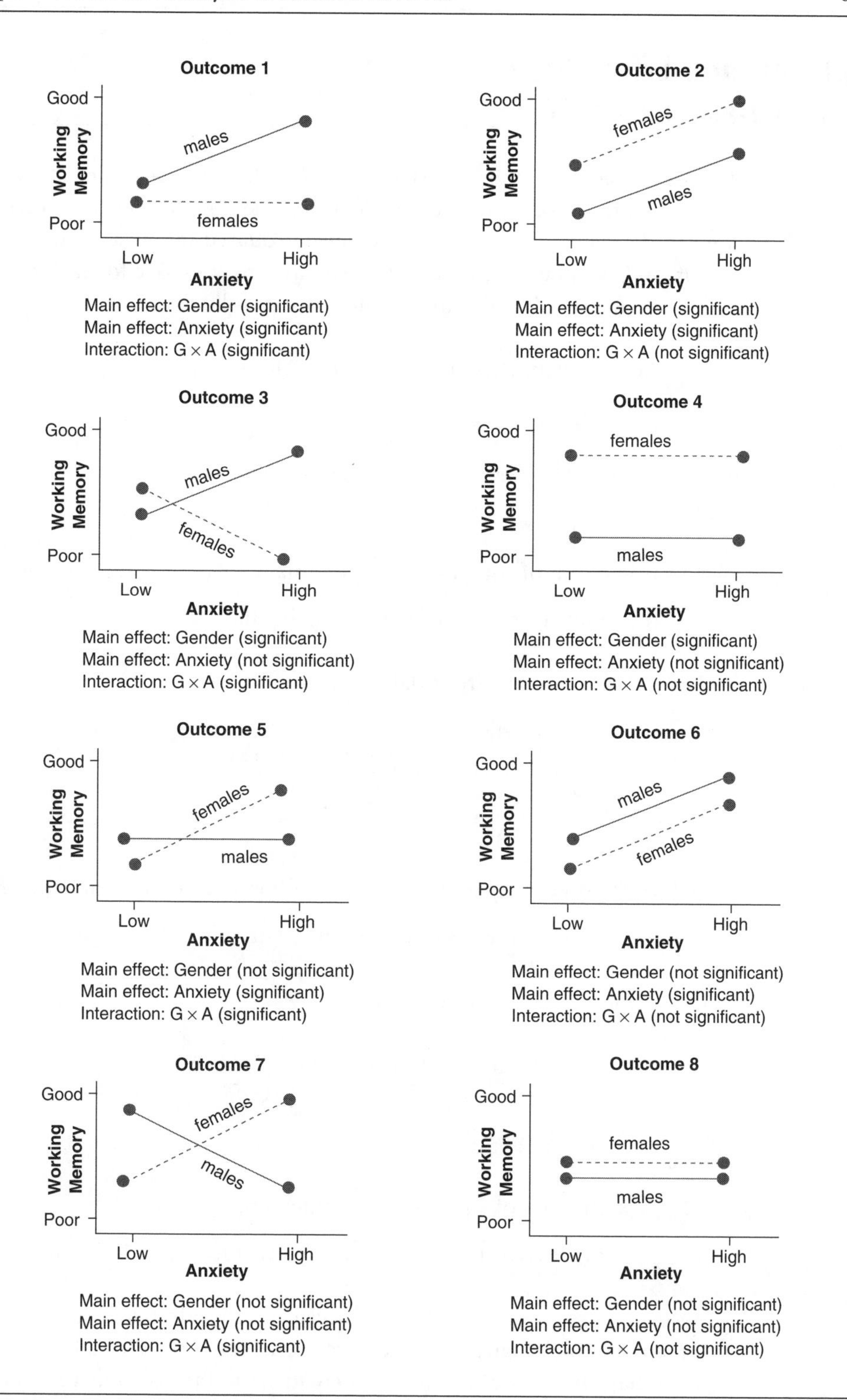

Figure 13.1 Eight Possible Main Effects and Interaction Outcomes of a 2 × 2 ANOVA

NOTE: In this hypothetical example of a two-factor ANOVA, gender (males vs. females) and anxiety level (low or high) were the main effects. The dependent variable was working memory (short-term memory capacity). The solid line represents the males, and the dotted line represents the females.

Determining Effect Size in a Two-Factor ANOVA

Because a significant or nonsignificant F value still does not reveal the strength of the effect of the independent variables on the dependent variable, we can determine the effect size by **omega squared** (ω^2) just as in the one-factor ANOVA, but it requires three separate analyses, one for each F value. For a two-factor ANOVA, the formulas are as follows:

For the first main effect (labeled factor A):

$$\omega_A^2 = \frac{\text{SSA} - (k - 1)(\text{MSW})}{\text{SST} + \text{MSW}}$$

where

SSA = the sum of squares for the first main effect or main effect A,

k = the number of levels of the first main effect.

For the second main effect (labeled factor B):

$$\omega_B^2 = \frac{\text{SSB} - (k - 1)(\text{MSW})}{\text{SST} + \text{MSW}}$$

where

SSB = the sum of squares for the second main effect or main effect B,

k = the number of levels of the second main effect.

For the interaction (labeled $A \times B$):

$$\omega_{A\times B}^2 = \frac{\text{SS}_{A\times B} - (df_{A\times B})(\text{MSW})}{\text{SST} + \text{MSW}}$$

where

$\text{SS}_{A \times B}$ = the sum of squares for the interaction,

$df_{A \times B}$ = degrees of freedom for the interaction.

In most cases, we will only be concerned with effect size determination when an F value is significant. However, when the power of an ANOVA is low (perhaps because of too few participants), and the obtained F value did not reach significance but was close (often called a trend), you may be interested in determining effect size to see if the independent variable did have an impact on the dependent variable despite the lack of significance.

In the previous computational example, for the gender main effect:

$$\omega_A^2 = \frac{38.521 - (2 - 1)(5.616)}{813.979 + 5.616} = .04$$

In the previous computational example, for the age level main effect:

$$\omega_B^2 = \frac{329.292 - (3 - 1)(5.616)}{813.979 + 5.616} = .39$$

In the previous computational example, for the interaction:

$$\omega_{A \times B}^2 = \frac{210.291 - (2)(5.616)}{813.979 + 5.616} = .24$$

According to omega-squared interpretation guidelines,

$$\omega^2 > .15 = \text{large effect}$$
$$\omega^2 > .06 = \text{medium effect}$$
$$\omega^2 > .01 = \text{small effect}$$

Thus, we can see that for the gender main effect, there was between a small and medium effect size. The main effect for age level had a very large effect size, with $\omega^2 = .39$. Remember that ω^2 may also be interpreted as the percentage of variance in the dependent variable accounted for by the independent variable. ω^2 is a measure of how much variation is accounted for by the treatment and, by default, how much of the total variation in scores is accounted for by random factors. For the age level main effect, 39% of the total variance in the scores can be accounted for by the age effect, and 61% of the total variance is accounted for by unknown or random factors. For the interaction, $\omega^2 = .24$, which is indicative of a large effect size, and thus, the interaction between the two main effects accounts for about 24% of the variance in the dependent variable.

History Trivia

Fisher and Smoking

In 1958, at the age of 68, Ronald Fisher, as noted earlier, published three papers in prestigious journals questioning the relationship between cancer and smoking. As a long-term pipe smoker, he was adamant that studies of the relationship between lung cancer and smoking were correlational in nature and could not imply a causal relationship. According to statistics historian David Salsburg (2001), Fisher even accused some leading researchers of making up their data. In 1959, statisticians and cancer experts wrote a 30-page paper that reviewed all the studies available and Fisher's objections in addition to objections raised by the tobacco companies. Their conclusions were that the evidence was overwhelmingly in favor of a causative relationship between smoking and lung cancer. It is ironic that Fisher took such a stand in his career, when earlier he emphasized so strongly the need for replication and a sequence of experiments. However, it was consistent with his strong stand that nothing could ever be proven without a randomized, controlled experiment.

In 1962, he was invited to India to attend the 30th anniversary of the founding of the Indian Statistical Institute. Jerzy Neyman, Egon Pearson, and other famous statisticians from all over the world came for the conference, and it is reported that the attendees were excited to see the science of statistics influencing virtually every discipline. Ronald Fisher died of a heart attack at the age of 72 on a boat when he was returning from the conference. Despite his last faux pas with regard to smoking, his articles and books have probably had the single most profound influence on the modern practice of statistics.

Key Terms, Symbols, and Definitions

Highly significant—A term to be avoided because (a) it has the connotation of a value judgment (i.e., that the results are important and valuable), and (b) it suggests that the null hypothesis was rejected at a lower *p* level than .05. Most statisticians argue that the null hypothesis is rejected at $p = .05$ or less, and the *p* level is simply reported to the lowest *p* level possible.

Omega squared (ω^2)—The percentage of variance in the dependent variable accounted for by the independent variable.

The null hypothesis was accepted—A phrase to be avoided because most statisticians do not like the implication that the null hypothesis has been endorsed in any way when one simply fails to reject the null hypothesis. Indeed, the null hypothesis could be correct, but the experiment may lack power (too few subjects), the independent variable may have too weak an effect, or the dependent variable may be inappropriate or insensitive to the independent variable.

Chapter 13 Practice Problems

1. When is a multiple comparison test necessary for a significant main effect?
2. When is a multiple comparison test necessary for a significant interaction?
3. Can you name the conditions in factorial ANOVA when a multiple comparison test is not required or appropriate?
4. For some statisticians, why is the phrase "highly significant" annoying?
5. Why do some statisticians object to the phrase "retain the null hypothesis"?
6. Perform a post hoc analysis for problem 6 in Chapter 12.

Chapter 13 Test Questions

1. When checking for the significance of a main effect, the degrees of freedom for the numerator are the
   a. number of levels of the main effect
   b. number of levels of the main effect minus 1
   c. number of levels of the main effect times 2
   d. degrees of freedom for the denominator minus 1
2. When checking for the significance of a main effect, the degrees of freedom for the denominator are equal to the
   a. error degrees of freedom (*df* for SSW)
   b. error degrees of freedom minus 1
   c. total sum of squares' degrees of freedom
   d. number of levels of the main effect minus 1
3. A significant main effect always requires a post hoc analysis
   a. if there are more than two levels
   b. if there are only two levels
   c. even if there is only one level of the main effect
   d. when the interaction is not significant
4. In the calculation of Tukey's HSD value, the calculation can be made at either $p < .05$ or .01
   a. but those values are chosen based on the significance of the F value
   b. depending on which level explains your data better
   c. depending on the effect size analysis
   d. or even $p < .001$
5. With a df_{error} of 16, $r = 5$, and $\alpha = .05$, Tukey's q value would be
   a. 5.49
   b. 4.33
   c. 4.05
   d. 5.19
6. Which of the following is not true:
   a. post hoc analyses of main effects are minimized or sometimes not conducted at all if the interaction is significant
   b. a post hoc analysis is not required of a significant main effect whose $df = 1$
   c. a post hoc analysis is not required of a significant main effect whose $df = 2$
   d. a post hoc analysis is not required of a nonsignificant interaction
7. In a 3×5 factorial ANOVA, there will be ____ means in the interaction.
   a. 4
   b. 8
   c. 15
   d. 2
8. Which of the following statements would probably have the fewest objections from statisticians?
   a. highly significant
   b. retain the null hypothesis

c. proved the alternative hypothesis
d. fail to reject the null hypothesis

9. Because there are only two possible hypothesis testing outcomes (significant or not) and there are two main effects and an interaction in a 2 × 2 factorial ANOVA, then there are ____ possible outcomes.

a. 2
b. 4
c. 6
d. 8

10. In the graphic plot of an interaction in factorial ANOVA, parallel lines indicate

a. a significant interaction
b. a nonsignificant interaction
c. two significant main effects and a significant interaction
d. two nonsignificant main effects and a nonsignificant interaction

Problems 11–19 (same as Problems 6–15 in Chapter 12). A psychologist wishes to determine whether short-term memory scores are affected by the type of medication in hyperactive children. Sixty hyperactive children (30 boys and 30 girls) were randomly assigned to receive either Cylert or Ritalin (two kinds of amphetamines that are central nervous system stimulants). Their short-term memory was tested on a 20-point scale (where 0 = *no memory* and 20 = *perfect memory*).

Boys			*Girls*		
Cylert	*Placebo*	*Ritalin*	*Cylert*	*Placebo*	*Ritalin*
12	5	3	5	4	12
10	6	2	6	6	9
13	7	5	2	7	14
10	6	6	4	6	13
9	4	1	3	5	9
12	5	4	2	4	12
13	6	2	4	5	8
10	7	6	6	6	15
11	5	1	2	7	12
12	7	5	4	7	11

11. The q value at $p = .05$ for Tukey's HSD test for the main effect for drug levels is

a. 3.44
b. 4.37
c. 3.79
d. 4.70

12. Tukey's HSD value at $p = .05$ for the main effect for drug levels is

a. 3.44
b. 2.628
c. 20
d. 1.25

13. The results of Tukey's test for the main effect of drug levels reveal that
    a. short-term memory is not different between Cylert and Ritalin conditions, but both are better than the placebo
    b. short-term memory is better for Cylert than the placebo or Ritalin
    c. short-term memory for Ritalin is better than for Cylert or the placebo
    d. short-term memory is poorer for Cylert and Ritalin than for the placebo
14. The q value at $p = .05$ for Tukey's HSD for the interaction between gender and drug levels is
    a. 3.44
    b. 4.23
    c. 5.11
    d. 5.24
15. Tukey's HSD value for the interaction between gender and drug levels (at $p = .05$) is
    a. 3.44
    b. 4.23
    c. 2.17
    d. 1.17
16. For the boys, the results of Tukey's test in the interaction revealed
    a. Ritalin worked best
    b. Cylert worked best
    c. the placebo worked best
    d. none of the above
17. For the girls, the results of Tukey's test for in the interaction revealed
    a. Ritalin worked best
    b. Cylert worked best
    c. The placebo worked best
    d. none of the above
18. Overall, for the interaction, which of the following is true?
    a. boys' short-term memory is better on Ritalin
    b. girls' short-term memory is better on Cylert
    c. boys' short-term memory is better on Cylert
    d. none of the above
19. When comparing boys and girls in the interaction, which of the following is true?
    a. boys do better on Cylert and girls do better on Ritalin
    b. boys do better on Ritalin and girls do better on Cylert
    c. there is a significant difference in short-term memory between girls on Cylert and boys on Ritalin
    d. none of the above
20. In an effect size analysis in a two-factor ANOVA for a main effect, k is equal to
    a. the number of levels of the main effect
    b. the number of levels times 2
    c. the number of levels minus 2
    d. the number of levels for the interaction

14 Factorial Analysis of Variance

Additional Designs

Chapter 14 Goals

- Learn how to analyze a two-factor split-plot ANOVA design
- Learn how to analyze a two-factor ANOVA where both factors are repeated measures

The Split-Plot Design

The previous chapter introduced the completely randomized factorial ANOVA. The present chapter will present two additional two-factor ANOVA designs: the **split-plot ANOVA** and the **repeated-measures ANOVA.** Chapter 12 introduced the two-factor completely randomized ANOVA where both of the independent variables were considered to be between-groups or **between-subjects variables** (i.e., different participants were assigned to each level of each factor). The split-plot design is a combination of a completely randomized design and a repeated-measures design. It requires one between-groups variable (different participants in each level of the independent variable) and one **within-subjects variable** (the same participant serves in each level of the other independent variable). Frequently, the within-subjects variable consists of taking measurements of participants across time or over successive trials. Interestingly, the split-plot design takes its name from the historical roots of statistics in agriculture. Remember that Gosset and Fisher both worked on statistics related to the growing of crops (in Gosset's case, wheat, barley, and malt used in the brewing of beer). The notion of splitting a plot of land and trying different growing techniques (e.g., fertilizers, watering cycles, etc.) on each plot led to the statistical designs being labeled *split-plot.*

A two-factor repeated-measures ANOVA has two factors with two or more levels of each factor, and the same group of participants serves in every level of both factors. The design has the advantage of using fewer participants than the two-factor ANOVA, without a major loss of power. However, there are many experimental designs where the repeated-measures design is unfeasible. The two-factor repeated-measures design will be discussed in greater detail later in this chapter.

Overview of the Split-Plot ANOVA

The split-plot ANOVA has three F values for the two main effects and the interaction just as in the two-factor completely randomized ANOVA. The difference between the two designs lies in the error term. In the two-factor completely randomized ANOVA, there was only one error term for both main effects and the interaction. The denominator in the F ratio was the same for all three effects. In the two-factor split-plot ANOVA, we will derive two different error terms: one error term for the between-subjects variable and the other error term for the within-subjects variable and the interaction. Otherwise, the assumptions are the same for both designs as are the null and alternative hypotheses, significance testing, and the interpretation.

Computational Example

A social facilitation experiment was conducted where there were two levels of a social facilitation factor: a participant working alone in a room and a participant working with another participant (a confederate) present in a room. The task required the participant to judge different weights and order them according to magnitude. The other person present in the room only watched and did not speak or participate. Each participant served for three trials, and three different ranges of weights were randomized across participants, so that each participant received a different weight range on each trial. Six participants were randomly assigned to each of the two social facilitation conditions for a total of 12 participants. Time and number of errors were combined in a special manner for one overall score per trial for each participant, and lower scores are better scores. The data are tabled as follows:

	Participants Alone				*Participants Observed*		
	Trials				*Trials*		
	1	*2*	*3*		*1*	*2*	*3*
P_1	9	7	7	P_7	6	7	5
P_2	8	9	7	P_8	8	6	5
P_3	10	6	8	P_9	5	5	4
P_4	7	8	8	P_{10}	6	6	5
P_5	9	8	8	P_{11}	7	5	4
P_6	8	9	8	P_{12}	6	6	4

Step 1. Obtain $\Sigma\ x$ and $\Sigma\ x^2$ for each treatment combination and the overall $\Sigma\ x$ and $\Sigma\ x^2$.

	Alone			*Observed*			
	Trial 1	*Trial 2*	*Trial 3*	*Trial 1*	*Trial 2*	*Trial 3*	*Overall*
$\Sigma\ x$	51	47	46	38	35	27	244
$\Sigma\ x^2$	439	375	354	246	207	123	1744

Step 2. The correction for the mean (CM) is obtained by squaring the overall $\Sigma\ x$ and dividing by the total number of observations.

$$\text{CM} = \frac{(\text{overall } \sum x)^2}{N}$$

$$\text{CM} = \frac{(244)^2}{36} = 1653.78$$

Step 3. The sum of squares total (SST) is obtained by taking the overall $\Sigma\ x^2$ and subtracting the CM.

$$\text{SST} = \text{overall } \sum x^2 - \text{CM}$$

$$\text{SST} = 1744 - 1653.78$$

$$\text{SST} = 90.22$$

Step 4. To calculate the between-subjects sum of squares (SSB), add the three trial scores together for each participant.

Alone		*Observed*	
P_1	23	P_7	18
P_2	24	P_8	19
P_3	24	P_9	14
P_4	23	P_{10}	17
P_5	25	P_{11}	16
P_6	25	P_{12}	16

Next, square each of these sums and add them together.

$$23^2 + 24^2 + 24^2 + 23^2 + 25^2 + 25^2 + 18^2 + 19^2 + 14^2 + 17^2 + 16^2 + 16^2 = 5142$$

Divide this obtained value by the number of trials summed across and subtract the CM to obtain SSB.

$$SSB = \frac{5142}{3} - CM$$
$$SSB = 1714 - 1653.78$$
$$SSB = 60.22$$

Step 5. To calculate the sum of squares for the social facilitation factor (SSF), obtain $\Sigma\, x$ for each level of the social facilitation factor, square each and divide each by the number of scores within each group, and add the quotients.

Sum of participants alone: $51 + 47 + 46 = 144$

Sum of participants with observer: $38 + 35 + 27 = 100$

$$\frac{(144)^2}{18} + \frac{(100)^2}{18} = 1707.56$$

From this value, subtract the CM to obtain SSF.

$$SSF = 1707.56 - CM$$
$$SSF = 1707.56 - 1653.78$$
$$SSF = 53.78$$

Step 6. To compute the sum of squares error term for the between-subjects factor (SSW_B), subtract SSF from SSB.

$$SSF_B = SSB - SSF$$
$$SSW_B = 60.22 - 53.78$$
$$SSW_B = 6.44$$

Step 7. To compute the overall within-subjects sum of squares (SSW), subtract SSB from SST.

$$SSW = SST - SSB$$
$$SSW = 90.22 - 60.22$$
$$SSW = 30.00$$

Step 8. To compute the sum of squares for trials (SSTR), add the scores of Trial 1 across both social facilitation conditions, and do the same for Trial 2 and Trial 3.

	Trials		
	1	*2*	*3*
$\Sigma\, x$	89	82	73

Next, square each of these sums and divide these values by the number of scores upon which each sum was based and add the quotients.

$$\frac{89^2}{12} + \frac{82^2}{12} + \frac{73^2}{12} = 1664.50$$

From this value, subtract the CM to obtain SSTR.

$$\text{SSTR} = 1664.50 - \text{CM}$$
$$\text{SSTR} = 1664.50 - 1653.78$$
$$\text{SSTR} = 10.72$$

Step 9. To compute the interaction sum of squares between trials and social facilitation ($SS_{F \times TR}$), obtain the sums for each combination of the trials and social facilitation factors.

	Alone			*Observed*		
	Trial 1	*Trial 2*	*Trial 3*	*Trial 1*	*Trial 2*	*Trial 3*
Σx	51	47	46	38	35	27

Next, square each of these values and divide each by the number of scores upon which each sum was based and add the quotients.

$$\frac{51^2}{6} + \frac{47^2}{6} + \frac{46^2}{6} + \frac{38^2}{6} + \frac{35^2}{6} + \frac{27^2}{6} = 1720.67$$

From this value, subtract the CM, SSF, and SSTR to obtain $SS_{F \times TR}$.

$$SS_{F \times TR} = 1720.67 - \text{CM} - \text{SSF} - \text{SSTR}$$
$$SS_{F \times TR} = 1720.67 - 1653.78 - 53.78 - 10.72$$
$$SS_{F \times TR} = 2.39$$

Step 10. To compute the within-subjects sum of squares error term (SSW_E) for the within-subjects main effect and the interaction, take SSW and subtract SSTR and $SS_{F \times TR}$.

$$SSW_E = \text{SSW} - \text{SSTR} - \text{SSF} \times \text{TR}$$
$$SSW_E = 30.00 - 10.72 - 2.39$$
$$SSW_E = 16.89$$

Step 11. The *df* are now calculated.

$$df_{\text{SST}} = \text{total number of scores} - 1$$
$$df_{\text{SST}} = 36 - 1 = 35$$
$$df_{\text{SSB}} = \text{total number of participants} - 1$$
$$df_{\text{SSB}} = 12 - 1 = 11$$
$$df_{\text{SSF}} = \text{total number of levels of the social facilitation factor} - 1$$
$$df_{\text{SSF}} = 2 - 1 = 1$$
$$df_{\text{SSW}_\text{B}} = df_{\text{SSB}} - df_{\text{SSF}}$$
$$df_{\text{SSW}_\text{B}} = 11 - 1 = 10$$
$$df_{\text{SSW}} = df_{\text{SST}} - df_{\text{SSB}}$$
$$df_{\text{SSW}} = 35 - 11 = 24$$
$$df_{\text{SSTR}} = \text{the number of trials (for each participant)} - 1$$
$$df_{\text{SSTR}} = 3 - 1 = 2$$
$$df_{\text{SS}_{\text{F}\times\text{TR}}} = df_{\text{SSTR}} \times df_{\text{SSF}}$$
$$df_{\text{SS}_{\text{F}\times\text{TR}}} = (2) \times (1) = 2$$
$$df_{\text{SSW}_\text{E}} = df_{\text{SSW}} - df_{\text{SSTR}} - df_{\text{SS}_{\text{F}\times\text{TR}}}$$
$$df_{\text{SSW}_\text{E}} = 24 - 2 - 2 = 20$$

Step 12. The mean squares (MS) are computed by dividing the appropriate sum of squares by the appropriate *df* for each sum of squares.

$$\text{MSF} = \frac{\text{SSF}}{df_{\text{SSF}}} = \frac{53.78}{1} = 53.78$$

$$\text{MSW}_\text{B} = \frac{\text{SSW}_\text{B}}{df_{\text{SSW}_\text{B}}} = \frac{6.44}{10} = 0.64$$

$$\text{MSTR} = \frac{\text{SSTR}}{df_{\text{SSTR}}} = \frac{10.72}{2} = 5.36$$

$$\text{MS}_{\text{F}\times\text{TR}} = \frac{\text{SS}_{\text{F}\times\text{TR}}}{df_{\text{SS}_{\text{F}\times\text{TR}}}} = \frac{2.39}{2} = 1.20$$

$$\text{MSW}_\text{E} = \frac{\text{SSW}_\text{E}}{df_{\text{SSW}_\text{E}}} = \frac{16.89}{20} = .85$$

Step 13. The appropriate *F* values are computed by the following formulas.

$$\text{Main effect: Social facilitation } F = \frac{\text{MSF}}{\text{MSW}_\text{B}} = \frac{53.78}{.64} = 84.03$$

$$\text{Main effect: Trials } F = \frac{\text{MSTR}}{\text{MSW}_\text{E}} = \frac{5.36}{.85} = 6.31$$

$$\text{Interaction: Social Facilitation} \times \text{Trials } F = \frac{\text{MS}_{\text{F}\times\text{TR}}}{\text{MSW}_\text{E}} = \frac{1.20}{.85} = 1.41$$

Step 14. To check these obtained F values for significance, refer to Appendix E for the F distribution. To enter the table, the df_1 is the df for the appropriate numerator sum of squares, and df_2 is the df for the error term used to test the respective sum of squares.

Thus,

Social facilitation factor	$df_1 = 1, df_2 = 10$
Trials factor	$df_1 = 1, df_2 = 20$
Social facilitation and trials	$df_1 = 1, df_2 = 20$

Step 15. After checking for significance, table the data as follows:

Source	*SS*	*df*	*MS*	*F*	*p*
Social facilitation	53.78	1	53.78	84.03	$< .001$
Error (SSW_B)	6.44	10	0.64		
Trials	10.72	2	5.36	6.31	$< .01$
Social Facilitation × Trials	2.39	2	1.20	1.41	$> .05$
Error (SSW_E)	16.89	20	0.85		
Total	90.22	35			

Step 16. As in all two-factor ANOVA designs, there are three null and alternative hypotheses.

Main Effect: Social Facilitation

In this example, the null and alternative hypotheses for the social facilitation factor are as follows:

$$H_0: \mu_{\text{alone}} = \mu_{\text{observed}}$$

$$H_a: \mu_{\text{alone}} \neq \mu_{\text{observed}}$$

Since the F value is significant for the social facilitation factor, H_0 is rejected. There are only two means associated with the rejected H_0, and no multiple comparison test is necessary. An inspection of the means reveals that the mean error score (5.56) of the participants who were observed is significantly lower than the mean error score (8.00) of the unobserved participants, $F(1, 10) = 84.03$, $p < .01$.

Main Effect: Trials

The trials factor was also significant. The H_0 and H_a are as follows:

$$H_0: \mu_{\text{Trial 1}} = \mu_{\text{Trial 2}} = \mu_{\text{Trial 3}}$$

H_a: H_0 is not true.

Since the trials factor null hypothesis was rejected, and because it contains more than two means, a multiple comparison test is necessary to determine which of the means are significantly different from each other. From the ANOVA, it can only be concluded that at least one trial mean is significantly different from at least one other trial mean. The pattern of mean differences can only be revealed with a multiple comparison test. The F may be reported as $F\ (2, 20) = 6.31, p < .01$.

Interaction: Social Facilitation × Trials

The interaction between the social facilitation and the trials factors was not significant. The null and alternative hypotheses are as follows:

H_0: There is no interaction between the social facilitation and trials factors on the error score.

H_a: There is an interaction between the social facilitation and trials factors on the error score.

Since the obtained F for the interaction did not exceed the critical value of F at $p = .05$, H_0 is retained, and no multiple comparison test is necessary or appropriate. It can be concluded that there is no significant interaction between the social facilitation main effect and the trials main effect on the number of errors. The F can be reported as $F(2, 20) = 1.41, p > .05$.

Two-Factor ANOVA: Repeated Measures on Both Factors Design

The two-factor repeated-measures design is appropriate for the analysis of two factors with two or more levels of each factor and where a single group of participants serves in every level of every factor. One advantage of this design is that it uses fewer participants than in a two-factor completely randomized ANOVA. However, there are many circumstances where this design cannot be used—for example, where the two conditions of one factor are inherently fixed (such as in situ designs), such as in the factor of gender. If the participants must remain naive under all conditions, then different participants and a between-subjects design will be necessary. Also, it is important to note whether one factor may influence the outcome of the other factor. In this case (as in drug studies and other medical studies), the assumption of independence between factors will be violated, and the two-factor repeated-measures ANOVA is completely inappropriate. The assumptions for the repeated-measures factorial ANOVA are the same as for the split-plot ANOVA.

Overview of the Repeated-Measures ANOVA

The repeated-measures ANOVA is similar to the completely randomized and split-plot ANOVA designs. There will be three F values for the two main effects and the interaction. The assumptions also remain the same, but the assumption of independence between the factors is essential. While the assumptions of normality and homogeneity of variance may be violated to some extent because of the robustness of ANOVA, the assumption that scores under one factor are independent of the other factor is sacrosanct.

Computationally, one interesting feature of the two-factor repeated-measures ANOVA is that each main effect and the interactions have their own unique error term. Thus, the denominators of the three F values are all different.

Computational Example

An experimenter was interested to see whether chimpanzees varied in learning ability from morning to evening. A maze learning task was employed, and the number of errors (false turns) was recorded for three trials. Different mazes were used in the morning and evening conditions, and prior testing indicated that they were of equal difficulty. The data are tabled as follows:

	Morning			*Evening*		
	Trial 1	*Trial 2*	*Trial 3*	*Trial 1*	*Trial 2*	*Trial 3*
Chimp 1	6	5	6	3	7	5
Chimp 2	8	4	4	5	5	6
Chimp 3	6	6	7	4	6	8
Chimp 4	7	8	7	4	4	1
Chimp 5	7	6	7	4	6	6
Chimp 6	6	5	6	4	6	6
Chimp 7	8	5	9	5	7	4
Chimp 8	7	6	6	4	6	5
Chimp 9	9	5	8	5	3	5
Chimp 10	6	7	6	4	5	3

Step 1. Obtain Σx and Σx^2 for each treatment combination and the overall Σx and Σx^2 for each participant's scores.

	Morning				*Evening*				
	Trial 1	*Trial 2*	*Trial 3*	*Row* Σx	*Trial 1*	*Trial 2*	*Trial 3*	*Row* Σx	*Total Row* Σx
S_1	6	5	6	17	3	7	5	15	32
S_2	8	4	4	16	5	5	6	16	32
S_3	6	6	7	19	4	6	8	18	37
S_4	7	8	7	22	4	4	1	9	31
S_5	7	6	7	20	4	6	6	16	36
S_6	6	5	6	17	4	6	6	16	33
S_7	8	5	9	22	5	7	4	16	38
S_8	7	6	6	19	4	6	5	15	34
S_9	9	5	8	22	5	3	5	13	35
S_{10}	6	7	6	19	4	5	3	12	31
Σx	70	57	66		42	55	49		
Σx^2	500	337	452		180	317	273		

$$\text{Overall} \sum x = 339$$

$$\text{Overall} \sum x^2 = 2059$$

Step 2. To obtain the correction for the mean (CM), square the overall $\Sigma\, x$ and divide by the total number of observations.

$$\text{CM} = \frac{\left(\text{overall} \sum x\right)^2}{N}$$

$$\text{CM} = \frac{(339)^2}{60} = 1915.35$$

Step 3. To obtain the sum of squares total (SST), subtract the CM from the overall $\Sigma\, x^2$.

$$\text{SST} = \text{overall} \sum x^2 - \text{CM}$$

$$\text{SST} = 2059 - 1915.35$$

$$\text{SST} = 143.65$$

Step 4. To obtain the sum of squares for subjects (SSS), obtain the sum of each participant's scores, square them, and add these squared values together. Divide this summed value by the number of scores summed across (how many scores went into an individual's sum) and subtract the CM.

$$\frac{32^2}{6} + \frac{32^2}{6} + \frac{37^2}{6} + \frac{31^2}{6} + \frac{36^2}{6} + \frac{33^2}{6} + \frac{38^2}{6} + \frac{34^2}{6} + \frac{35^2}{6} + \frac{31^2}{6} = \frac{11,549}{6} = 1924.83$$

$$\text{SSS} = 1924.83 - \text{CM}$$

$$\text{SSS} = 1924.83 - 1915.35$$

$$\text{SSS} = 9.48$$

Step 5. To calculate sum of squares for the circadian (morning vs. evening) factor (SSC), sum the scores for all the morning trials and then sum all the scores for the evening trials.

Sum for morning trials: $70 + 57 + 66 = 193$

Sum for evening trials: $42 + 55 + 49 = 146$

Square these sums, add the resulting values together, and divide by the number of scores summed across to obtain an individual sum.

$$\frac{(193)^2}{30} + \frac{(146)^2}{30} = \frac{58,565}{30} = 1952.17$$

To obtain SSC, subtract the CM from this resulting value.

$$\text{SSC} = 1952.17 - \text{CM}$$
$$\text{SSC} = 1952.17 - 1915.35$$
$$\text{SSC} = 36.82$$

Step 6. To obtain the sum of squares for the trials factor (SSTR), sum all the scores for Trial 1, sum all the scores for Trial 2, and sum the scores for Trial 3.

Sum for Trial 1: $70 + 42 = 112$

Sum for Trial 2: $57 + 55 = 112$

Sum for Trial 3: $66 + 49 = 115$

Now square each of these sums and add them together, and divide this resulting value by the number of scores summed across for an individual trial.

$$\frac{(112)^2}{20} + \frac{(112)^2}{20} + \frac{(115)^2}{20} = \frac{38,313}{20} = 1915.65$$

Now, subtract the CM from this resulting value to obtain SSTR.

$$\text{SSTR} = 1915.65 - \text{CM}$$
$$\text{SSTR} = 1915.65 - 1915.35$$
$$\text{SSTR} = 0.30$$

Step 7. To obtain the interaction sum of squares ($SS_{C \times TR}$), take the $\Sigma\, x$ for each treatment combination, square each of these values, divide each by the number of scores that went into $\Sigma\, x$, and add the quotients.

$$\frac{(70)^2}{10} + \frac{(57)^2}{10} + \frac{(66)^2}{10} + \frac{(42)^2}{10} + \frac{(55)^2}{10} + \frac{(49)^2}{10} = 1969.50$$

Step 8. To obtain $SSS_{C \times TR}$, subtract from the value obtained in Step 7, the CM, SSC, and SSTR.

$$SS_{C\times TR} = 1969.50 - \text{CM} - \text{SSC} - \text{SSTR}$$
$$SS_{C\times TR} = 1969.50 - 1915.35 - 36.82 - 0.30$$
$$SS_{C\times TR} = 17.03$$

Step 9. To calculate the error term for the circadian factor (SSWC), obtain $\Sigma\, x$ for each participant for each level of the circadian factor (ignoring the trials factor), square these sums, add them together, and divide by the number of scores added to get each of the individual sums.

	Circadian Factor	
	Morning	*Evening*
S_1	6 + 5 + 6 = 17	3 + 7 + 5 = 15
S_2	16	16
S_3	19	18
S_4	22	9
S_5	20	16
S_6	17	16
S_7	22	16
S_8	19	15
S_9	22	13
S_{10}	19	12

Thus,

$$\frac{17^2}{3} + \frac{16^2}{3} + \frac{19^2}{3} + \frac{22^2}{3} + \frac{20^2}{3} + \frac{17^2}{3} + \frac{22^2}{3} + \frac{19^2}{3} + \frac{22^2}{3} + \frac{19^2}{3}$$
$$+ \frac{15^2}{3} + \frac{16^2}{3} + \frac{18^2}{3} + \frac{9^2}{3} + \frac{16^2}{3} + \frac{16^2}{3} + \frac{16^2}{3} + \frac{15^2}{3} + \frac{13^2}{3} + \frac{12^2}{3}$$
$$= \frac{5961}{3} = 1987.00$$

Step 10. To obtain the within-subjects sum of squares error term for the circadian factor (SSWC), take the value obtained in Step 9 and subtract the CM, SSS, and SSC.

$$SSW_C = 1987.00 - CM - SSC - SSC$$
$$SSW_C = 1987.00 - 1915.35 - 9.48 - 36.82$$
$$SSW_C = 25.35$$

Step 11. To calculate the error term for the trials factor (SSWTR), first obtain $\Sigma\, x$ for each participant, for each level of the trials factor (ignoring

	Trials		
	1	*2*	*3*
S_1	6 + 3 = 9	5 + 7 = 12	6 + 5 = 11
S_2	13	9	10
S_3	10	12	15
S_4	11	12	8
S_5	11	12	13
S_6	10	11	12
S_7	13	12	13
S_8	11	12	11
S_9	14	8	13
S_{10}	10	12	9

the circadian factor), square these sums, add them together, and divide by the number of scores added to get each of the individual sums.

Thus,

$$\frac{9^2}{2}+\frac{12^2}{2}+\frac{11^2}{2}+\frac{13^2}{2}+\frac{9^2}{2}+\frac{10^2}{2}+\frac{10^2}{2}+\frac{12^2}{2}+\frac{15^2}{2}+\frac{11^2}{2}+\frac{12^2}{2}$$
$$+\frac{8^2}{2}+\frac{11^2}{2}+\frac{12^2}{2}+\frac{13^2}{2}+\frac{10^2}{2}+\frac{11^2}{2}+\frac{12^2}{2}+\frac{13^2}{2}$$
$$+\frac{12^2}{2}+\frac{13^2}{2}+\frac{11^2}{2}+\frac{12^2}{2}+\frac{11^2}{2}+\frac{14^2}{2}$$
$$+\frac{8^2}{2}+\frac{13^2}{2}+\frac{10^2}{2}+\frac{12^2}{2}+\frac{9^2}{2}$$
$$=\frac{3915}{2}=1957.50$$

Step 12. To obtain the error sums of squares for trials (SSW_{TR}), take the value obtained in Step 11 and subtract the CM, SSS, and SSTR.

$$SSW_{TR} = 1957.50 - CM - SSS - SSTR$$
$$SSW_{TR} = 1957.50 - 1915.35 - 9.48 - 0.30$$
$$SSW_{TR} = 32.37$$

Step 13. To obtain the error term for the interaction ($SSW_{C \times TR}$), take SST and subtract from it SSS, SSC, SSTR, $SS_{C \times TR}$, SSW_{TR}, and SSW_C.

$$SSW_{C\times TR} = SST - SSS - SSC - SSTR - SS_{C\times TR} - SSW_{TR} - SSW_C$$
$$SSW_{C\times TR} = 143.65 - 9.48 - 36.82 - 0.30 - 17.03 - 32.37 - 25.35$$
$$SSW_{C\times TR} = 22.30$$

Step 14. The *df* are now calculated.

$$df_{SST} = \text{total number of scores} - 1$$
$$df_{SST} = 60 - 1 = 59$$
$$df_{SSS} = \text{the number of participants} - 1$$
$$df_{SSS} = 10 - 1 = 9$$
$$df_{SSC} = \text{the number of levels of the circadian factor} - 1$$
$$df_{SSC} = 2 - 1 = 1$$
$$df_{SSTR} = \text{the number of levels of the trials factor} - 1$$
$$df_{SSTR} = 3 - 1 = 2$$
$$df_{SS_{C\times TR}} = df_{SSC} \times df_{SSTR}$$
$$df_{SS_{C\times TR}} = 1 \times 2 = 2$$
$$df_{SSW_C} = df_{SSS} \times df_{SSC}$$
$$df_{SSW_C} = 9 \times 1 = 9$$

$$df_{\text{SSW}_{\text{TR}}} = df_{\text{SSS}} \times df_{\text{SSTR}}$$
$$df_{\text{SSW}_{\text{TR}}} = 9 \times 2 = 18$$
$$df_{\text{SSW}_{\text{C}\times\text{TR}}} = df_{\text{SSS}} \times df_{\text{SSC}} \times df_{\text{SSTR}}$$
$$df_{\text{SSW}_{\text{C}\times\text{TR}}} = 9 \times 1 \times 2 = 18$$

Step 15. The mean squares (MS) are now calculated by dividing the sum of squares by its appropriate *df* (see Step 14).

$$\text{MSC} = \frac{\text{SSC}}{df_{\text{SSC}}} = \frac{36.82}{1} = 36.82$$
$$\text{MSTR} = \frac{\text{SSTR}}{df_{\text{SSTR}}} = \frac{0.30}{2} = 0.15$$
$$\text{MS}_{\text{C}\times\text{TR}} = \frac{\text{SS}_{\text{C}\times\text{TR}}}{df_{\text{SS}_{\text{C}\times\text{TR}}}} = \frac{17.03}{2} = 8.52$$
$$\text{MSW}_{\text{C}} = \frac{\text{SSW}_{\text{C}}}{df_{\text{SSW}_{\text{C}}}} = \frac{25.35}{9} = 2.82$$
$$\text{MSW}_{\text{TR}} = \frac{\text{SSW}_{\text{TR}}}{df_{\text{SSW}_{\text{TR}}}} = \frac{32.37}{18} = 1.80$$
$$\text{MSW}_{\text{C}\times\text{TR}} = \frac{\text{SSW}_{\text{C}\times\text{TR}}}{df_{\text{SSW}_{\text{C}\times\text{TR}}}} = \frac{22.30}{18} = 1.24$$

Step 16. The appropriate *F* values are now calculated by the following formulas:

$$\text{Main effect: Circadian factor } F = \frac{\text{MSC}}{\text{MSW}_{\text{C}}} = \frac{36.82}{2.82} = 13.06$$
$$\text{Main effect: Trials factor } F = \frac{\text{MSTR}}{\text{MSW}_{\text{TR}}} = \frac{.15}{1.80} = 0.08$$
$$\text{Interaction: Circadian} \times \text{Trials } F = \frac{\text{MS}_{\text{C}\times\text{TR}}}{\text{MSW}_{\text{C}\times\text{TR}}} = \frac{8.52}{1.24} = 6.87$$

Step 17. To check these obtained *F* values for significance, refer to the *F* distribution in Appendix E. To enter the table, df_1 is the *df* associated with the numerator or sum of squares in the obtained *F* value, and df_2 is the *df* for the error term or the denominator in the obtained *F* value.

Thus,

Circadian factor $df_1 = 1, df_2 = 9$

Trials factor $df_1 = 2, df_2 = 18$

Interaction factor $df_1 = 2, df_2 = 18$

Step 18. After checking for significance, table the data as follows:

Source	*SS*	*df*	*MS*	*F*	*p*
Circadian (SSC)	36.82	1	36.82	13.06	$< .01$
Error (SSW_C)	25.35	9	2.82		
Trials (SSTR)	0.30	2	0.15	0.08	$> .05$
Error (SSW_{TR})	32.37	18	1.80		
Circadian × Trials (C × TR)	17.03	2	8.52	6.87	$< .01$
Error ($SSW_{C \times TR}$)	22.30	18	1.24		
Subjects	9.48	9	1.05		
Total	143.65	59			

Step 19. As in all two-factor ANOVA designs, there are three null and alternative hypotheses. In this example, there is an H_0 and H_a for the circadian factor:

$$H_0: \mu_{\text{morning}} = \mu_{\text{evening}}$$
$$H_a: \mu_{\text{morning}} \neq \mu_{\text{evening}}$$

Because the *F* value is significant for the circadian factor, H_0 is rejected. There are only two means associated with the rejected H_0, so no multiple comparison test is necessary. An inspection of the means reveals that the mean number of errors by the chimps (6.43) in the morning was significantly higher than their mean errors (4.87) in the evening. In a report, the *F* may be reported as $F(1, 9) = 13.06, p < .01$.

The trials factor was not significant. The H_0 and H_a are as follows:

$$H_0: \mu_{\text{Trial 1}} = \mu_{\text{Trial 2}} = \mu_{\text{Trial 3}}$$
$$H_a: H_0 \text{ is not true.}$$

Because the trials factor H_0 was retained, no multiple comparison test is necessary. It may be concluded that the means for the three trials do not differ significantly from one another. The *F* may be reported as $F(2, 18) = 0.08$, $p > .05$.

The interaction between the circadian factor and the trials factor was significant. The null and alternative hypotheses are as follows:

H_0: There is no interaction between the circadian and trials factors on the number of errors.

H_a: There is an interaction between the circadian and trials factors on the number of errors.

Because the derived *F* for the interaction exceeded the critical value of *F* at $p < .05$, H_0 is rejected, and it may be concluded that at least one of these means is significantly different from one other. To determine the pattern of differences between the six means, a multiple comparison test is necessary. The F is reported as $F(2, 18) = 8.52, p < .01$.

Key Terms and Definitions

Between-subjects variable—A factor in a completely randomized ANOVA where the participants are randomly assigned to each level of a factor, resulting in different participants in each level. It also refers to an in situ independent variable where different participants serve in each level of the variable, but they were not randomly assigned to the levels.

Repeated-measures ANOVA—An experimental design where the same participant serves under every level of at least one factor in the ANOVA.

Split-plot ANOVA—A two-factor ANOVA where one of the independent variables has the participants randomly assigned to the different levels or different participants on each level, in situ, and the other independent variable is a repeated measure, where the same participant serves under all levels.

Within-subjects variable—A factor where the same group of participants is measured on the dependent variable at every level of the independent variable.

Chapter 14 Practice Problems

1. A school psychologist wishes to determine whether school bullies' anger changes from the beginning of a school year to the end of the school year. School bullies (as identified by multiple disciplinary actions) and age- and gender-matched controls were measured on an anger scale (*T* score) at the beginning of the school year and at the end of the year. Perform a split-plot ANOVA, create an ANOVA source table, and write up the results.

	Time 1	*Time 2*		*Time 1*	*Time 2*
Bully 1	62	72	Control 1	45	45
Bully 2	75	80	Control 2	40	43
Bully 3	83	85	Control 3	38	44
Bully 4	66	90	Control 4	55	45
Bully 5	74	81	Control 5	57	51
Bully 6	78	78	Control 6	49	50
Bully 7	70	74	Control 7	42	45
Bully 8	81	85	Control 8	50	49

2. Seven schizophrenic patients in a state mental hospital are being evaluated to see if a new antipsychotic drug improves long-term memory. Without any medication, the schizophrenics are tested each day for 3 days on three different, but equivalent, long-term memory tasks (0 = *no memory*, 20 = *perfect memory*). Next, they are given the new antipsychotic drug and tested again on three additional memory tasks on 3 consecutive days. Perform a repeated-measures ANOVA, create an ANOVA source table, and write up the results.

	Unmedicated Trials			Medicated Trials		
	1	*2*	*3*	*1*	*2*	*3*
Patient 1	8	8	9	12	13	15
Patient 2	4	6	5	10	11	17
Patient 3	5	4	6	9	12	13
Patient 4	11	12	13	15	18	17
Patient 5	7	6	6	11	19	20
Patient 6	9	10	11	14	13	15
Patient 7	12	11	11	15	20	20

Chapter 14 Test Questions

Problems 1–10. A researcher wishes to determine whether the controversial healing touch therapy is successful when the patients are unaware that the therapist is using "healing touch." Twenty patients with acute back pain are randomly assigned to either an unaware condition, where the patients sleep through the sessions, or an aware condition, where they are awake. The patients then receive three sessions each of healing touch therapy. After each session, the patients are asked to rate their pain on a 1 to 10 scale, where 1 = *no pain* and 10 = *maximum pain.*

	Unaware Sessions				Aware Sessions		
	1	*2*	*3*		*1*	*2*	*3*
Patient 1	8	8	9	Patient 11	5	3	2
Patient 2	7	7	8	Patient 12	6	7	4
Patient 3	5	4	3	Patient 13	5	3	2
Patient 4	4	6	4	Patient 14	4	3	1
Patient 5	7	9	6	Patient 15	4	2	1
Patient 6	5	5	4	Patient 16	3	1	1
Patient 7	8	9	10	Patient 17	3	3	3
Patient 8	10	9	9	Patient 18	4	3	2
Patient 9	9	8	9	Patient 19	2	4	2
Patient 10	8	9	9	Patient 20	2	2	1

1. The SST is
   a. 1440.600
   b. 232.067
   c. 24.700
   d. 3.033
   e. 463.400

2. The SS for the awareness factor is
   a. 1440.600
   b. 232.067
   c. 24.700
   d. 3.033
   e. 463.400

3. The SS for the sessions factor is
    a. 1440.600
    b. 232.067
    c. 24.700
    d. 3.033
    e. 463.400

4. The SS for the interaction is
    a. 1440.600
    b. 232.067
    c. 24.700
    d. 3.033
    e. 463.400

5. The correction for the mean (CM) is
    a. 1440.600
    b. 232.067
    c. 24.700
    d. 3.033
    e. 463.400

6. The *df* for the awareness factor is
    a. 1
    b. 2
    c. 3
    d. 54
    e. 59

7. The *df* for the sessions factor is
    a. 1
    b. 2
    c. 3
    d. 54
    e. 59

8. The *df* for the interaction is
    a. 1
    b. 2
    c. 3
    d. 54
    e. 59

9. The *df* for SST is
    a. 1
    b. 2
    c. 3
    d. 54
    e. 59

10. The interaction was
    a. significant
    b. nonsignificant
    c. insignificant

Problems 11–20. An experimental neuropsychologist wishes to determine whether a new drug (X-9) enhances the growth of new, functional brain tissue after maximal brain growth has already been attained after a period of 1 month and 6 months. Five rats were run in a series of three different mazes. All of the mazes were judged to be of equal difficulty. The time in seconds to solve each maze is recorded for each rat across its three mazes. X-9 was given 2 weeks before testing. A prior study statistically demonstrated that rats not given X-9 did not improve their times. Is there any evidence X-9 works?

	1 Month			*6 Months*		
	Trial 1	*Trial 2*	*Trial 3*	*Trial 1*	*Trial 2*	*Trial 3*
Rat 1	10	12	5	4	5	4
Rat 2	15	14	8	6	5	4
Rat 3	14	14	7	6	7	7
Rat 4	9	10	5	7	8	8
Rat 5	12	11	4	7	6	7

11. The SST is
    a. 324.967
    b. 1936.033
    c. 116.033
    d. 64.067
    e. 68.467

12. The CM is
    a. 324.967
    b. 1936.033
    c. 116.033
    d. 64.067
    e. 68.467

13. The SS for time is
    a. 324.967
    b. 1936.033
    c. 116.033
    d. 64.067
    e. 68.467

14. The SS for trials is
    a. 324.967
    b. 1936.033
    c. 116.033
    d. 64.067
    e. 68.467

15. The SS for the interaction is

    a. 324.967
    b. 1936.033
    c. 116.033
    d. 64.067
    e. 68.467

16. The *df* for the time main effect is

    a. 1
    b. 2
    c. 3
    d. 24
    e. 29

17. The *df* for the trials main effect is

    a. 1
    b. 2
    c. 3
    d. 24
    e. 29

18. The *df* for the interaction is

    a. 1
    b. 2
    c. 3
    d. 24
    e. 29

19. The *df* for SST is

    a. 1
    b. 2
    c. 3
    d. 24
    e. 29

20. The interaction was

    a. significant
    b. nonsignificant
    c. insignificant

15 Nonparametric Statistics

The Chi-Square Test

Chapter 15 Goals

- Learn the difference between parametric and nonparametric statistics
- Learn the fundamentals of the chi-square distribution
- Learn how to conduct a 2 × 2 chi-square statistical analysis
- Learn how to interpret a 2 × 2 chi-square test

A recent study found that if two sports teams were evenly matched, the team that wore red was significantly more likely to win. Besides waiting for the numerous replications of this provocative finding, the data pose an interesting design and analysis problem. One variable appears to be color, and its two levels are red or not red. The second variable appears to be outcome, victory or defeat. The parametric statistical tests, such as *t* tests and ANOVA designs, require means of individual scores and two or more groups. Both *t* tests and ANOVA are considered parametric statistics because the underlying distributions are assumed to be normal, and the distributions may be consequently described by common statistical parameters such as the arithmetic mean, standard deviation, and variance. In the present example, we have two dichotomous variables, and we wish to see whether levels of one variable are consistently associated with one of the levels of the other variable (i.e., is red significantly more associated with victory). For designs such as this, statisticians have developed a class of statistics known as **nonparametric statistics.** Nonparametric statistics are used when the data are measured by nominal or categorical scales or when the underlying distributions violate the assumptions of the parametric statistics (where underlying distribution is nonnormal or unknown). The most common use of nonparametric statistics is probably not the latter case but the former: when the data are in name only (e.g., gender) and cannot be given any numerical value except arbitrarily (e.g., 1 = male, 2 = female or 1 = red, 2 = not red).

Nonparametric statistics are usually not the first choice of most statisticians. In nearly every experimental situation, a parametric statistic has more power (the ability to detect genuine relationships, genuine change, or real differences) than a nonparametric statistic. Yet some types of research, because of the nature of the data, must be analyzed nonparametrically. The most popular of the nonparametric statistics is chi-square or χ^2 (from the Greek letter χ, pronounced "k-eye" but in one syllable).

The **chi-square tests** are used most often to analyze data that consist of counts or frequencies. For example, frequency data establish how many people or events qualify for a particular category, such as how many people choose Coke or Pepsi in a taste test. Notice that the data in a taste test end up to be simple counts, such as 110 people chose Coke and 123 chose Pepsi. A frequency analysis answers the question of how many, and frequency analyses can compare frequency differences between two or more categories.

Overview of the Purpose of Chi-Square

Chi-square statistics are designed to determine whether an observed number differs either from chance or from what was expected. For example, the simplest chi-square design uses a single independent variable on a nominal or categorical scale with two levels, such as gender. If we theorized that nothing other than pure chance should affect the gender of babies, then 50% of all babies should be male and 50% should be female. The experiment or study would consist of coding the gender of all births at a hospital or hospitals in a city, and a chi-square test would determine whether the observed or actual birth rates differ from the theoretical expectation. Interestingly, in this example, they do. In the United States, male babies account for about 49% and female babies account for 51% of all births. When Karl Pearson created the chi-square test in the 1890s, he termed it a **goodness-of-fit test** in order to describe how well theoretically the observed frequencies fit the expected frequencies.

If there was no clear preference for either product in a taste test between Coke and Pepsi of 1000 people, then it would be expected that 500 people would prefer Pepsi and 500 would prefer Coke. In this case, the expected outcome would be an even split if we initially thought (the null hypothesis) there should be no preference for one over the other. A chi-square test could also test whether other probabilities could fit the observed frequencies. For example, if the study were a Third World country where Coke had been sold for the past 10 years and Pepsi had only been sold for 2 years, the hypothesis could be tested whether the observed frequencies "fit" a prior assumption (the null hypothesis) that Coke should outsell Pepsi 10 to 1, or a hypothesis could be tested where the null hypothesis might be that Coke should outsell Pepsi by 3 to 1.

Overview of Chi-Square Designs

The previous example of a taste test between two products is considered a two-cell chi-square design because there is only one independent variable (in this case, the product with two levels, Pepsi and Coke) and a single dependent variable (frequency or number of times chosen). The levels are called *cells* in chi-square analyses. Thus, as ANOVA was an extension of *t* tests (from a two-group analysis to more than two groups to be analyzed), a two-cell chi-square can consist of a single independent variable with two cells, or it can consist of more than two cells. Chi-square can also be similar to a factorial ANOVA, in that chi-square can analyze two or more independent variables at once. The most commonly used chi-square design is probably the 2×2 (pronounced "two-by-two") design where there are two independent variables, and each has two levels. To build on the birth example, we might test to see whether the birth ratio of males to females differs in America from India. One independent variable is gender (male or female), and the other is a geographical country (America or India). The null hypothesis states that these two independent variables are independent of each other (the ratio of girl to boy births is the same in both countries even if the overall number of births is higher in India). The alternative hypothesis states that the two independent variables are dependent (which means that the ratios of girls to boys are different in the two countries).

Chi-Square Test: Two-Cell Design (Equal Probabilities Type)

The simplest form of a chi-square test is a two-cell experiment where the outcome of the experiment is that each participant or event falls into only one of the two cells. Let us continue with our gender examples. An experimenter wishes to know whether male deep-sea divers have more female children than male children. The experimenter asks 10 consecutive long-time deep-sea divers who come to a dive store about the gender of their children. The null hypothesis would be that the probability of having a boy is 1/2 or .5, and the probability of having a girl is 1/2 or .5, and the sum of the two probabilities is 1.0. Let us say that the 10 divers reported having a total of 25 children: 10 boys and 15 girls. Do these observed values differ from what we expected to be a 50-50 split?

Computation of the Two-Cell Design

In a chi-square test of this experiment, the null and alternative hypotheses would be

$$H_0: p_B = p_G$$
$$H_a: p_B \neq p_G$$

where

p_B = the probability of having a boy, and

p_G = the probability of having a girl.

Step 1. Construct a table of the observed and expected cell frequencies.

	Boy	*Girl*
Observed cell frequency	10	15
Expected cell frequency	12.5	12.5

Note that the expected cell frequency is obtained by multiplying the theoretical probability of having a boy (1/2 or .5) times the total number of participants in the study (10 + 15 = 25). The expected cell frequency for both cells is .5 times 25 = 12.5. In this case, the expected cell frequencies are equal to each other because it was stated that way in the null hypothesis—hence the title "Equal Probabilities Type."

Step 2. The chi-square value is obtained by the following formula:

$$\chi^2 = \sum_{i=1}^{k} \frac{(\text{Observed Cell Frequency}_i - \text{Expected Cell Frequency}_i)^2}{\text{Expected Cell Frequency}_i}$$

where i = the first cell, then the second cell, and so on up to k, where k is the last cell. In this example, we have just two cells contributing to the overall chi-square value.

So,

$$\chi^2 = \frac{(10 - 12.5)^2}{12.5} + \frac{(15 - 12.5)^2}{12.5}$$

$$\chi^2 = \frac{(-2.5)^2}{12.5} + \frac{(2.5)^2}{12.5}$$

$$\chi^2 = \frac{6.25}{12.5} + \frac{6.25}{12.5}$$

$$\chi^2 = .50 + .50$$

$$\chi^2 = 1.00$$

Step 3. Next, the degrees of freedom (*df*) are obtained by the following formula:

$$df = \text{number of cells} - 1$$
$$df = 2 - 1 = 1$$

Step 4. Compare the derived chi-square value to the critical value in Appendix D containing the chi-square distribution. The critical value of chi-square at $p = .05$ with $df = 1$ is 3.84. Our informal write-up might appear as follows:

> The derived value of $\chi^2 = 1.00$ does not exceed the tabled critical value of $\chi^2 = 3.84$ at $p = .05$ with $df = 1$. Therefore, H_0 is retained, and it is concluded that a deep-sea diver's probability of having a boy is not different from the probability of having a girl, χ^2 $(1, N = 10) = 1.00$, $p > .05$. Thus, although it was observed that divers had more girls than boys in the study, the differences were not statistically different beyond what might be expected by chance.

A formal write-up in a journal article might appear as follows:

> A chi-square test revealed that the observed ratio of 1.5:1 (girls to boys) among children of 10 deep-sea divers did not vary significantly from the expected ratio of 1:1, χ^2 $(1, N = 10) = 1.00$, $p > .05$. In terms of the research question, deep-sea divers did not have significantly more girls than boys.

If we divide the number of girls (the larger observed value) by the number of boys, we get a ratio of girls to boys of 1.5:1 (it reads, "1.5 *to* 1"). We could also view this as a 1.5 times greater rate of girls to boys, although we have already performed a statistical test (the chi-square test) that determined that this rate may be within the realm of chance; that is, it is not significantly different from a 1:1 ratio.

A chi-square test can suffer from the same lack of statistical power, or abuse of power, as the parametric tests, such as correlation, *t* tests, or ANOVA. Thus, too few observations in this example (i.e., 10 divers and their 25 children) may have led us to underuse the power of the test. Indeed, had we polled 50 divers and found that they had 100 children in the same ratio (1.5:1, girls to boys), we would have obtained statistical significance with the chi-square test.

The Chi-Square Distribution

Like the *t* distribution or *F* distribution, the chi-square distribution is actually a family of theoretical distributions. Like the *F* value in ANOVA, the chi-square value is a one-tailed test, but unlike the *F* distribution, chi-square values have *df* based on the number of cells or categories and not on the number of observations or participants.

Assumptions of the Chi-Square Test

Although it was stated that the chi-square test makes no assumption about the shape of the underlying population, there are a few important assumptions when using the chi-square test.

1. The scores in each cell should be independent of one another. This means that a score in one cell should have no effect on a score in another cell. A violation of this assumption could occur if participants became aware that Cell 1 was being chosen more than Cell 2, and as a result, Cell 1 was chosen at even greater frequencies later in the experiment because the participants thought that the choice of Cell 1 was somehow better than Cell 2. Independence also means that each score (count or frequency) represents only one unique participant or event. The same participant may not be represented by more than one score. In our first example, independence might have been violated if all of the divers and their wives decided to keep having children only until they had one boy.

2. There should be a minimum of five participants or events in any one cell. Violations of this assumption do not automatically disqualify the use of the chi-square statistic, but we have seen that power is already reduced by the choice of a nonparametric statistic, and the chi-square test requires substantial numbers of scores to be sensitive to real or genuine differences in the data.

3. The dependent variable in the chi-square test is assumed to be a frequency or count, such as numbers of participants. It is not appropriate to analyze continuous variables unless they have been dichotomized. One method of dichotomizing a continuous variable is to split the data into two halves by the median. Those above the median become Group 1, and those below the median become Group 2. However, it would be unwise to use a nonparametric statistic on a dichotomized continuous variable if the distribution was normal or mound-shaped since the equivalent parametric test would be much more powerful and sensitive to real differences in the data. Dichotomizing a continuous variable lumps all scores along an interval into just two categories, and thus much unique information is lost in the process.

Chi-Square Test: Two-Cell Design (Different Probabilities Type)

There is a less common variation of the simple two-cell chi-square test where the theoretical probabilities are expected to be different for the two cells instead of being equal. Let us use the example of handedness. In the general population, 95% are right-handed people and 5% are non-right-handed people. An experimenter wishes to determine whether there are significantly more non-right-handed people in a sample of 100 people with epilepsy, and the experimenter found 85 right-handed people and 15 non-right-handed people. Do these observed differences vary significantly from the theoretical expectation in the population?

Computation of the Two-Cell Design

The null and alternative hypotheses are as follows:

$$H_0: p_R = .95 \text{ and } p_{NR} = .05$$
$$H_a: H_0 \text{ is not true.}$$

where

p_R = the probability of being a right-handed person, and

p_{NR} = the probability of being a non-right-handed person.

Step 1. Construct a table of the observed and expected cell frequencies.

	Right-Handed	*Non-Right-Handed*
Observed cell frequency	85	15
Expected cell frequency	95	5

Note that the expected cell frequency is obtained by multiplying the theoretical probability of being right-handed (95/100 or .95) times the total number of participants in the study ($85 + 15 = 100$). Thus, the expected cell frequency for the right-handers is $.95 \times 100 = 95$, and for non-right-handers, it is $.05 \times 100 = 5$.

Step 2. The chi-square value is obtained by the following formula:

$$\chi^2 = \sum_{i=1}^{k} \frac{(\text{Observed Cell Frequency}_i - \text{Expected Cell Frequency}_i)^2}{\text{Expected Cell Frequency}_i}$$

where i = the first cell, then the second cell, and so on up to k, where k = the last cell.

So,

$$\chi^2 = \frac{(85 - 95)^2}{95} + \frac{(15 - 5)^2}{5}$$

$$\chi^2 = \frac{(-10)^2}{95} + \frac{(10)^2}{5}$$

$$\chi^2 = \frac{100}{95} + \frac{100}{5}$$

$$\chi^2 = 1.05 + 20.00$$

$$\chi^2 = 21.05$$

Step 3. Next, the degrees of freedom (*df*) are obtained by the following formula:

$$df = \text{number of cells} - 1$$
$$df = 2 - 1 = 1$$

Step 4. The critical value of chi-square is obtained from Appendix D, which contains the chi-square distribution. The critical value of chi-square at $p = .05$ with $df = 1$ is 3.84. Our informal write-up would appear as follows:

The derived value of $\chi^2 = 21.05$ does exceed this tabled critical value of chi-square. Therefore, H_0 is rejected, and it is concluded that the observed cell frequencies do significantly vary from the theoretical cell frequency beyond what might be expected by chance, $\chi^2(1, N = 100) = 21.05$, $p < .001$. In terms of the research question, it was observed that there were significantly more non-right-handed people in a sample of people with epilepsy than would be expected in the population of people without epilepsy.

A formal write-up in a journal article might appear as follows:

The incidence of non-right-handedness (.15) in a group of 100 epileptic patients was significantly greater than the rate of non-right-handedness in a nonepileptic sample, $\chi^2(1, N = 100) = 21.05$, $p < .001$. In terms of the research question, it appears that non-right-handedness occurs at a three times greater rate in this epileptic sample than would be expected in a nonepileptic population.

Interpreting a Significant Chi-Square Test for a Newspaper

It is important throughout all of statistics to remind ourselves of the grand meaning of what we are doing. It is fine to report the study as significant, $\chi^2(1) = 21.05$, $p < .01$, to a sophisticated group of statisticians, but what would we tell our mothers or fathers we were doing (assuming they were not statisticians)? The contemporary psychologist-statistician Abelson asks his students, "If your study were reported in the newspaper, what would the headline be?"(Abelson, 1995, p. xiii).

I have long told my students to write a final paragraph of their study that summarizes the statistical findings in a way that the curious-but-unsophisticated reader could comprehend. While we did not directly test the significance of the percentages or the rates, it is completely valid to report the percentages or rates, along with the statement of their statistical significance, because the average person can understand differences in percentages or greater rates in a particular group. Thus, in the previous example, it is relatively easy to understand the findings of a three times greater rate of non-right-handedness in epileptic patients. In fact, it is easier for nonstatisticians *and* statisticians to comprehend the results of a chi-square test in this manner.

Chi-Square Test: Three-Cell Experiment (Equal Probabilities Type)

Another common form of a chi-square test is a three-cell design (and all of the following discussion and formulas can be applied to more than three cells) where the outcome of the experiment is that each observation (people or event) can fall into one of three cells. In this design, it is assumed at the outset that the probability of ending up in any cell is equal to the probability of ending up in any other cell. For example, an experimenter wishes to determine whether rats have a preference in a maze for a door on the right (D1), a middle door (D2), or a door on the left (D3). Let us suppose that 30 rats ran the maze, and the observed frequencies for the three doors were as follows: Door 1 = 13, Door 2 = 10, and Door 3 = 7.

Computation of the Three-Cell Design

In a chi-square test of this experiment, the null and alternative hypotheses would be

$$H_0 : p_{D1} = p_{D2} = p_{D3}$$
$$H_a : H_0 \text{ is not true.}$$

where

p_{D1} = the probability of choosing Door 1,

p_{D2} = the probability of choosing Door 2, and

p_{D3} = the probability of choosing Door 3.

Step 1. Construct a table of the observed and expected cell frequencies:

	Door 1	*Door 2*	*Door 3*
Observed cell frequencies	13	10	7
Expected cell frequencies	10	10	10

Note that the expected cell frequency is obtained by multiplying the theoretical probability of choosing Door 1 (1/3 or .33) times the total number of participants (rats) in the study (30). The expected cell frequency for all three cells is obtained the same way. The expected cell frequency is 1/3 × 30 = 10.

Step 2. Obtain the chi-square value by the following formula:

$$\chi^2 = \sum_{i=1}^{k} \frac{(\text{Observed Cell Frequency}_i - \text{Expected Cell Frequency}_i)^2}{\text{Expected Cell Frequency}_i}$$

where i = the first cell, then the second cell, and so on up to k, where k = the last cell.

So,

$$\chi^2 = \frac{(13-10)^2}{10} + \frac{(10-10)^2}{10} + \frac{(7-10)^2}{10}$$

$$\chi^2 = \frac{(3)^2}{10} + \frac{(0)^2}{10} + \frac{(-3)^2}{10}$$

$$\chi^2 = \frac{9}{10} + \frac{0}{10} + \frac{9}{10}$$

$$\chi^2 = 0.90 + 0.00 + 0.90$$

$$\chi^2 = 1.80$$

Step 3. Next, calculate the degrees of freedom (*df*) by the following formula:

$$df = \text{number of cells} - 1$$
$$df = 3 - 1 = 2$$

Step 4. The critical value of chi-square is obtained from the chi-square distribution in Appendix D. The critical value of chi-square at $p = .05$ with $df = 2$ is 5.99. An informal write-up appears as follows:

> The derived value of $\chi^2 = 1.80$ does not exceed the tabled critical value of chi-square at $p = .05$ with $df = 2$. Therefore, H_0 is retained, and it is concluded that the observed cell frequencies do not significantly vary from the theoretical cell frequencies beyond what might be expected by chance, $\chi^2(2, N = 30) = 1.80$, $p > .05$. In terms of the research question, it appears that the rats have no greater preference for a particular door over any other door. Although it was observed that the door on the right was chosen with the greatest frequency and the door on the left was chosen with the least frequency, the chi-square test revealed that none of the doors was chosen with significantly greater or lesser frequency than what would be expected by chance.

A formal write-up in a journal article appears as follows:

> A chi-square test revealed that the rats had no greater preference for any of the three doors in the maze, $\chi^2(2, N = 30) = 1.80$, $p > .05$.

Chi-Square Test: Two-by-Two Design

A more sophisticated and much more common chi-square design is the two-by-two design. It is actually used frequently in statistical analyses and is useful in many experimental situations. It is analogous to the two-factor analysis of variance designs in some respects. The experimenter is interested in knowing whether two factors are independent of each other. If they are independent, then knowing a participant's score on one factor tells the experimenter nothing about that participant's score on the second factor. If the factors are dependent on one another, then knowledge about one factor can help predict the second factor. If the two independent variables are dependent and the null hypothesis has been rejected, then there is a significant and predictive relationship between the two variables.

For example, let us return to the factors of gender and handedness. Are these two factors independent of each other? Or would knowing a person's gender help predict his or her handedness? Let us suppose that a professor notes the frequencies of the gender and handedness of 107 students. There are 22 right-handed males, 5 non-right-handed males, 76 right-handed females, and 4 non-right-handed females. Although it may not be obvious by observing the frequencies, the experimental question is whether the rate of non-right-handedness in males is greater than the rate of non-right-handedness in females. Thus, if the experimenter's hunch is true, and the two factors are dependent, then knowing a person's gender helps predict the greater likelihood of being right-handed or non-right-handed.

Computation of the Chi-Square Test: Two-by-Two Design

The null and alternative hypotheses are as follows:

H_0: Gender and handedness are independent.

or

H_0: $p_{r-m} = p_{r-f}$ (the probability of being right-handed for males is the same as the probability of being right-handed for females).

H_a: Gender and handedness are dependent.

or

H_a: $p_{r-m} \neq p_{r-f}$ (the probability of being right-handed for males is different from the probability of being right-handed for females).

Step 1. Construct a table of the observed cell frequencies.

	Male	*Female*	*Total*
Right-handed	22	76	98
Non-right-handed	5	4	9
Total	27	80	107

Step 2. Obtain the expected cell frequencies. For ease of computation, let us label the cells as follows:

	Male	*Female*	*Total*
Right-handed	22[a]	76[b]	98
Non-right-handed	5[c]	4[d]	9
Total	27	80	107

The expected frequency of Cell A is obtained by multiplying the column total for Cell A (27) times the row total for Cell A (98) and dividing by the overall total (107). Thus,

$$\text{Expected Cell A Frequency} = \frac{(\text{Column Total for Cell A} \times \text{Row Total for Cell A})}{\text{Overall Total}}$$

$$\text{Expected Cell A Frequency} = \frac{(27 \times 98)}{107}$$

$$\text{Expected Cell A Frequency} = 24.73$$

The expected frequency of Cell B is obtained by multiplying the column total for Cell B (80) times the row total for Cell B (98) and dividing by the overall total (107).

Thus,

$$\text{Expected Cell B Frequency} = \frac{(\text{Column Total for Cell B} \times \text{Row Total for Cell B})}{\text{Overall Total}}$$

$$\text{Expected Cell B Frequency} = \frac{(80 \times 98)}{107}$$

$$\text{Expected Cell B Frequency} = 73.27$$

The expected frequency for Cell C is obtained by multiplying the column total for Cell C (27) times the row total for Cell C (9) and dividing by the overall total (107). Thus,

$$\text{Expected Cell C Frequency} = \frac{(\text{Column Total for Cell C} \times \text{Row Total for Cell C})}{\text{Overall Total}}$$

$$\text{Expected Cell C Frequency} = \frac{(27 \times 9)}{107}$$

$$\text{Expected Cell C Frequency} = 2.27$$

The expected frequency of Cell D is obtained by multiplying the column total for Cell D (80) times the row total for Cell D (9) and dividing by the overall total (107).

Thus,

$$\text{Expected Cell D Frequency} = \frac{(\text{Column Total for Cell D} \times \text{Row Total for Cell D})}{\text{Overall Total}}$$

$$\text{Expected Cell D Frequency} = \frac{(80 \times 9)}{107}$$

$$\text{Expected Cell D Frequency} = 6.73$$

Now let us table the observed and expected cell frequencies:

	Male		*Female*	
	Observed	*Expected*	*Observed*	*Expected*
Right-handed	22	24.73	76	73.27
Non-right-handed	5	2.27	4	6.73

Step 3. The chi-square test formula remains the same:

$$\chi^2 = \sum_{i=1}^{k} \frac{(\text{Observed Cell Frequency}_i - \text{Expected Cell Frequency}_i)^2}{\text{Expected Cell Frequency}_i}$$

where i = the first cell, then the second cell, and so on up to k, where k is the last cell.

So,

$$\chi^2 = \frac{(22 - 24.73)^2}{24.73} + \frac{(76 - 73.27)^2}{73.27} + \frac{(5 - 2.27)^2}{2.27} + \frac{(4 - 6.73)^2}{6.73}$$

$$\chi^2 = \frac{(-2.73)^2}{24.73} + \frac{(2.73)^2}{73.27} + \frac{(2.73)^2}{2.27} + \frac{(-2.73)^2}{6.73}$$

$$\chi^2 = .30 + .10 + 3.28 + 1.11$$

$$\chi^2 = 4.79$$

Step 4. Next, the degrees of freedom (df) are obtained by the following formula: df = (number of levels of Factor 1 – 1) × (number of levels of Factor 2 – 1)

$$df = (2-1)\,(2-1) = 1 \times 1 = 1$$

Step 5. The critical value of chi-square is obtained from the chi-square distribution in Appendix D. The critical value of chi-square at $p = .05$ with $df = 1$ is 3.84. The informal write-up appears as follows:

> The derived value of $\chi^2 = 4.79$ exceeds the tabled critical value of $\chi^2 = 3.84$ at $p = .05$ with $df = 1$. Therefore, H_0 is rejected, and it is concluded that the two factors, gender and handedness, are dependent, $\chi^2(1, N = 107) = 4.79$, $p < .05$. In terms of the research question, it appears that in this sample, males have more than a four times greater rate of non-right-handedness (23% of the males were non-right-handed) than females (5%).

A formal write-up for a journal article might appear as follows:

> Males have more than a four times greater rate of non-right-handedness (23% of the males were non-right-handed) than females (5%), $\chi^2(1, N = 107) = 4.79$, $p < .05$.

What to Do After a Chi-Square Test Is Significant

There is an equivalent to the multiple comparison tests of analysis of variance in chi-square, and it is referred to as the **test of the standardized residuals.** This test examines each cell and determines whether the cell makes a substantial contribution to the overall significance of the test. The formula, which is applied to each cell, is as follows:

$$R = \frac{\text{Observed Cell Frequency} - \text{Expected Cell Frequency}}{\text{Expected Cell Frequency}}$$

where R = the absolute value of the standardized residual.

Note that the absolute value of the standardized residual must exceed a value of 2.00 in order for that cell to be considered a major contributor to the overall significance of the chi-square value. Interestingly, in the previous 2×2 example, none of the cells exceeds the critical value of 2.00. In this case, it appears that the test of standardized residuals is too conservative. Since post hoc analyses in chi-square are relatively rare, and in a theoretical sense, all of the cells must contribute to the overall significance of the chi-square statistic, one post hoc strategy (besides the test of standardized residuals) is to look simply at the individual chi-square values of each cell in the original chi-square formula or in the test of standardized residuals and note their relative rankings. In the previous examples (Step 3), it was observed that Cell C and Cell D contributed the most to the overall chi-square value. Thus, it can be concluded that the relative *high* rate of non-right-handed males and the relative *low* rate of non-right-handed females had the most substantial contributions to the overall chi-square significance. It should also be noted that there would be a perfect correlation between the rankings of the cells in terms of their contributions to the original chi-square formula and the rankings of the cells in the test of standardized residuals.

When Cell Frequencies Are Less Than 5 Revisited

Chi-square designs can become more complicated than a simple two-by-two design. There can be two or more levels of any number of factors such as 2 × 3, 2 × 10, or 2 × 2 × 3 designs. Many times in these more complicated designs, more than one cell might fall below the minimum observed cell frequency of 5. One solution in these designs is to collapse one of the factors into only two cells or three cells to increase the cell frequencies. Some information is invariably lost in this procedure, but the chi-square test will be more sensitive to real differences among the cells with the increase in cell size.

For example, examine the following frequency table:

	Handedness by Education (in years)					
	9	*10*	*11*	*12*	*Some College*	*B.A. or B.S.*
Right-handed	2	2	3	10	5	3
Non-right-handed	1	1	1	3	3	2

In this 2 × 6 chi-square design, only 2 of the 12 cells meet the minimum recommended observed cell frequency. However, if the education factor is collapsed into a dichotomous variable, such as 12th grade and below versus some college work or greater, then the following table results:

	Handedness by Education (in years)	
	12th Grade or Less	*Some College or Greater*
Right-handed	17	8
Non-right-handed	6	5

Now, all of the cells meet the minimum criterion.

History Trivia

Pearson and Biometrika

Karl Pearson (1857–1936) developed the chi-square test in the 1890s. He termed it a test of goodness-of-fit by measuring the discrepancy between an actual distribution of numbers and a mathematical model based on probability theory. His son, Egon Pearson, later wrote that chi-square was a "new powerful weapon in the hands of one who sought to battle with the myths of a dogmatic world" (Peters, 1987, p. 105). He also wrote that chi-square was "one of Pearson's greatest single contributions to statistical theory." Helen Walker, a statistician and historian, wrote of Karl Pearson that few people "in all of the history of science have stimulated so many other people to cultivate and enlarge the fields they have planted" (Walker, 1929/1975).

In 1900, Karl Pearson had written a paper on the similarities and differences in plants and submitted it to the Royal Society in England for publication. Pearson felt that the paper's referees were overly critical of his paper, and he thought that the mainstream biologists did not fully appreciate statistical methods. Although the paper did finally get published, eminent zoologist W. F. R. Weldon, who was sympathetic to Pearson's ideas, suggested that Pearson establish a journal devoted to the mathematical and statistical contributions to the biological sciences. This new journal was called *Biometrika,* and Pearson was its editor for almost 40 years.

Key Terms, Symbols, and Definitions

Chi-square (χ^2) test—One of the most popular nonparametric tests that involves the assessment of one or more independent variables, each with two or more levels of nominal or categorical data.

Goodness-of-fit test—Karl Pearson's original name for the chi-square statistic. A goodness-of-fit test measures how well observed frequencies match theoretically expected frequencies.

Nonparametric statistics—A variety of statistical tests that make no assumptions about the shape of the underlying population distribution. The word *nonparametric* implies that standard statistical parameters, such as mean and standard deviation, do not apply or are not appropriate to these types of data.

Test of the standardized residuals—A post hoc analysis in chi-square, analogous to multiple comparison tests in ANOVA, which determines which cells make substantial contributions to the overall significance of the chi-square test.

Chapter 15 Practice Problems

1. The Humane Society has to set up a budget for the coming year. The concern is whether they need to buy more dog food or more cat food. Over the past year, they have housed 637 cats

and 725 dogs. If the population of dogs and cats is theoretically equal, was the sample they cared for last year different from the norm?

2. A psychologist wishes to determine whether Tourette's syndrome occurs more in males than in females. She noted the gender and the absence and presence of Tourette's syndrome in a large hospital's consecutive admissions for the past 5 years. Are gender and Tourette's syndrome independent?

   1024 men without Tourette's
   27 men with Tourette's
   986 women without Tourette's
   9 women with Tourette's

Chapter 15 Test Questions

1. A nonparametric statistical test has _____ power compared to a parametric statistical test.
   a. more
   b. less
   c. the same
   d. about the same

2. Chi-square tests are designed to determine whether an observed frequency differs from
   a. chance or an expected value
   b. the parametric value
   c. the mean in the null hypothesis
   d. the mean and standard deviation of the population

3. Karl Pearson originally called the chi-square test a
   a. parametrically challenged test
   b. normal distribution test
   c. "jolly good" test
   d. goodness-of-fit test

4. A chi-square test can be similar to a factorial ANOVA in that it can analyze two or more __________ at once.
   a. independent variables
   b. normal distributions
   c. people or subjects
   d. mean scores

5. Which of the following is not an assumption of the chi-square test?
   a. the data come from normal distributions
   b. independent scores
   c. minimum of five scores per cell
   d. dependent variable is a frequency or count

Problems 6–8. A statistician in a small town wishes to determine if persons treated for an accident involving a joint injury were more likely to go to an orthopedist or a chiropractor. In the small town, there is only one orthopedist and one chiropractor. The orthopedist saw 118 people for joint injuries in one year. The chiropractor saw 83.

6. The degrees of freedom for this chi-square problem are
   a. 1
   b. 2
   c. 5
   d. 200

7. The derived chi-square value for this problem is
   a. 3.94
   b. 1.09
   c. 5.69
   d. 6.09

8. The conclusion for this problem would be
   a. people preferred the orthopedist significantly more at $p < .02$
   b. people preferred the orthopedist significantly more at $p < .001$
   c. people preferred the chiropractor more at $p < .05$
   d. there was no significant difference between the two according to people's preference

Problems 9–11. A consumer psychologist was testing a new coffee blend. She randomly chose 230 people to try a taste of three samples (Sample A, Sample B, and Sample C). The results were that 98 people chose Sample A, 42 people chose Sample B, and 90 chose Sample C.

9. The degrees of freedom for this chi-square problem are
   a. 1
   b. 2
   c. 5
   d. 200

10. The derived chi-square value for this problem is
    a. 23.93
    b. 16.55
    c. 17.16
    d. 14.14

11. The conclusion for this problem would be
    a. there were significant differences in blend preferences at $p < .05$
    b. there were significant blend preferences at $p < .01$
    c. there were significant blend preferences at $p < .001$
    d. there were no significant blend preference differences ($p > .05$)

Problems 12–15. A heart surgeon wishes to determine whether an atherectomy (cutting out plaque in arteries) or balloon angioplasty (inserting a balloon in an artery) is a safer procedure. The heart surgeon randomly assigns 1024 heart attack patients to one of the two procedures. After 1 year, the outcome is as follows:

495 patients alive after atherectomy
503 patients alive after balloon angioplasty

17 patients died after atherectomy
9 patients died after balloon angioplasty

12. The degrees of freedom for this chi-square problem are
    a. 1
    b. 2
    c. 5
    d. 200

13. The derived chi-square value for this problem is
    a. 2.53
    b. 17.16
    c. 6.09
    d. 3.84

14. The expected cell frequency for the people living after atherectomy is
    a. 562.62
    b. 499.00
    c. 387
    d. 56.62

15. The conclusion for this problem would be
    a. an atherectomy is significantly more dangerous than a balloon angioplasty
    b. a balloon angioplasty is significantly more dangerous than an atherectomy
    c. an atherectomy is not significantly more dangerous than a balloon angioplasty
    d. there is an insignificant difference between a balloon angioplasty and an atherectomy

Problems 16–20. A psychologist wanted to determine what was more important in the choice of a spouse: love or money. The psychologist surveyed 290 unmarried psychology students and asked them whether they would choose a prospective mate based on love or money. The professor was also interested to see if the choice would vary by gender. The professor found that 68 men preferred a spouse based on love, 33 men preferred a spouse based on money, 79 women preferred a spouse based on love, and 110 women preferred a spouse based on money.

16. The degrees of freedom for this chi-square problem are
    a. 1
    b. 2
    c. 5
    d. 200

17. The expected cell frequency for the men who preferred a spouse for love is
    a. 51.197
    b. 52.197
    c. 53.197
    d. 54.197

18. The critical value of the chi-square for this problem at $p = .05$ is
    a. 2.71
    b. 3.84

c. 5.41
d. 6.64

19. The derived chi-square value for this problem is

    a. 69.37
    b. 17.16
    c. 6.09
    d. 3.84

20. The conclusion for this problem would be

    a. a greater percentage of men than women married for love
    b. a greater percentage of women than men married for love
    c. more women married for love
    d. more men married for money

16 Other Statistical Parameters and Tests

Chapter 16 Goals

- Learn common statistical parameters of the health sciences and epidemiology
- Appreciate multivariate statistics philosophically
- Appreciate multivariate statistical techniques strategically

There is a plethora of other statistical parameters, tests, and investigative procedures besides the ones presented in this book. Most scientific disciplines share a general body of statistical knowledge like basic statistical parameters such as the arithmetic mean, standard deviation, frequency distributions, hypothesis testing, and correlational procedures. Yet most disciplines also favor particular statistical tests and investigative procedures, and most have statistics that are unique to or favored by their discipline. The purpose of the present chapter is twofold: first, to present an introduction to some of the statistical parameters that are frequently used in medicine, nursing, epidemiology, and health science disciplines, and second, to present an overview of some of the more popular yet more complicated statistical tests and procedures known as **multivariate statistics.** Multivariate statistical procedures have formulas so complicated that their computation can only be performed by computers, and their interpretation is not without controversy.

Health Science Statistics

Test Characteristics

While the health sciences share most of the statistical parameters, tests, and experimental investigative procedures presented in this book, these sciences also employ some discipline-specific parameters. Many of them come from the discipline of epidemiology, which is the study of disease. One of the most essential concepts in this discipline is also part of signal detection theory, which was discussed earlier in this book.

Signal detection theory is reconfigured in epidemiology and can be graphically represented by a **receiver operating curve** (**ROC**). Before panic sets in, allow me to explain at the outset that these are relatively simple ideas, parts of which you have undoubtedly used or have understood intuitively already in your lives. At its heart, we are concerned with whether a disease is truly present or absent. For example, the early detection of cancer is extremely important to us all since cancer treatment remains as primitive as ever: burn it, poison it, or cut it out. And statistics reveal that the earliest treatment of cancer is associated with a better prognosis. Many types of cancers have tests that are designed to predict their earliest presence or precancerous conditions. However, the tests are not perfect. Ideally, we would prefer that the test works perfectly and is never wrong.

There are two ways a test can be right and two ways a test can be wrong. One way for the test to be right is for the test to indicate the presence of cancer only when the patient really does have cancer. This situation is said to be a **true positive.** The first word in this tandem refers to the genuine truth or falsity of the test's decision. The second word refers to whether the test indicated the presence of the disease (positive) or absence of the disease (negative). A true positive indicates that the test determined that cancer was present (the test results were positive) and, indeed, the patient really did have cancer (true positive). The other way the test can be right is by a **true negative.** In this situation, the test has determined that the patient does not have cancer (the test results were negative), and it is subsequently found that the patient did not have cancer (true negative).

One way the test can be wrong is by a **false positive.** A false positive indicates that the test determined that cancer was present; that is, the test came out positive, but the test was wrong, and the patient did not really have cancer. The other way the test can be wrong is by a **false negative.** In this situation, the test indicates that the patient does not have cancer (the test results are negative) when the patient really does have cancer (false negative). While the latter mistake has mortal consequences, so may the former mistake. With a false negative, the cancer may spread until the cancer cannot be treated because the test failed to pick up the earliest signs of the disease. We would always hope that the test we use to detect the earliest signs of cancer would be sensitive to the earliest signs of cancer. Thus, a characteristic of a good test is the test's **sensitivity.**

A test possesses good sensitivity if it has a high true-positive rate. In practice, this rate may come at some cost, and that is, it may also have a few false positives to achieve a high true-positive rate. In other words, the test is a good one if it identifies nearly all of the people who really have the disease, plus a few people who do not. While a false-positive indication does not have all of the dire consequences of a false negative, there are consequences: A patient with a false-positive indication may unnecessarily be subjected to unneeded treatment or surgery. The patient may even be harmed or killed as a result of these treatments. The adverse effects of the treatment of a disease are referred to as **iatrogenic effects.** With the disease of cancer, either false positives or false negatives can have mortal consequences because many cancer treatments have mortality rates associated with them. Refer to Table 16.1 for a summary of these decisions.

Table 16.1

		In reality, the patient	
		. . . has the disease	*. . . does not have the disease*
This test indicates:	Positive	True positive a	False positive b
	Negative	False negative c	True negative d

This model of disease detection does assume that the patients have been followed up long term, and in the final analysis, there is no question as to whether the patient did or did not have the disease. This assumption is obviously questionable, and we will discuss later in the chapter how test results and outcomes may be specious. Nevertheless, epidemiologists use the model to explain the characteristics of good and bad tests, and the model is certainly useful in disease prevention and treatment.

As discussed earlier, a test's sensitivity is defined as its ability to detect most of the people with the disease and maybe a few patients who do not have the disease. Sensitivity may be calculated as a proportion by taking the number of true positives and dividing by the total number of people with the disease. Thus,

$$\text{Sensitivity} = \frac{a}{(a + c)}$$

If a test is reported to have a sensitivity of 87%, it means that of 100 patients who are really known to have the disease, 87 of them tested positively for the presence of the disease (true positives). The other 13 are considered to be false negatives because the test said they do not have the disease, but they do.

Another characteristic of a good test is its **specificity,** and this means that it should also correctly rule out people who do not have the disease, although it may include a few people who do not have the disease, but the test said they do (false positives). A test's specificity may be calculated as follows:

$$\text{Sensitivity} = \frac{d}{(b + d)}$$

Another way of thinking about a test's specificity is to answer this question: If a person does not have the disease, what is the probability the test will be negative? If a test has a specificity of 79%, it means that out of 100 people who are really known not to have the disease, 79 of them did not have the disease according to the test (true negatives). The other 21 are considered to be false positives because the test said they have the disease, but they do not.

The overall **diagnostic accuracy** may be calculated by combining the total number correct (true positives + true negatives) and dividing by the total combination of all outcomes (true positives + false positives + true negatives + false negatives). The diagnostic accuracy of a test is sometimes called the *efficiency* of a diagnostic test. Hence,

$$\text{Diagnostic Accuracy} = \frac{(a + d)}{(a + b + c + d)}$$

Another increasingly popular measure of diagnostic accuracy is the likelihood ratio. The **likelihood ratio for a positive test** combines the sensitivity and specificity of a test into a single parameter. This ratio expresses the likelihood that the test will be positive in a person with the disease compared to a positive test in a person without the disease. The formula consists of dividing the test's sensitivity by 1 minus the test's specificity. Thus,

$$\text{Likelihood Ratio for a Positive Test} = \frac{\frac{a}{(a+c)}}{1 - \frac{d}{(b+d)}}$$

The likelihood ratio of a positive test might be reported as 8.5, which would be interpreted that a positive result of the test is 8.5 times more likely to be seen in a person with the disease compared to a person without the disease.

The **prevalence** of a disease may also be derived by the previous characteristics. Prevalence of a disease is defined as the number of people in a population who have the disease at a particular point in time divided by the total number of people in the entire population (those with and without the disease). Thus,

$$\text{Prevalence} = \frac{(a + c)}{(a + b + c + d)}$$

The **incidence** of a disease is defined as the number of verified new cases of the disease in a population for a period of time (usually 1 year) divided by the total number of people in the population. The parameters of prevalence and incidence are useful in the prediction of diseases and helpful in decision making regarding the control and treatment of diseases. For example, although autistic disorders in children are well known and stories about the disease are common in the media, the prevalence of the disease is actually very low, and its incidence is extremely stable. In contrast, the prevalence of AIDS in some countries used to be the same as autistic disorders, but the incidence has increased dramatically every year. In the same fashion, in some countries, while the prevalence of AIDS reached higher and higher levels, the incidence began to drop. Some epidemiologists inferred that the drop in incidence indicated that prevention programs had begun to work.

Earlier, in Chapter 3, it was noted that sometimes test results are not so clear, and sometimes the final determination may even be uncertain or equivocal. One common example of **equivocal results** comes from home pregnancy tests. In some of these kits, the test is supposed to turn red if the woman is pregnant or stay white if she is not pregnant. But what happens if the indicator turns pink? This is an example of an **intermediate result** where the result is neither positive nor negative but falls between these two poles.

A second type of equivocal result is the **indeterminate result.** This means that the test result is neither positive nor negative and does not fall between the two poles. In this case, there may be a variety of reasons related or unrelated to the test that precludes a valid reading. An example of an indeterminate result is when an X-ray cannot determine whether a person has a bone fracture or not, or an MRI cannot tell whether a person has sustained a closed head injury.

A third type of equivocal result is the **uninterpretable result,** in which the test was not performed according to the standardized procedures. Uninterpretable results can occur because the person using the test is not completely familiar with the testing procedures. Equivocal results may be annoying, but they are common throughout the research realm. It may or may not be the fault of the experimenter. In cases where it may not be the experimenter's fault, the test may be state of the art, but it may also be that the "art" is simply not that good at this point in time. In the latter case, it would behoove anyone involved with a scientific experiment to be fully informed of the experimental and standardized procedures. In dealing with equivocal results, it is the responsibility of the researcher to report the frequency or proportion of equivocal results since their publication may motivate others to improve upon inadequate tests, testing procedures, and so on.

Risk Assessment

A common method of determining which procedure should be chosen in health-related professions is the statistical assessment of risk. As noted earlier, diagnostic procedures and treatments may themselves have harmful or even deadly (iatrogenic) effects, so their choice becomes a very important decision. **Absolute risk reduction** refers to the reduction in risk between two groups treated differently when their outcomes are compared. For example, imagine if, after a first heart attack, 100 women took Vitamin E supplements daily for 3 years while 100 similar women did not. The incidence of a second heart attack is determined for both groups over the 3-year period. If the incidence of a second heart attack in the Vitamin E group is 11% and the incidence for the unsupplemented group is 14%, then the absolute risk reduction is 3 percentage points (or 14%–11%). The absolute risk reduction is formulaically obtained by subtracting the proportion of people with the outcome of interest who were untreated from the proportion of those who were treated. Thus,

Absolute Risk Reduction = Proportion of Treated Group's Outcome – Proportion of Untreated Group's Outcome

An absolute risk reduction of 3 percentage points indicates that Vitamin E supplements may reduce the risk of heart attack by 3 percentage points. A separate statistical procedure, significance testing, determines whether these results could have occurred by chance, and even under those circumstances, replication of the findings by other experimenters would be necessary for the findings to be trustworthy. Absolute risk reduction is often coupled with significance testing, making it a highly valuable and easily interpreted parameter of general statistical interest.

The relative risk reduction is a similar parameter, except it is expressed as a percentage. **Relative risk reduction** is a measure of how much reduction in risk (as a percent) has occurred between two groups treated differently (most commonly, the experimental and control groups) on an outcome of interest when compared to the untreated group's percentage of the outcome of interest. The formula is as follows:

$$\text{Relative Risk Reduction} = \frac{(P_C - P_E)}{P_C} \times 100\%$$

where

P_C = the proportion of the control group with the outcome of interest,

P_E = the proportion of the experimental group with the outcome of interest.

In a recent study of 1100 potato chip eaters, experimenters were interested in the incidence of gastric troubles and flatulence in a diet potato chip compared to regular potato chips. It was found that the incidence (of these outcomes for 48 hours) for the diet chip was 16%, while the incidence was 18% for the regular chip. Thus, the relative risk reduction (RRR) would be calculated as RRR = [(18% – 16%) / 18%] × 100% or 11%. The experimenters observed an 11% reduction in gastric problems and flatulence with the new diet potato chips compared to regular potato chips. Once again, however, only significance testing would reveal whether the observed relative risk reduction was attributable to chance.

Parameters of Mortality and Morbidity

Mortality refers to the proportion of deaths relative to the population, while morbidity refers to the incidence of disease or sickness in a sample or population. Since the specific causes of death (morbidity) are more difficult to measure than death itself (mortality), there are more indicators of mortality than morbidity. Nevertheless, there are a wide array of rates, proportions, and percentages based on mortality and morbidity throughout the health sciences. One of the simplest and most common measures of the general health of a population is the annual crude death rate. The **annual crude death rate** is defined as the number of deaths in a year divided by the population (as of midyear or July 1 of that same year). Frequently, the quotient in this determination is multiplied by a factor of 100 or a higher multiple (1000, 1,000,000, etc.) to make more interpretable sense of the statistic. The multiple is chosen based on the magnitude of the quotient. If the quotient is very small (e.g., .000001), it would be senseless to use a base of 1000, while with a larger quotient (.067), it might make perfect sense to use a base of 1000. The formula is as follows:

Annual Crude Death Rate = [Number of Deaths in a Calendar Year/The Population at Midyear] × [A Multiple of 100 or Higher]

For example, approximately 2,523,000 died in the United States last year. The population at midyear was approximately 290,000,000 (these two numbers are not official estimates). So the calculation would be

$$\text{Annual Crude Death Rate} = \frac{2{,}523{,}000}{290{,}000{,}000} \times 1000$$

$$\text{Annual Crude Death Rate} = 8.7 \text{ deaths per 1000 people}$$

The annual crude death rate is used frequently to compare the general health of countries across the world. It is acknowledged that it is not an exact indicator of the health status of a country since it sums across such factors as age, gender, ethnicity, and conditions such as war and natural disasters. In 1992, the annual crude death rate in Costa Rica was 4.0 deaths per 1000, one of the lowest in the world, while in Afghanistan, the annual crude death rate was 22.1 deaths per 1000, one of the highest in the world.

Death rates may also be reported as age specific, cause specific, age-cause specific, or other combinations. In **age-specific death rates,** an age range is specified (e.g., 16–24 years), and the number of deaths in that age range (for a given year) is divided by the total population of that age range at midyear. Factor-specific death rates are useful in determining populations of people at risk as a function of their age, gender, ethnicity, or other variables and are very frequently used as arguments before political groups for health intervention and disease prevention. In cause-specific death rates, a cause of death (e.g., motor vehicle, accidents, suicide, etc.) is specified, and the number of deaths due to that specific cause is divided by the population at midyear. For example, motor vehicle accidents have the greatest cause-specific death rates of all accidental deaths in the United States at approximately 17 deaths per 100,000.

In the prevention and treatment of disease, the incidence of morbidity is a highly useful parameter even though morbidity can be much more difficult to determine than mortality. Morbidity is concerned with the measurement of disease, and many times the stigma and biases associated with diseases prevent their accurate report. For example, Freddy Mercury, the lead singer of the rock group Queen, publicly admitted he had AIDS the day before he died of the disease.

Probably the most common measure of morbidity is the incidence rate. Incidence rate is defined as the number of new cases of a particular disease in a year divided by the population at midyear. The quotient is multiplied by an interpretable factor such as 1000 or higher. For example, in 1992, there were 28,215 new cases of AIDS reported. The total population of the United States at that time was about 257,000,000. Thus, the incidence of AIDS in 1992 was 11.0 cases per 100,000.

The prevalence proportion is another useful statistical parameter associated with morbidity. The **prevalence proportion** is defined as the total number of cases of a disease known to exist at a given point in time divided by the total population at the same time. This quotient is also multiplied by an interpretable multiple such as 1000, 100,000, and so on. The prevalence proportion can also be reported specific to an age, gender, or other demographic combination. For example, in 1992, there were 1991 known cases of AIDS in Denver, Colorado. The population in Denver at that time was about 1,700,000. The prevalence proportion was 11.7 per 10,000. The prevalence proportion of AIDS in Miami, Florida, at this same time was 38.4 per 10,000.

The case-fatality proportion is used as an indicator of how serious a particular disease might be. The **case-fatality proportion** is defined as the number of deaths attributable to a particular disease during a period of time divided by the prevalence proportion of the disease at the same time. The quotient is reported as an interpretable multiple. For example, in 1992, there were 23,411 known deaths from AIDS in the United States. The prevalence of AIDS patients at that time was approximately 82,000. Thus, the case-fatality proportion was 28.5 per 100 cases. Obviously, the latter figure would indicate that AIDS was a very serious and fatal disease in 1992. With the advent of new drug combinations for the treatment of AIDS, the case-fatality proportion for people with AIDS has fallen dramatically. Despite its usefulness and popularity, the case-fatality proportion is still fraught with interpretation and measurement problems. Because AIDS, in particular, is known to mutate, it is possible that the newer cases of the disease may not have the same mortality rates as earlier cases of the disease. Also, because of successful new treatments of AIDS, newer cases may have lower mortality rates than older or more advanced cases. At best, the case-fatality proportion is a useful but crude index of the seriousness of a disease.

Analysis of Covariance

An **analysis of covariance (ANCOVA)** is similar to a standard analysis of variance. The focus is on determining differences among group means, either for a single independent variable with two or more levels or multiple independent variables with two or more levels each. There is also a single dependent variable. Thus, ANCOVA can have the same designs as in ANOVA; however, an ANCOVA statistically controls for a confounding or nuisance variable that has an effect on the dependent variable. For example, imagine a researcher was interested in age differences in working memory (short-term memory). Fifteen younger people and 15 older people were tested for their working memory (the maximum number of digits that could be repeated in a listening paradigm). Consistent with the researcher's hypothesis, the younger people showed a small but statistically significant advantage. During the preparation of the manuscript for publication, the researcher reads that working memory appears to have a strong positive correlation with IQ ($r = .50$). Perhaps the younger people in the study simply have higher IQs than the older people and IQ, not age, accounts for the difference in working memory. The researcher happens to have an estimate of the participants' IQs because of an elaborate pretesting procedure. An ANCOVA will remove or covary out the effect of IQ on working memory and still allow for the assessment of the effect of age on working memory.

ANCOVA has all of the assumptions of ANOVA, and in addition, ANCOVA assumes that the effect of the covariate (the nuisance variable) is the same on the dependent variable under all levels of the independent variable. It is also important to note that the covariate should be measured prior to the measurement of the dependent variable. A violation of this latter assumption may mean that the ANCOVA will remove important and useful variance from the dependent variable if they are gathered at the same time.

Multivariate Analysis of Variance

A **multivariate analysis of variance (MANOVA)** is similar to all design forms of ANOVA with the exception that two or more dependent variables are measured at once. A MANOVA can give a better picture of the complex effects of independent variables on behavior because it allows for the measurement of more than just one dependent variable. However, the cost of this picture is complexity both in computation (it can only be performed by computer) and in interpretation. When a simple one-factor MANOVA is statistically significant, it indicates that the combination of dependent variables varies among the levels of the independent variable. The interpretation, however, is restricted to univariate post hoc interpretations of each dependent variable by itself. A significant MANOVA does not reveal the pattern of significant differences.

Furthermore, there is a controversy about whether a MANOVA provides protection against the Type I error when using multiple dependent variables. It was initially thought that it does provide protection against the Type I error, but it seems a consensus is forming against a MANOVA providing Type I error protection. Whether it does or not, the inclusion of a number of dependent variables in a MANOVA that are not thought to vary as a function of the independent variable may increase the probability of the Type II error. In this situation, the combination of irrelevant dependent variables hinders the detection of a potentially significant dependent variable. One moral of this tale is that one should always be thoroughly familiar with a statistical procedure, able to defend its use, and know of its abuses. Otherwise, it is better to employ statistical procedures with which one is comfortable. The ultimate criterion on a statistical choice should be whether the statistic helps the experimenter to better understand the phenomenon in question. Again, however, a significant MANOVA would only tell us that the vector of means of the dependent variables was significantly different between the groups. The post hoc analyses would be traditional ANOVAs. For example, I once performed a MANOVA on 12 personality disorders as the dependent variables and a diagnosis of closed head injury or not as the two levels of a single independent variable. I found a nonsignificant overall MANOVA, but post hoc analyses did reveal a few significant individual personality disorders differing between the two groups.

Multivariate Analysis of Covariance

A **multivariate analysis of covariance** (**MANCOVA**) consists of analyzing the effects of one or more independent variables on more than one dependent variable while controlling for the effect of one or more covariates (nuisance or confounding variables) on these dependent variables. Again, the interpretation of a MANCOVA can be difficult. One of my last classes in graduate school was a course in multivariate statistics, and I used MANCOVAs in my dissertation. During my final oral defense, one member of my dissertation examining committee sat unusually quiet. After my successful defense, I asked him why he did not ask me any questions. "Because I didn't understand a darn thing you did," was his reply. In my dissertation, I was interested in measuring two forms of memory, verbal and visual (the dependent variables), as a function of varying amounts of rapid eye movement (REM) sleep and slow-wave sleep (independent variables). I found that the time of day confounded the dependent measures, so I used time of day as the covariate in the MANCOVA.

Factor Analysis

There are a number of purposes of the family of statistical techniques known as **factor analysis.** Factor analysis (often used interchangeably with the term *principal components analysis*) explores the underlying conceptual structure of a set of dependent variables by examining the correlations between each variable in the set with every other variable in the set. Factor analysis helps to determine which variables in a set are highly correlated. Thus, factor analysis can be used to reduce a set of variables to a smaller set by removing redundant variables. Factor analysis can also be used to identify underlying factors in a large set of variables so that the entire set may be better understood conceptually. For example, factor analysis played an important role in the concept of IQ. A factor analysis performed on sets of various intellectual tasks most often yielded a single underlying factor (i.e., most of the variables in the set of tasks were intercorrelated, and all loaded on a single underlying factor). Even modern computer programs do not label the underlying factor. It must be named by the statistician by looking at which individual items have the highest single loadings. In the case of the factor analysis of IQ, the single underlying factor for the set of intellectual tasks was subsequently named the "*g*" factor for general intelligence.

The experimental design in a factor analysis does not consist of levels of an independent variable. There is only one large sample of participants measured on a number of dependent variables. It is recommended that the number of participants should be greater than 10 times the number of dependent variables. If there are 15 items, questions, or dependent variables, then there should be at least 150 participants in the factor analysis to yield replicable results.

For example, one major focus of personality research in psychology has been the number of factors underlying people's personalities. One initial step in this research was to gather hundreds and hundreds of adjectives that could be used to describe personality. Next, a summary list is compiled of these adjectives, and a self-rating scale is developed, such as 1 = not like me, 2 = sometimes like me, and 3 = exactly like me. If 50 adjectives were compiled on the list, then 500 subjects would be required for an appropriate factor analysis. The results of the factor analysis, of course, would be highly dependent on the adjectives in the original list. Thus, the survey of adjectives should be comprehensive and exhaustive. Once these data were actually subjected to factor analyses, the results revealed five underlying factors, with each of the adjectives loading on at least one factor. It is a mathematical goal of some types of factor analysis to attempt to find a solution where each adjective only loads on one factor. By examining the highest loadings (which are actually read like correlation coefficients), each factor can be named. Personality researchers found, for example, that one group of adjectives clustered on what appeared to be a continuum of extraversion to introversion. This personality factor has been one of the most consistent findings in personality.

One interesting aspect of factor analysis is that there are an infinite number of correct solutions. The "correct" solution is the one that answers the research question the best according to the opinion of the person conducting the research.

Multiple Regression

Multiple regression, as noted in Chapter 6, is an extension of bivariate correlation. The word *regression* generally implies a concern with prediction, while the word *correlation* implies that the concern is with the relationship among variables. In **multiple regression,** a single dependent variable (or criterion variable) is predicted from several independent variables (or predictor variables). The research concern is often twofold: To what extent can the independent variables predict the dependent variable, and what is the strength of each independent variable in the prediction of the dependent variable? There are many different types of multiple regression, such as stepwise regression, forward regression, and backward regression. All of these techniques vary the way in which the independent variables are introduced into the regression equation. The different techniques may give different pictures of the interrelationships between the independent variables and the dependent variable. Interestingly, the different techniques often yield similar results.

Canonical Correlation

Canonical correlation is essentially a correlation-regression technique in combination with a kind of factor analysis that yields different estimates of how two different sets of variables may be related to each other. The two sets of variables can be specified as the criterion set and the predictor set, and the experimental question is to what extent a weighted combination of the criterion set correlates with a weighted combination of the predictor set. The two sets of variables may also be specified as two sets of independent variables. The resulting correlation is called the canonical correlation coefficient. A canonical analysis also derives an amount of variance that is accounted for in the criterion set by the predictor set. While canonical correlation is not exceptionally popular, it is useful in determining complex relationships (as in a test's construct validity) between two personality tests. For example, I recently compared the results of a canonical correlation between a measure of 14 personality disorders and a three-dimensional measure of personality based on the work of American psychoanalyst Karen Horney (Horney-Coolidge Tridimensional Inventory, or HCTI) (Coolidge, 1998). I found that the first canonical variate accounted for about 45% of the total variance between the two tests and that it consisted of positive contributions between nearly all the 14 personality disorders, with a negative contribution from the

Compliance scale of the HCTI and a positive contribution from the Aggression scale of the HCTI. These results were interpreted as demonstrating that most people with personality disorders have a substantial amount of aggression and that they are not very compliant, which in part explains their great difficulty in therapy. Multiple canonical variates can be extracted, but each successive variate will account for a decreasing amount of variance.

Linear Discriminant Function Analysis

Multiple regression was used to determine how a criterion variable could be predicted by a set of independent variables or predictor variables. **Linear discriminant function analysis (LDFA)** is similar in that independent variables are used to predict group membership. In this design, the criterion variable is dichotomous instead of continuous. There is also a family of LDFA techniques that differs on the entry of the predictor variables in the LDFA equation.

LDFA could also be used on the data in the previous example, but dichotomous group membership in the criterion variable would have to be established initially. For example, we might identify two groups of people, those with personality disorders and those without, and then these groups might be arbitrarily labeled Group 1 and Group 2. Next, each person's three HCTI scores (Compliance, Aggression, Detachment) would be used as predictor or independent variables. Essentially, it would be like three independent *t* tests would be conducted between each of the two groups. The highest *t* value of the three predictor variables would be chosen first for the formation of a linear discriminant function equation. Then, the next highest *t* value for the remaining two predictor variables would enter the linear discriminant function equation. This process would repeat until a nonsignificant *t* value was encountered (in stepwise linear discriminant function analysis) or until all three predictors were entered (normal linear discriminant function). At this point, a linear discriminant function equation would be generated for each group, and group membership for each person would be predicted and then compared to his or her actual preassigned group membership (personality disordered or not). The success of these predictions would then be reported.

Note that LDFA requires that the two criterion groups be specified before the LDFA. Group differences (as in personality disordered or not) are an a priori assumption. The subsequent LDFA will decide the extent to which these groups may be successfully classified according to the predictor variables. One problem associated with LDFA is the problem of shrinkage. Shrinkage refers to the phenomenon where an LDFA equation, derived on a sample of data, will have substantially less predictive power when applied to a new sample. This is often seen where LDFA on a specific sample has a high classification rate (usually around 90% or better) yet falls near 50% when the same LDFA equation is used to classify another sample. Shrinkage can be minimalized by having a large sample (the number of participants in the smallest of the two groups should be at least 10 times the number of predictor variables). An LDFA that is unsuccessful may be due to poor predictor variables; that is, the predictor variables may have no relationship to group membership. An LDFA may also be unsuccessful because the group membership is spurious.

Cluster Analysis

The object of **cluster analysis** is to classify a large set of objects (people, plants, kinds of beer, etc.) into distinct subgroups based on predictor variables. These subgroups will be homogeneous with respect to the group's scores on the predictor variables if the cluster analysis is successful. Cluster analysis is also referred to as hierarchical cluster analysis or taxonomy analysis. The statistical object of cluster analysis is to minimize the within-group variation in each group while maximizing the between-group variation among the groups. The cluster analyses' success will be limited by the ability of the predictor variables to yield identifiable clusters. Most clustering procedures also yield coefficients of group similarity that give relative estimates of how alike the members of each group are. For example, imagine a cluster analysis of 100 types of beer. The beers are first rated on predictor variables such as color (reds, yellows, ambers, browns, etc.); alcohol content (1% to 10% alcohol by volume); clarity (clear to opaque); and so on. A cluster analysis might show that Group 1, consisting of Beers 1, 7, 23, 33, 39, 42, 67, and 88, has a group similarity of $r = .58$, while Group 2, consisting of Beers 2, 3, 5, 6, 19, and 55, had a similarity profile of $r = .89$. Examination of graphic plots of each group's means on the predictor variables allows a labeling of the groups—such as the high-alcohol, clear, red group; the low-alcohol, opaque, brown group; and so on—depending on the nature of the predictor variables. Hierarchical cluster analysis allows the researcher to determine what number of groups is optimal since the procedure continues to combine the objects until there are only two groups (but the profiles of group similarity will be low and sometimes senseless at that point).

A Summary of Multivariate Statistics

As the contemporary Yale University statistician Robert P. Abelson has noted, "Don't use Greek, if you don't speak the language." While multivariate techniques are impressive, their interpretation can be complex and controversial. Anyone who chooses to perform a multivariate statistical analysis should understand its assumptions, understand the repercussions of the violations of the assumptions, understand the computer printout, and be able to defend the choice of the statistic. As the late Vanderbilt University statistician Jum Nunnally, wrote, "Don't let the tail wag the dog." If univariate statistics will answer your experimental question just as well, or if you will be able to understand and explain a univariate analysis of your data better than a multivariate statistic, then your choice is clear. One should also not shy away from new techniques, as they may yield new and interesting relationships in your data heretofore unrecognized. Also, remember that few people in the whole world are fully and completely comfortable with their knowledge of statistics. Statisticians and researchers are continually contacting one another, asking for advice and guidance. So do not hesitate, during the rest of your statistical careers, to seek help, advice, and guidance on the choice of your experimental design and the choice of your statistical analyses. Also, be guided by your ever-growing statistical intuition.

Coda

Two wonderful statistics books I heartily recommend (not books of statistics but books about the science of statistics), now that you have finished your first course in statistics, are *The Lady Tasting Tea* by D. Salsburg and *Statistics as Principled Argument* by R. Abelson (1995).

Key Terms and Definitions

Absolute risk reduction—The reduction in risk between two groups treated differently when their outcomes are compared.

Age-specific death rate—The number of deaths in a given age range (for a given year) divided by the total population of that age range at midyear.

Analysis of covariance (ANCOVA)—A method of determining differences among means, on a single dependent variable, while statistically adjusting for the effects of a confounding or nuisance variable.

Annual crude death rate—The number of deaths in a year divided by the population as of midyear of that same year.

Canonical correlation—A statistical technique that yields estimates of how two different sets of variables may be related to each other. The two sets of variables are often specified as the criterion set and the predictor set, and the experimental question is, To what extent is a weighted combination of the criterion set correlated with a weighted combination of the predictor set?

Case-fatality proportion—The number of deaths attributable to a particular disease during a period of time divided by the prevalence proportion of the disease at the same time. The quotient is reported as an interpretable multiple such as 1000.

Cluster analysis—A technique that allows classification of a large set of objects into distinct subgroups based on predictor variables. These subgroups will be homogeneous with respect to the group's scores on the predictor variables.

Diagnostic accuracy—A proportion of the number correct (true positives + true negatives) divided by the total combination of all outcomes (true positives + true negatives + false positives + false negatives).

Equivocal results—A condition when a test's outcome is unclear, uncertain, or unreadable.

Factor analysis—A statistical procedure that explores the underlying conceptual structure of a set of dependent variables by examining the correlations between each variable in the set with every other variable in the set. Factor analysis can also be used to reduce a set of variables by identifying redundant or unnecessary items.

False negative—The condition where a test does not indicate the presence of a disease, and the patient really does have the disease.

False positive—The condition where a test indicates the presence of a disease, and the patient really does not have the disease.

Iatrogenic effect—A condition that arises when the diagnosis of a disease or condition or the treatment of a disease or condition is harmful in and of itself.

Incidence—The number of verified new cases of a disease in a period of time (usually 1 year) divided by the total number of people in the population (with and without the disease).

Indeterminate results—A condition when a test's outcome is neither positive nor negative and does not fall between these two conditions.

Intermediate results—A condition when a test's outcome is neither positive nor negative but falls between these two conditions.

Likelihood ratio for a positive test—A parameter that combines the sensitivity and specificity of a test into a single parameter. The ratio expresses the likelihood that the test will be positive in a person with the disease compared to a positive test in a person without the disease.

Linear discriminant function analysis (LDFA)—A multivariate statistical technique that uses a set of predictor or independent variables to predict group membership on a dichotomous criterion variable. The interest is in whether the predictor variables can accurately classify the a priori specified dichotomous criterion variable.

Multiple regression—A technique that allows a single dependent variable (or criterion variable) to be predicted from several independent variables (or predictor variables). The research concern is often twofold: To what extent can the independent variables predict the dependent variable, and what is the strength of each independent variable in the prediction of the dependent variable?

Multivariate analysis of covariance (MANCOVA)—A technique that analyzes the effects of one or more independent variables on more than one dependent variable while controlling for the effect of one or more covariates (nuisance or confounding variables).

Multivariate analysis of variance (MANOVA)—A technique that analyzes the effects of one or more independent variables on more than one dependent variable. There is a consensus that MANOVA does not control for the Type I error arising from the analysis of multiple dependent variables, and its use with a large group of indiscriminately chosen dependent variables may obscure true significant relationships.

Multivariate statistics—A family of different statistical designs and procedures involving the analysis of more than one dependent variable at the same time. Multivariate statistics are almost always performed by computer programs, and their interpretation can be complex and sometimes controversial.

Prevalence—The number of people who have a specified disease at a given point in time divided by the total number of people in the population (with and without the disease).

Prevalence proportion (of a disease)—The total number of cases of a disease known to exist at a given point in time divided by the total population at the same time. This quotient is also multiplied by an interpretable multiple such as 1000.

Receiver operating curve (ROC)—A graphical representation of signal detection theory where the sensitivity and specificity of tests can be visually analyzed.

Relative risk reduction—A measure of how much reduction in risk has occurred between two groups treated differently when compared to the untreated group's percentage on the outcome of interest.

Sensitivity—A characteristic of a good test where it is very likely to indicate the presence of a disease when the patient really does have the disease.

Specificity—A characteristic of a good test where it is very likely to be correct when it does not indicate the presence of a disease.

True negative—The condition where a test does not indicate the presence of a disease, and the patient really does not have the disease.

True positive—The condition where a test indicates the presence of a disease, and the patient really does have the disease.

Uninterpretable results—A condition when a test's outcome is invalid because it has not been given according to its correct instructions or directions.

Chapter 16 Practice Problems

1. Understand signal detection theory and the receiver operating curve.
2. Understand true and false positives and negatives.
3. Understand the difference between sensitivity and specificity.
4. Understand the definitions of and differences between prevalence and incidence.
5. Understand the four kinds of results.
6. Understand the parameters of mortality and morbidity.
7. Be able to recognize and describe the major kinds of multivariate statistics.

Chapter 16 Test Questions

1. Signal detection theory can be graphically represented by a
   a. scatterplot
   b. receiver operating curve

c. histogram
d. bar graph

2. A true positive is defined as
   a. a test result is positive, and the patient actually has the condition
   b. a test is negative, and the patient actually has the condition
   c. a test is positive, and the patient doesn't have the condition
   d. a test is negative, and the patient doesn't have the condition

3. A false positive is defined as
   a. a test result is positive, and the patient actually has the condition
   b. a test is negative, and the patient actually has the condition
   c. a test is positive, and the patient doesn't have the condition
   d. a test is negative, and the patient doesn't have the condition

4. A true negative is defined as
   a. a test result is positive, and the patient actually has the condition
   b. a test is negative, and the patient actually has the condition
   c. a test is positive, and the patient doesn't have the condition
   d. a test is negative, and the patient doesn't have the condition

5. A false negative is defined as
   a. a test result is positive, and the patient actually has the condition
   b. a test is negative, and the patient actually has the condition
   c. a test is positive, and the patient doesn't have the condition
   d. a test is negative, and the patient doesn't have the condition

6. If a patient is harmed or killed just by the treatment for a disease, it is called the ___________ effect.
   a. not too good
   b. bad
   c. iatrogenic
   d. polygenic

7. A test possesses good sensitivity if it has a
   a. high true-positive rate
   b. high true-negative rate
   c. high false-positive rate
   d. high false-negative rate

8. A test possesses good specificity if it has a
   a. high true-positive rate
   b. high true-negative rate
   c. high false-positive rate
   d. high false-negative rate

9. The diagnostic accuracy of a test is sometimes called the ________ of a diagnostic test.
   a. likelihood
   b. sensitivity

c. specificity
d. efficiency

10. The number of new verified cases of a disease in a year is called
    a. prevalence
    b. incidence
    c. likelihood
    d. number of new cases

11. Which of the following is true about a multivariate analysis of variance?
    a. one or more independent variables, one or more dependent variables
    b. two or more independent variables, with one or more levels of each independent variable
    c. one or more independent variables, two or more dependent variables
    d. all of the above are true

12. Which of the following is true of multivariate statistics?
    a. almost always computed by computer programs
    b. can be easily computed by the average calculator
    c. there are few debates about their use and interpretation
    d. all of the above are true

13. Which of the following is not true of ANCOVA?
    a. covaries the effect of one or more nuisance variables
    b. covaries the effect of one or more nuisance variables on the independent variable
    c. covaries the effect of one or more nuisance variables on the dependent variable
    d. covaries the effect of one or more confounding variables

14. Which of the following is not true about MANCOVA?
    a. handles more than one dependent variable
    b. covaries one or more nuisance variables
    c. handles one or more independent variables
    d. covaries only one confounding variable but one or more nuisance variables

15. Which of the following is not true of factor analysis?
    a. there are an infinite number of correct solutions
    b. it can be used to reduce the number of items on a test or measure
    c. factor analysis designs are similar to ANOVA designs
    d. in IQ research, it found a single "g" factor

16. To have a reliable factor analysis of 57 items, the study should be conducted on at least ________ people.
    a. 57
    b. 285
    c. 570
    d. 1140

17. Canonical correlation comes closest to being a technique related to
    a. correlation
    b. regression

c. factor analysis
d. all of the above

18. In linear discriminant function analysis, the criterion variable
    a. is usually dichotomous
    b. is usually continuous
    c. is an a priori assumption by the experimenter
    d. all of the above
    e. a and c

19. Shrinkage in linear discriminant function analysis can be minimalized by
    a. having the number of participants in the smallest group be 10 times the number of predictor variables
    b. having the number of participants in the largest group be 10 times the number of predictor variables
    c. having the number of participants in the smallest group be at least 30
    d. having the number of participants in both groups be at least 150

20. The statistical object of cluster analysis is to minimize ______ variation and maximize variation between ________.
    a. within-group; groups
    b. group; subjects
    c. subject; confounding variables
    d. trend; cycles

Appendix A

z Distribution

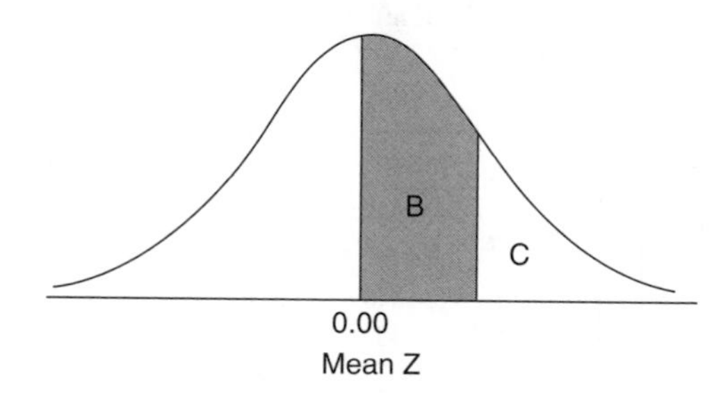

(A) z	(B) Area Between Mean and z	(C) Area Beyond z
.00	.0000	.5000
.01	.0040	.4960
.02	.0080	.4920
.03	.0120	.4880
.04	.0160	.4840
.05	.0199	.4801
.06	.0239	.4761
.07	.0279	.4721
.08	.0319	.4681
.09	.0359	.4641
.10	.0398	.4602
.11	.0438	.4562
.12	.0478	.4522
.13	.0517	.4483
.14	.0557	.4443
.15	.0596	.4404
.16	.0636	.4364
.17	.0675	.4325
.18	.0714	.4286
.19	.0753	.4247

(A) z	(B) Area Between Mean and z	(C) Area Beyond z
.20	.0793	.4207
.21	.0832	.4168
.22	.0871	.4129
.23	.0910	.4090
.24	.0948	.4052
.25	.0987	.4013
.26	.1026	.3974
.27	.1064	.3936
.28	.1103	.3897
.29	.1141	.3859
.30	.1179	.3821
.31	.1217	.3783
.32	.1255	.3745
.33	.1293	.3707
.34	.1331	.3669
.35	.1368	.3632
.36	.1406	.3594
.37	.1443	.3557
.38	.1480	.3520
.39	.1517	.3483

(A) z	(B) Area Between Mean and z	(C) Area Beyond z
.40	.1554	.3446
.41	.1591	.3409
.42	.1628	.3372
.43	.1664	.3336
.44	.1700	.3300
.45	.1736	.3264
.46	.1772	.3228
.47	.1808	.3192
.48	.1844	.3156
.49	.1879	.3121
.50	.1915	.3085
.51	.1950	.3050
.52	.1985	.3015
.53	.2019	.2981
.54	.2054	.2946
.55	.2088	.2912
.56	.2123	.2877
.57	.2157	.2843
.58	.2190	.2810
.59	.2224	.2776

(Continued)

(Continued)

(A) z	(B) Area Between Mean and z	(C) Area Beyond z
.60	.2257	.2743
.61	.2291	.2709
.62	.2324	.2676
.63	.2357	.2643
.64	.2389	.2611
.65	.2422	.2578
.66	.2454	.2546
.67	.2486	.2514
.68	.2517	.2483
.69	.2549	.2451
.70	.2580	.2420
.71	.2611	.2389
.72	.2642	.2358
.73	.2673	.2327
.74	.2704	.2296
.75	.2734	.2266
.76	.2764	.2236
.77	.2794	.2206
.78	.2823	.2177
.79	.2852	.2148
.80	.2881	.2119
.81	.2910	.2090
.82	.2939	.2061
.83	.2967	.2033
.84	.2995	.2005

(A) z	(B) Area Between Mean and z	(C) Area Beyond z
.85	.3023	.1977
.86	.3051	.1949
.87	.3078	.1922
.88	.3106	.1894
.89	.3133	.1867
.90	.3159	.1841
.91	.3186	.1814
.92	.3212	.1788
.93	.3238	.1762
.94	.3264	.1736
.95	.3289	.1711
.96	.3315	.1685
.97	.3340	.1660
.98	.3365	.1635
.99	.3389	.1611
1.00	.3413	.1587
1.01	.3438	.1562
1.02	.3461	.1539
1.03	.3485	.1515
1.04	.3508	.1492
1.05	.3531	.1469
1.06	.3554	.1446
1.07	.3577	.1423
1.08	.3599	.1401
1.09	.3621	.1379

(A) z	(B) Area Between Mean and z	(C) Area Beyond z
1.10	.3643	.1357
1.11	.3665	.1335
1.12	.3686	.1314
1.13	.3708	.1292
1.14	.3729	.1271
1.15	.3749	.1251
1.16	.3770	.1230
1.17	.3790	.1210
1.18	.3810	.1190
1.19	.3830	.1170
1.20	.3849	.1151
1.21	.3869	.1131
1.22	.3888	.1112
1.23	.3907	.1093
1.24	.3925	.1075
1.25	.3944	.1056
1.26	.3962	.1038
1.27	.3980	.1020
1.28	.3997	.1003
1.29	.4015	.0985
1.30	.4032	.0968
1.31	.4049	.0951
1.32	.4066	.0934
1.33	.4082	.0918
1.34	.4099	.0901

Column A contains the *z* scores. Column B gives the proportion of the total area under the curve between the mean (i.e., $z = 0.00$) and the given *z* value. Column C contains the proportion of the total area under the curve above the *z* value.

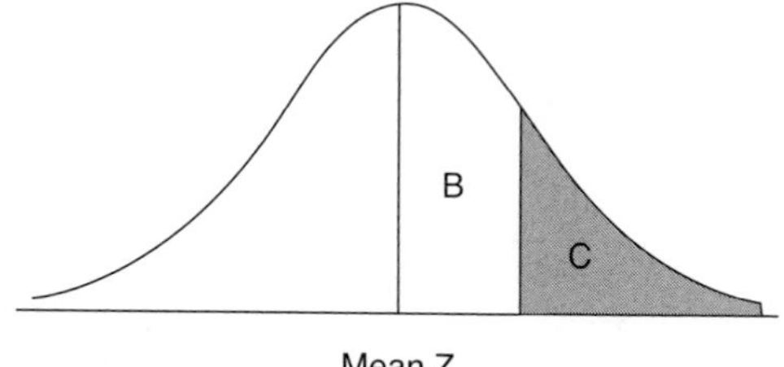

(A) z	(B) Area Between Mean and z	(C) Area Beyond z
1.35	.4115	.0885
1.36	.4131	.0869
1.37	.4147	.0853
1.38	.4162	.0838
1.39	.4177	.0823
1.40	.4192	.0808
1.41	.4207	.0793
1.42	.4222	.0778
1.43	.4236	.0764
1.44	.4251	.0749
1.45	.4265	.0735
1.46	.4279	.0721
1.47	.4292	.0708
1.48	.4306	.0694
1.49	.4319	.0681
1.50	.4332	.0668
1.51	.4345	.0655
1.52	.4357	.0643
1.53	.4370	.0630
1.54	.4382	.0618
1.55	.4394	.0606
1.56	.4406	.0594
1.57	.4418	.0582
1.58	.4429	.0571
1.59	.4441	.0559
1.60	.4452	.0548
1.61	.4463	.0537
1.62	.4474	.0526
1.63	.4484	.0516
1.64	.4495	.0505
1.65	.4505	.0495
1.66	.4515	.0485
1.67	.4525	.0475
1.68	.4535	.0465
1.69	.4545	.0455

(A) z	(B) Area Between Mean and z	(C) Area Beyond z
1.70	.4554	.0446
1.71	.4564	.0436
1.72	.4573	.0427
1.73	.4582	.0418
1.74	.4591	.0409
1.75	.4599	.0401
1.76	.4608	.0392
1.77	.4616	.0384
1.78	.4625	.0375
1.79	.4633	.0367
1.80	.4641	.0359
1.81	.4649	.0351
1.82	.4656	.0344
1.83	.4664	.0336
1.84	.4671	.0329
1.85	.4678	.0322
1.86	.4686	.0314
1.87	.4693	.0307
1.88	.4699	.0301
1.89	.4706	.0294
1.90	.4713	.0287
1.91	.4719	.0281
1.92	.4726	.0274
1.93	.4732	.0268
1.94	.4738	.0262
1.95	.4744	.0256
1.96	.4750	.0250
1.97	.4756	.0244
1.98	.4761	.0239
1.99	.4767	.0233
2.00	.4772	.0228
2.01	.4778	.0222
2.02	.4783	.0217
2.03	.4788	.0212
2.04	.4793	.0207

(A) z	(B) Area Between Mean and z	(C) Area Beyond z
2.05	.4798	.0202
2.06	.4803	.0197
2.07	.4808	.0192
2.08	.4812	.0188
2.09	.4817	.0183
2.10	.4821	.0179
2.11	.4826	.0174
2.12	.4830	.0170
2.13	.4834	.0166
2.14	.4838	.0162
2.15	.4842	.0158
2.16	.4846	.0154
2.17	.4850	.0150
2.18	.4854	.0146
2.19	.4857	.0143
2.20	.4861	.0139
2.21	.4864	.0136
2.22	.4868	.0132
2.23	.4871	.0129
2.24	.4875	.0125
2.25	.4878	.0122
2.26	.4881	.0119
2.27	.4884	.0116
2.28	.4887	.0113
2.29	.4890	.0110
2.30	.4893	.0107
2.31	.4896	.0104
2.32	.4898	.0102
2.33	.4901	.0099
2.34	.4904	.0096
2.35	.4906	.0094
2.36	.4909	.0091
2.37	.4911	.0089
2.38	.4913	.0087
2.39	.4916	.0084

(A) z	(B) Area Between Mean and z	(C) Area Beyond z
2.40	.4918	.0082
2.41	.4920	.0080
2.42	.4922	.0078
2.43	.4625	.0075
2.44	.4927	.0073
2.45	.4929	.0071
2.46	.4931	.0069
2.47	.4932	.0068
2.48	.4934	.0066
2.49	.4936	.0064

(A) z	(B) Area Between Mean and z	(C) Area Beyond z
2.50	.4938	.0062
2.51	.4940	.0060
2.52	.4941	.0059
2.53	.4943	.0057
2.54	.4945	.0055
2.55	.4946	.0054
2.56	.4948	.0052
2.57	.4949	.0051
2.58	.4951	.0049
2.59	.4952	.0048

(A) z	(B) Area Between Mean and z	(C) Area Beyond z
2.60	.4953	.0047
2.61	.4955	.0045
2.62	.4956	.0044
2.63	.4957	.0043
2.64	.4959	.0041
2.65	.4960	.0040
2.66	.4961	.0039
2.67	.4962	.0038
2.68	.4963	.0037
2.69	.4964	.0036

For a negative z, column B contains the proportion of the area under the curve between the given negative z and the mean ($z = 0.00$). Column C contains the proportion of the area under the curve below the given negative z value.

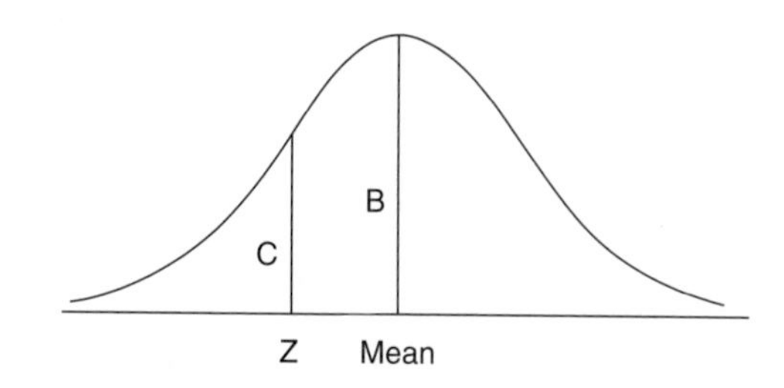

(A) z	(B) Area Between Mean and z	(C) Area Beyond z
2.70	.4965	.0035
2.71	.4966	.0034
2.72	.4967	.0033
2.73	.4968	.0032
2.74	.4969	.0031
2.75	.4970	.0030
2.76	.4971	.0029
2.77	.4972	.0028
2.78	.4973	.0027
2.79	.4974	.0026
2.80	.4974	.0026
2.81	.4975	.0025
2.82	.4976	.0024
2.83	.4977	.0023
2.84	.4977	.0023
2.85	.4978	.0022
2.86	.4979	.0021
2.87	.4979	.0021
2.88	.4980	.0020
2.89	.4981	.0019
2.90	.4981	.0019
2.91	.4982	.0018
2.92	.4982	.0018
2.93	.4983	.0017
2.94	.4984	.0016

(A) z	(B) Area Between Mean and z	(C) Area Beyond z
2.95	.4984	.0016
2.96	.4985	.0015
2.97	.4985	.0015
2.98	.4986	.0014
2.99	.4986	.0014
3.00	.4987	.0013
3.01	.4987	.0013
3.02	.4987	.0013
3.03	.4988	.0012
3.04	.4988	.0012
3.05	.4989	.0011
3.06	.4989	.0011
3.07	.4989	.0011
3.08	.4990	.0010
3.09	.4990	.0010
3.10	.4990	.0010
3.11	.4991	.0009
3.12	.4991	.0009
3.13	.4991	.0009
3.14	.4992	.0008
3.15	.4992	.0008
3.16	.4992	.0008
3.17	.4992	.0008
3.18	.4993	.0008
3.19	.4993	.0007

(A) z	(B) Area Between Mean and z	(C) Area Beyond z
3.20	.4993	.0007
3.21	.4993	.0007
3.22	.4994	.0006
3.23	.4994	.0006
3.24	.4994	.0006
3.30	.4995	.0005
3.40	.4997	.0003
3.50	.4998	.0002
3.60	.4998	.0002
3.70	.4999	.0001
3.80	.49993	.00007
3.90	.49995	.00005
4.00	.49997	.00003

Appendix B

t Distribution

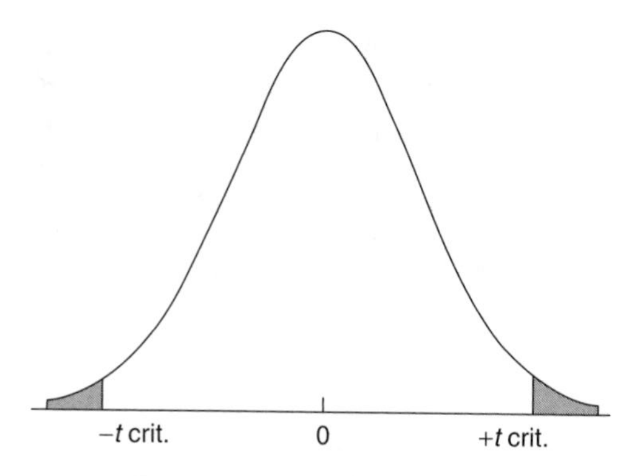

Two-Tailed or Nondirectional Test

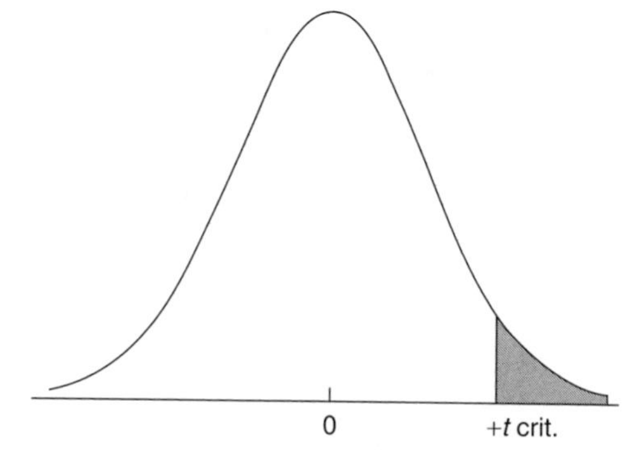

One-Tailed or Directional Test

	Level of Significance				*Level of Significance*		
df	*.05*	*.01*	*.001*	*df*	*.05*	*.01*	*.001*
1	12.706	63.657	636.62	1	6.314	31.821	318.31
2	4.303	9.925	31.598	2	2.920	6.965	22.326
3	3.182	5.841	12.924	3	2.353	4.541	10.213
4	2.776	4.604	8.610	4	2.132	3.747	7.173
5	2.571	4.032	6.869	5	2.015	3.365	5.893
6	2.447	3.707	5.959	6	1.943	3.143	5.308
7	2.365	3.499	5.408	7	1.895	2.998	4.785
8	2.306	3.355	5.041	8	1.860	2.896	4.501
9	2.262	3.250	4.781	9	1.833	2.821	4.297
10	2.228	3.169	4.587	10	1.812	2.764	4.144
11	2.201	3.106	4.437	11	1.796	2.718	4.025
12	2.179	3.055	4.318	12	1.782	2.681	3.930
13	2.160	3.012	4.221	13	1.771	2.650	3.852
14	2.145	2.977	4.140	14	1.761	2.624	3.787
15	2.131	2.947	4.073	15	1.753	2.602	3.733

(Continued)

(Continued)

	Level of Significance				Level of Significance		
df	*.05*	*.01*	*.001*	*df*	*.05*	*.01*	*.001*
16	2.120	2.921	4.015	16	1.746	2.583	3.686
17	2.110	2.898	3.965	17	1.740	2.567	3.646
18	2.101	2.878	3.922	18	1.734	2.552	3.610
19	2.093	2.861	3.883	19	1.729	2.539	3.579
20	2.086	2.845	3.850	20	1.725	2.528	3.552
21	2.080	2.831	3.819	21	1.721	2.518	3.527
22	2.074	2.819	3.792	22	1.717	2.508	3.505
23	2.069	2.807	3.767	23	1.714	2.500	3.485
24	2.064	2.797	3.745	24	1.711	2.492	3.467
25	2.060	2.787	3.725	25	1.708	2.485	3.450
26	2.056	2.779	3.707	26	1.706	2.479	3.435
27	2.052	2.771	3.690	27	1.703	2.473	3.421
28	2.048	2.763	3.674	28	1.701	2.467	3.408
29	2.045	2.756	3.659	29	1.699	2.462	3.396
30	2.042	2.750	3.646	30	1.697	2.457	3.385
40	2.021	2.704	3.551	40	1.684	2.423	3.307
60	2.000	2.660	3.460	60	1.671	2.390	3.232
120	1.980	2.617	3.373	120	1.658	2.358	3.160
∞	1.960	2.576	3.291	∞	1.645	2.326	3.090

Appendix C

Spearman's Correlation

Critical Values for the Spearman Rank-Order Correlation Coefficient

	Level of Significance for a One-Tailed Test			
	.05	*.025*	*.01*	*.005*
	Level of Significance for a Two-Tailed Test			
Number of Pairs (n)	*.10*	*.05*	*.02*	*.01*
5	0.900	1.000	1.000	—
6	0.829	0.886	0.943	1.000
7	0.714	0.786	0.893	0.929
8	0.643	0.738	0.833	0.881
9	0.600	0.683	0.783	0.833
10	0.564	0.648	0.746	0.794
12	0.506	0.591	0.712	0.777
14	0.456	0.544	0.645	0.715
16	0.425	0.506	0.601	0.665
18	0.399	0.475	0.564	0.625
20	0.377	0.450	0.534	0.591
22	0.359	0.428	0.508	0.562
24	0.343	0.409	0.485	0.537
26	0.329	0.392	0.465	0.515
28	0.317	0.377	0.448	0.496
30	0.306	0.364	0.432	0.478

NOTE: Reject the null hypothesis if the derived Spearman coefficient is equal to or greater than the tabled Spearman coefficient value

Appendix D

The Chi-Square Distribution

Critical Values of Chi-Square

df	α Level of Significance			
	.05	*.02*	*.01*	*.001*
1	3.84	5.41	6.64	10.38
2	5.99	7.82	9.21	13.82
3	7.82	9.84	11.34	16.27
4	9.49	11.67	13.28	18.46
5	11.07	13.39	15.09	20.52
6	12.59	15.03	16.81	22.46
7	14.07	16.62	18.48	24.32
8	15.51	18.17	20.09	26.12
9	16.92	19.68	21.67	27.88
10	18.31	21.16	23.21	29.59
11	19.68	22.62	24.72	31.26
12	21.03	24.05	26.22	32.91
13	22.36	25.47	27.69	34.53
14	23.68	26.87	29.14	36.12
15	25.00	28.26	30.58	37.70
16	26.30	29.63	32.00	39.25
17	27.59	31.00	33.41	40.79
18	28.87	32.35	34.80	42.31
19	30.14	33.69	36.19	43.82
20	31.41	35.02	37.57	45.32
21	32.67	36.34	38.93	46.80
22	33.92	37.66	40.29	48.27
23	35.17	38.97	41.64	49.73
24	36.42	40.27	42.98	51.18
25	37.65	41.57	44.31	52.62

(Continued)

(Continued)

26	38.88	42.86	45.64	54.05
27	40.11	44.14	46.96	55.48
28	41.34	45.42	48.28	56.89
29	42.56	46.69	49.59	58.30
30	43.77	47.96	50.89	59.70

SOURCE: This table is taken from Table IV of Fisher and Yates (1995), *Statistical Tables for Biological, Agricultural, and Medical Research,* published by Longman Group Ltd., London (previously published by Oliver and Boyd, Ltd., Edinburgh).

NOTE: Reject the null hypothesis if the derived chi-square value is equal to or greater than the tabled chi-square value.

Appendix E

F Distribution

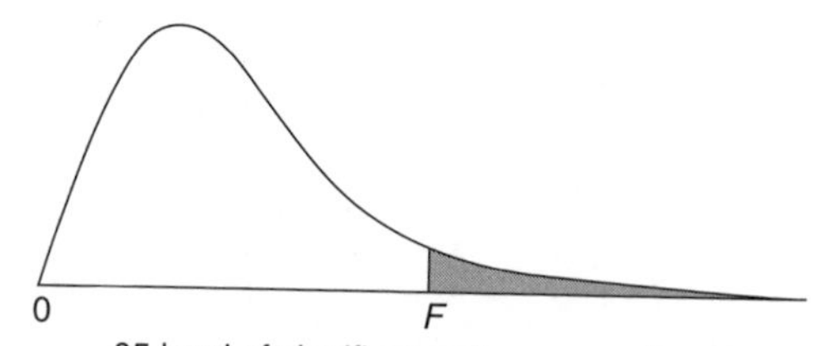

.05 level of significance (upper numbers)
.01 level of significance (lower numbers)

	df_1									
df_2	*1*	*2*	*3*	*4*	*5*	*6*	*7*	*8*	*9*	*10*
11	4.84	3.98	3.59	3.36	3.20	3.09	3.01	2.95	2.90	2.86
	9.65	7.20	6.22	5.67	5.32	5.07	4.88	4.74	4.63	4.54
12	4.75	3.88	3.49	3.26	3.11	3.00	2.92	2.85	2.80	2.76
	9.33	6.93	5.95	5.41	5.06	4.82	4.65	4.50	4.39	4.30
13	4.67	3.80	3.41	3.18	3.02	2.92	2.84	2.77	2.72	2.67
	9.07	6.70	5.74	5.20	4.86	4.62	4.44	4.30	4.19	4.10
14	4.60	3.74	3.34	3.11	2.96	2.85	2.77	2.70	2.65	2.60
	8.86	6.51	5.56	5.83	4.69	4.46	4.28	4.14	4.03	3.94
15	4.54	3.68	3.29	3.06	2.90	2.79	2.70	2.64	2.59	2.55
	8.68	6.36	5.42	4.89	4.56	4.32	4.14	4.00	3.89	3.80
16	4.49	3.63	3.24	3.01	2.85	2.74	2.66	2.59	2.54	2.49
	8.53	6.23	5.29	4.77	4.44	4.28	4.03	3.89	3.78	3.69
17	4.45	3.59	3.20	2.96	2.81	2.70	2.62	2.55	2.50	2.45
	8.40	6.11	5.18	4.67	4.34	4.10	3.93	3.79	3.68	3.59
18	4.41	3.55	3.16	2.93	2.77	2.66	2.58	2.51	2.46	2.41
	8.28	6.01	5.08	4.58	4.25	4.01	3.85	3.71	3.60	3.51
19	4.38	3.52	3.13	2.90	2.74	2.63	2.55	2.48	2.43	2.38
	8.18	5.83	5.01	4.50	4.17	3.94	3.77	3.63	3.52	3.43
20	4.35	3.49	3.10	2.87	2.71	2.60	2.52	2.45	2.40	2.35
	8.10	5.85	4.94	4.43	4.10	3.87	3.71	3.56	3.45	3.37
21	4.32	3.47	3.07	2.84	2.68	2.57	2.49	2.42	2.37	2.32
	8.02	5.78	4.87	4.37	4.04	3.81	3.65	3.51	3.40	3.31

(Continued)

(Continued)

					*df*₁					
df_2	*1*	*2*	*3*	*4*	*5*	*6*	*7*	*8*	*9*	*10*
22	4.30	3.44	3.05	2.82	2.66	2.55	2.47	2.40	2.35	2.30
	7.94	5.72	4.82	4.31	3.89	3.76	3.59	3.45	3.35	3.26
23	4.28	3.42	3.03	2.80	2.64	2.53	2.45	2.38	2.32	2.28
	7.88	5.66	4.76	4.26	3.94	3.71	3.54	3.41	3.30	3.21
27	4.21	3.35	2.96	2.73	2.57	2.46	2.37	2.30	2.25	2.20
	7.68	5.49	4.60	4.11	3.79	3.56	3.39	3.26	3.14	3.06
28	4.20	3.34	2.95	2.71	2.56	2.44	2.36	2.29	2.24	2.19
	7.64	5.45	4.57	4.07	3.76	3.53	3.36	3.23	3.11	3.03
29	4.18	3.33	2.93	2.70	2.54	2.43	2.35	2.28	2.22	2.18
	7.60	5.42	4.54	4.04	3.73	3.50	3.33	3.20	3.08	3.00
30	4.17	3.32	2.92	2.69	2.53	2.42	2.34	2.27	2.21	2.16
	7.56	5.39	4.51	4.02	3.70	3.47	3.30	3.17	3.06	2.98
32	4.15	3.30	2.90	2.67	2.51	2.40	2.32	2.25	2.19	2.14
	7.50	5.34	4.46	3.97	3.66	3.42	3.25	3.12	3.01	2.94
34	4.13	3.28	2.88	2.65	2.49	2.38	2.30	2.23	2.17	2.12
	7.44	5.29	4.42	3.93	3.61	3.38	3.21	3.08	2.97	2.89
36	4.11	3.26	2.86	2.63	2.48	2.36	2.28	2.21	2.15	2.10
	7.39	5.25	4.38	3.89	3.58	3.35	3.18	3.04	2.94	2.86
38	4.10	3.25	2.85	2.62	2.46	2.35	2.26	2.19	2.14	2.09
	7.35	5.21	4.34	3.86	3.54	3.32	3.15	3.02	2.91	2.82
40	4.08	3.23	2.84	2.61	2.45	2.34	2.25	2.18	2.12	2.07
	7.31	5.18	4.31	3.83	3.51	3.29	3.12	2.99	2.88	2.80
42	4.07	3.22	2.83	2.59	2.44	2.32	2.24	2.17	2.11	2.06
	7.27	5.15	4.29	3.80	3.49	3.26	3.10	2.96	2.86	2.77
44	4.06	3.21	2.82	2.58	2.43	2.31	2.23	2.16	2.10	2.05
	7.24	5.12	4.26	3.78	3.46	3.24	3.07	2.94	2.84	2.75
46	4.05	3.20	2.81	2.57	2.42	2.30	2.22	2.14	2.09	2.04
	7.21	5.10	4.24	3.76	3.44	3.22	3.05	2.92	2.82	2.73
48	4.04	3.19	2.80	2.56	2.41	2.30	2.21	2.14	2.08	2.03
	7.19	5.08	4.22	3.74	3.42	3.20	3.04	2.90	2.80	2.71
50	4.03	3.18	2.79	2.56	2.40	2.29	2.20	2.13	2.07	2.02
	7.17	5.06	4.20	3.72	3.41	3.18	3.02	2.88	2.78	2.70
55	4.02	3.17	2.78	2.54	2.38	2.27	2.18	2.11	2.05	2.00
	7.12	5.01	4.16	3.68	3.37	3.15	2.98	2.85	2.75	2.66
60	4.00	3.15	2.76	2.52	2.37	2.25	2.17	2.10	2.04	1.99
	7.08	4.98	4.13	3.65	3.34	3.12	2.95	2.82	2.72	2.63
65	3.99	3.14	2.75	2.51	2.36	2.24	2.15	2.08	2.02	1.98
	7.04	4.95	4.10	3.62	3.31	3.09	2.93	2.79	2.70	2.61
70	3.98	3.13	2.74	2.50	2.35	2.23	2.14	2.07	2.01	1.97
	7.01	4.92	4.08	3.60	3.29	3.07	2.91	2.77	2.67	2.59
80	3.96	3.11	2.72	2.48	2.33	2.21	2.12	2.05	1.99	1.95
	6.96	4.88	4.04	3.56	3.25	3.04	2.87	2.74	2.64	2.55
100	3.94	3.09	2.70	2.46	2.30	2.19	2.10	2.03	1.97	1.92
	6.90	4.82	3.98	3.51	3.20	2.99	2.82	2.69	2.59	2.51
125	3.92	3.07	2.68	2.44	2.29	2.17	2.08	2.01	1.95	1.90
	6.84	4.78	3.94	3.47	3.17	2.95	2.79	2.65	2.56	2.47
150	3.91	3.06	2.67	2.43	2.27	2.16	2.07	2.00	1.94	1.89
	6.81	4.75	3.91	3.44	3.14	2.92	2.76	2.62	2.53	2.44
200	3.89	3.04	2.65	2.41	2.26	2.14	2.05	1.98	1.92	1.87
	6.76	4.71	3.88	3.41	3.11	2.90	2.73	2.60	2.50	2.41

NOTE: Reject the null hypothesis if the derived F value is equal to or greater than the tabled F value. When $df_1 = \infty$ and $df_2 = \infty$, the tabled critical value of $F = 1.00$ at $p = .05$ and $.01$.

Appendix F

Tukey's Table

q Values

df_2	α	r = *Number of Means in the Null Hypothesis*[a]									
		2	3	4	5	6	7	8	9	10	11
5	.05	3.64	4.60	5.22	5.67	6.03	6.33	6.58	6.80	6.99	7.17
	.01	5.70	6.98	7.80	8.42	8.91	9.32	9.67	9.97	10.24	10.48
6	.05	3.46	4.34	4.90	5.30	5.63	5.94	6.12	6.32	6.49	6.65
	.01	5.24	6.33	7.03	7.56	7.97	8.32	8.61	8.87	9.10	9.40
7	.05	3.34	4.16	4.68	5.06	5.36	5.61	5.82	6.00	6.16	6.30
	.01	4.95	5.92	6.54	7.01	7.37	7.68	7.94	8.17	8.37	8.55
8	.05	3.26	4.04	4.53	4.89	5.17	5.40	5.60	5.77	5.92	6.05
	.01	4.75	5.64	6.20	6.62	6.96	7.24	7.47	7.68	7.86	8.03
9	.05	3.20	3.95	4.41	4.76	5.02	5.24	5.43	5.59	5.74	5.87
	.01	4.60	5.43	5.96	6.35	6.66	6.91	7.13	7.33	7.49	7.65
10	.05	3.15	3.88	4.33	4.65	4.91	5.12	5.30	5.46	5.60	5.72
	.01	4.48	5.27	5.77	6.14	6.43	6.67	6.87	7.05	7.21	7.36
11	.05	3.11	3.82	4.26	4.57	4.82	5.03	5.20	5.35	5.49	5.61
	.01	4.39	5.15	5.62	5.97	6.25	6.48	6.67	6.84	6.99	7.13
12	.05	3.08	3.77	4.20	4.51	4.75	4.95	5.12	5.27	5.39	5.51
	.01	4.32	5.05	5.50	5.84	6.10	6.32	6.51	6.67	6.81	6.94
13	.05	3.06	3.73	4.15	4.45	4.69	4.88	5.05	5.19	5.32	5.43
	.01	4.26	4.96	5.40	5.73	5.98	6.19	6.37	6.53	6.67	6.79
14	.05	3.03	3.70	4.11	4.41	4.64	4.83	4.99	5.13	5.25	5.36
	.01	4.21	4.89	5.32	5.63	5.88	6.08	6.26	6.41	6.54	6.66
15	.05	3.01	3.67	4.06	4.37	4.59	4.78	4.64	5.08	5.20	5.31
	.01	4.17	4.84	5.19	5.56	5.80	5.99	6.16	6.31	6.44	6.55
16	.05	3.00	3.65	4.05	4.33	4.56	4.74	4.90	5.03	5.15	5.26
	.01	4.13	4.79	5.19	5.49	5.72	5.92	6.08	6.22	6.35	6.46
17	.05	2.98	3.63	4.02	4.30	4.52	4.70	4.86	4.99	5.11	5.21
	.01	4.10	4.74	5.14	5.43	5.66	5.85	6.01	6.15	6.27	6.38
18	.05	2.97	3.61	4.00	4.28	4.49	4.67	4.82	4.96	5.07	5.17
	.01	4.07	4.70	5.09	5.38	5.60	5.79	5.94	6.08	6.20	6.31
19	.05	2.96	3.59	3.98	4.25	4.47	4.65	4.79	4.92	5.04	5.14
	.01	4.05	4.67	5.05	5.33	5.55	5.73	5.89	6.02	6.14	6.25
20	.05	2.95	3.58	3.96	4.23	4.45	4.62	4.77	4.90	5.01	5.11
	.01	4.02	4.64	5.02	5.29	5.51	5.69	5.84	5.97	6.09	6.19

(Continued)

		r = Number of Means in the Null Hypothesis[a]									
df_2	α	2	3	4	5	6	7	8	9	10	11
24	.05	2.92	3.53	3.90	4.17	4.37	4.54	4.68	4.81	4.92	5.01
	.01	3.96	4.55	4.91	5.17	5.37	5.54	5.69	5.81	5.92	6.02
30	.05	2.89	3.49	3.85	4.10	4.30	4.46	4.60	4.72	4.82	4.92
	.01	3.89	4.45	4.80	5.05	5.24	5.40	5.54	5.65	5.76	5.85
40	.05	2.86	3.44	3.79	4.04	4.23	4.39	4.52	4.63	4.73	4.82
	.01	3.82	4.37	4.70	4.93	5.11	5.26	5.39	5.50	5.60	5.69
60	.05	2.83	3.40	3.74	3.98	4.16	4.31	4.44	4.55	4.65	4.73
	.01	3.76	4.28	4.59	4.82	4.99	5.13	5.25	5.36	5.45	5.53
120	.05	2.80	3.36	3.68	3.92	4.10	4.24	4.36	4.47	4.56	4.64
	.01	3.70	4.20	4.50	4.71	4.87	5.01	5.12	5.21	5.30	5.37
∞	.05	2.77	3.31	3.63	3.86	4.03	4.17	4.29	4.39	4.47	4.55
	.01	3.64	4.12	4.40	4.60	4.76	4.88	4.99	5.08	5.16	5.23

a. For the interaction, r = (the number of levels in the first main effect) × (the number of levels in the second main effect).

References

Abelson, R. P. (1995). *Statistics as principled argument.* Hillsdale, NJ: Lawrence Erlbaum.

American Psychological Association. (2001). *Publication manual of the American Psychological Association* (5th ed.). Washington, DC: Author.

Bailey, J. M., Dunne, M. P., & Martin, N. G. (2000). Genetic and environmental influences on sexual orientation and its correlates in an Australian twin sample. *Journal of Personality and Social Psychology, 78,* 524–536.

Bailey, J. M., Pillard, R. C., Dawood, K., Miller, M. B., Farrer, L. A., Trivedi, S., & Murphy, R. L. (1999). A family history study of male sexual orientation using three independent samples. *Behavior Genetics, 29,* 79–86.

Bailey, J. M., Pillard, R. C., Neale, M. C., & Agyei, Y. (1993). Heritable factors influence sexual orientation in women. *Archives of General Psychiatry, 50,* 217–223.

Barnes, D. M. (1986). Promising results halt trial of anti-AIDS drug. *Science, 234,* 15–16.

Brown, W. A. (1998). The placebo effect. *Scientific American, 278*(1), 90–95.

Coolidge, F. L. (1983). WISC-R discrimination of learning-disabled and emotionally disturbed children: An intragroup and intergroup analysis. *Journal of Consulting and Clinical Psychology, 51,* 320.

Coolidge, F. L. (1998). *Horney-Coolidge Tridimensional Inventory: Manual.* Colorado Springs, CO: Author.

Coolidge, F. L., & Fish, C. E. (1983). Dreams of the dying. *Omega: Journal of Death and Dying, 14,* 1–8.

Coolidge, F. L., Thede, L. L., & Young, S. E. (2002). The heritability of gender identity disorder in a child and adolescent twin sample. *Behavior Genetics, 32,* 251–257.

Cronbach, L. J. (1957). The two disciplines of scientific psychology. *American Psychologist, 12,* 671–684.

Fisher, R. A. (1925). *Statistical methods for research workers.* Edinburgh, UK: Oliver and Boyd.

Fisher, R. A. (1935). *The design of experiments.* Edinburgh, UK: Oliver and Boyd.

Fisher, R. A., & Yates, F. (1995). *Statistical tables for biological, agricultural, and medical research.* London: Longman.

Herbert, J. D., Lilienfeld, S. O., Lohr, J. M., Montgomery, R. W., O'Donohue, W. T., Rosen, G. M., et al. (2000). Science and pseudoscience in the development of eye movement desensitization and reprocessing: Implications for clinical psychology. *Clinical Psychology Review, 20,* 945–971.

Kirk, R. E. (1995). *Experimental design: Procedures for the behavioral sciences* (3rd ed.). Pacific Grove, CA: Brooks/Cole.

Micceri, T. (1989). The unicorn, the normal curve, and other improbable creatures. *Psychological Bulletin, 105,* 156–166.

Needleman, H. L., Reiss, J. A., Tobin, M. J., Biesecker, G. E., & Greenhouse, J. B. (1996). Bone lead levels and delinquent behavior. *Journal of the American Medical Association, 275,* 363–369.

Park, R. (2000). *Voodoo science: The road from foolishness to fraud.* Oxford, UK: Oxford University Press.

Paulos, J. A. (1995). *A mathematician reads the newspaper.* New York: Basic Books.

Peters, W. S. (1987). *Counting for something: Statistical principles and personalities.* New York: Springer-Verlag.

Rosnow, R. L., & Rosenthal, R. (1989). Statistical procedures and the justification of knowledge in psychological science. *American Psychologist, 44,* 1276–1284.

Rosnow, R. L., & Rosenthal, R. (1996). *Beginning behavioral research: A conceptual primer.* Englewood Cliffs, NJ: Prentice Hall.

Sagan, C. (1996). *The demon-haunted world: Science as a candle in the dark.* New York: Ballantine.

Salsburg, D. (2001). *The lady tasting tea: How statistics revolutionized science in the twentieth century.* New York: Freeman.

Schmidt, F. L. (1996). Statistical significance testing and cumulative knowledge in psychology: Implications for training of researchers. *Psychological Methods, 1,* 115–120.

Sugiyama, M. S. (2001). Food, foragers, and folklore: The role of narrative in human subsistence. *Evolution and Human Behavior, 22,* 221–240.

Tabachnick, B. G., & Fidell, L. S. (2001). *Using multivariate statistics* (4th ed.). New York: HarperCollins.

Tufte, E. R. (1983). *The visual display of quantitative information.* Cheshire, CT: Graphics Press.

Tufte, E. R. (1997). *Envisioning information.* Cheshire, CT: Graphics Press.

Tufte, E. R. (2003). *The cognitive style of PowerPoint.* Cheshire, CT: Graphics Press.

Tufte, E. R. (2005). *Beautiful evidence.* Cheshire, CT: Graphics Press.

Tukey, J. W. (1969). Analyzing data: Sanctification or detective work? *American Psychologist, 24,* 83–91.

Walker, H. M. (1975). *Studies in the history of statistical method.* New York: Arno Press. (Original work published 1929)

Index

Absolute risk reduction, 360, 369
Abuse of power, 13, 29
Age-specific death rates, 361, 369
Alpha, 128, 147
Alternative hypothesis, 44, 122, 147
Analysis of covariance (ANCOVA), 363
Analysis of variance (ANOVA), 242, 256
Annual crude death rate, 361, 369
ANOVA assumptions, 248
ANOVA designs, 247
ANOVA repeated measures design, 273, 279
A posteriori tests, 263
Archival data, 155, 190
Assumptions of the dependent *t* test, 221

Bar graph, 60
Bell-shaped curve, 47, 61
Beta, 128, 147
Between-groups variance, 244, 256
Between-subjects variables, 315, 330
Bimodal distribution, 49, 56, 60
Binomial distribution, 144
Bonferroni correction, 242, 256

Canonical correlation, 366, 369
Case-fatality proportion, 362, 369
Categorical scales, 23
Causal relationship, 156, 190
Central limit theorem, 80, 83
Chance-is-lumpy problem, 161, 190
Changing scales midstream, 57
Chart junk, 57
Chi-square distribution, 188, 190
Chi-square tests, 336, 350
Cluster analysis, 368, 369
Coefficient of determination, 172, 190
Combinations theorem of probability, 141
Completely randomized ANOVA, 247, 256
Completely randomized factorial design, 283, 291
Confidence intervals, 209
Confounding variable, 29
Constant, 190
Continuous variable, 22, 29
Controlled experiment, 121, 147, 197, 214
Correcting for bias, 77
Correlation, 153, 190
Correlation and prediction, 161
Cumulative frequency distribution, 52, 60
Curvilinear relationship, 168, 190

Degrees of freedom, 190, 252
Demand characteristics, 137, 147
Dependent *t* test, 219, 235
Dependent variable, 20, 21, 30
Descriptive statistics, 7, 8, 30
Diagnostic accuracy, 358, 370
Dichotomous variable, 22, 30
Directional alternative hypothesis, 125, 130, 147
Discrete variable, 22, 30
Double-blind experiment, 12, 30
Drawbacks of the dependent *t* test designs, 223

Effect size, 208, 214, 227, 231, 234
Empirical rule, 80, 84
Equivocal results, 74, 84, 359, 370
Estimating the standard error, 212
Experimental control, 18
Experimental design, 18, 30
Experimental error, 244, 256

F ratio, 245
F test statistic, 244
Factor analysis, 365, 370
Factorial design, 283, 291
False negative, 356, 370
False positive, 356, 370
First main effect, 287
Fixed effects, 284
Frequency distribution, 46, 61
Frequency histogram, 47, 61
Frequency polygon, 47, 61

Goodness-of-fit test, 336, 350
Graphs, 38

Harmonic mean, 269
Heterogeneity of variance, 214
Heteroscedasticity, 172, 190
Homogeneity of variance, 200, 214
Homoscedasticity, 172, 190
Hypothesis, 30

Iatrogenic effects, 357, 370
Incidence, 359, 370
Increasing power of a *t* test, 222
Independent groups *t* test, 197, 214
Independent variable, 20, 30
Indeterminate results, 75, 84, 359, 370
Inferential statistics, 7, 8, 30, 119
Insignificant findings, 129, 148
In situ design, 199, 214
Interaction, 291
Intermediate results, 74, 84, 359, 370
Interpretation of interaction effect, 302
Intervals, 49
Interval scales, 24, 30

Klinkers, 73, 84
Kurtosis, 55, 61

Labeling the graph badly, 58
Least significant difference test (LSD), 263, 269
Least squares method, 190
Levene's test of homogeneity of variance, 202, 214
Leptokurtosis, 56, 61
Likelihood ratio for a positive test, 358, 370
Linear discriminant function analysis (LDFA), 367, 370
Linear regression, 174
Linear relationship, 168, 191
Low-density graphs, 57

Main effect, 291
Main effect interpretation, 297
Matched *t* test, 219, 235
Mean, 68, 84
Measurement, 21
Measurement error, 21, 30
Measurement scales, 22
Measures of central tendency, 67, 68, 84
Measures of variation, 84
Median, 69, 84
Mode, 71, 84
Multicolored graph, 58
Multiple comparison tests, 249, 256, 263, 269, 302
Multiple regression, 366, 370
Multiple regression analysis, 191
Multiple *t*s, 242, 256
Multiplication theorem of probability, 140
Multivariate, 61
Multivariate analysis of covariance (MANCOVA), 364, 370
Multivariate analysis of variance (MANOVA), 363, 370
Multivariate statistics, 355, 370

Negative *z* scores, 96
Negatively skewed, 49, 61
Nominal scales, 23, 30
Nondirectional alternative hypothesis, 125, 130, 148
Nonnormal frequency distributions, 55
Nonparametric statistics, 335, 350

Nonsignificant, 129, 148
Normal curve, 47, 61
Normality of the dependent variable, 200
Null hypothesis, 121, 148
Null hypothesis in two-factor ANOVA, 285
Null hypothesis accepted, 306, 310

Omega squared, 308, 310
Omnibus hypothesis, 257
Omnibus test, 257
One-tailed tests of significance, 223
Operational definition, 21, 30
Order of presentation, 221, 235
Ordinal scale, 23, 30
Outliers, 73, 84

Parameters, 8, 30
Pearson's *r*, 161
Pearson product moment correlation coefficient, 162, 191
Pearson product moment correlation (coefficient *r*), 162
Percentiles, 53, 61, 110, 113
Permutations theorem of probability, 142
Phi correlation, 162, 187, 188, 191
Placebo, 13
Placebo effect, 11, 30
p level, 148
Platykurtosis, 56, 61
Point-biserial correlation, 184, 191
Point biserial *r*, 162
Pooled variance, 201, 214
Population, 8, 30
Positively skewed, 49, 61
Post hoc tests, 263, 269
Power, 13, 30, 207, 215
Power analysis, 13, 30
PowerPoint graphs and presentations, 58
Prevalence, 358, 371
Prevalence proportion, 362, 371
Principle components analysis, 365
Probability, 139, 148

Qualitative scales, 23
Quartiles, 54, 61

Randomized block design, 274, 279
Random effects, 284
Range, 75, 84
Ratio scales, 25, 30
Receiver operating curve (ROC), 356, 371
Regression analysis, 180, 191
Regression effect, 235
Relative risk reduction, 360, 371
Repeated measures ANOVA, 315, 322, 330
Repeated measures *t* test, 219, 235
Replication, 31, 127, 148
Research hypothesis, 121, 148
Robustness, 215
Rounding, 25

Sample, 8, 31
Sampling distribution, 80, 84, 246, 257
Scatterplot, 167, 191
Second main effect, 287
Sensitivity, 356, 371
Shared variance, 173
Sigma, 31
Signal-to-noise ratio, 121, 148
Significance, 31, 129, 148
Significance levels, 129
Significant findings, 129
Simple effects, 291
Simple regression, 191
Skewed distribution, 49, 61
Spearman's correlation, 180, 191
Spearman's *r*, 161
Specificity, 371
Split plot, 315
Split-plot ANOVA, 315, 330
Standard deviation, 75, 76, 78, 84
Standard error of the mean, 80, 84, 246, 257
Standard score, 89, 113
Statistical control, 18
Statistical symbols, 26
Stem-and-leaf plot, 54, 61
Stratified samples, 8, 31
Strong negative relationship, 154, 191
Strong positive relationship, 154, 191
Subject error, 257

Tables, 38
t distribution, 164, 191
T score, 113
t test, 215
Test of the standardized
  residuals, 350
Ties in ranks, 182
Trend, 130, 148
True negative, 356, 371
True positive, 356, 371
Tukey's HSD test, 263, 269
Two-factor ANOVA, 322
Two-tailed tests of
  significance, 223
Type I error, 128, 148
Type II error, 128, 148

Unbiased estimator, 68, 84
Uncertain results, 74
Uncommon variance, 173
Unexplained variance, 173
Uninterpretable results, 75, 84,
  359, 371
Univariate, 61
Unshared variance, 173

Variance, 75, 79, 84

Within-groups variance, 244, 257
Within-subjects variables, 315, 330

z distribution, 91, 113
z score, 91, 113

About the Author

Frederick L. Coolidge received his B.A., M.A., and Ph.D. in psychology at the University of Florida. He completed a 2-year postdoctoral fellowship in clinical neuropsychology at Shands Teaching Hospital in Gainesville, Florida. He has been awarded three Fulbright Fellowships to India (1987, 1992, and 2005). He has also won three teaching awards at the University of Colorado (1984, 1987, and 1992), including the lifetime title of University of Colorado Presidential Teaching Scholar. In 2005, he received the University of Colorado at Colorado Springs College of Letters, Arts, and Sciences' Outstanding Research and Creative Works award.

Professor Coolidge conducts research in behavior genetics and has established the strong heritability of gender identity and gender identity disorder. He has published this work and others in *Behavior Genetics, Developmental Neuropsychology,* and *Journal of Personality Disorders.* He also conducts research in life span personality assessment and has established the reliability of posthumous personality evaluations, including his grandmother and Adolf Hitler. He has published this work in the *Journal of Clinical Geropsychology* and *Military Psychology.* He also applies cognitive models of thinking and language to explain evolutionary changes in the archaeological record and has published this work in the *Cambridge Archaeological Journal, Journal of Human Evolution,* and the *Journal of Anthropological Research.*

Professor Coolidge's hobbies include reading, traveling, collecting, mountain biking, playing in a rock band, and grandfathering.